AF339266

Green Trends in Insect Control

RSC Green Chemistry

Series Editors:
James H Clark, *Department of Chemistry, University of York, York, UK*
George A Kraus, *Department of Chemistry, Iowa State University, Iowa, USA*
Andrzej Stankiewicz, *Delft University of Technology, Delft, The Netherlands*

Titles in the Series:
 1: The Future of Glycerol: New Uses of a Versatile Raw Material
 2: Alternative Solvents for Green Chemistry
 3: Eco-Friendly Synthesis of Fine Chemicals
 4: Sustainable Solutions for Modern Economies
 5: Chemical Reactions and Processes under Flow Conditions
 6: Radical Reactions in Aqueous Media
 7: Aqueous Microwave Chemistry
 8: The Future of Glycerol: 2nd Edition
 9: Transportation Biofuels: Novel Pathways for the Production of Ethanol, Biogas and Biodiesel
10: Alternatives to Conventional Food Processing
11: Green Trends in Insect Control

How to obtain future titles on publication:
A standing order plan is available for this series. A standing order will bring delivery of each new volume immediately on publication.

For further information please contact:
Book Sales Department, Royal Society of Chemistry, Thomas Graham House, Science Park, Milton Road, Cambridge, CB4 0WF, UK
Telephone: +44 (0)1223 420066, Fax: +44 (0)1223 420247
Email: books@rsc.org
Visit our website at http://www.rsc.org/Shop/Books/

Green Trends in Insect Control

Edited By

Óscar López and José G. Fernández-Bolaños
Department of Organic Chemistry, Faculty of Chemistry,
University of Seville, Spain

RSC Publishing

RSC Green Chemistry No. 11

ISBN: 978-1-84973-149-2
ISSN: 1757-7039

A catalogue record for this book is available from the British Library

Published by The Royal Society of Chemistry,
Thomas Graham House, Science Park, Milton Road,
Cambridge CB4 0WF, UK

Registered Charity Number 207890

For further information see our web site at www.rsc.org

Preface

This book is part of the *Green Chemistry* series published by the *Royal Society of Chemistry*, and is designed to provide a modern overview of the current status of insecticides. We present the current approaches for insect pest control as *green* alternatives to classical agrochemicals, which should be of interest to a vast group of researchers: agrochemists, biochemists, chemists, toxicologists, *etc.* Throughout the book, the different approaches to pest control which involve "greener chemicals" in particular are emphasized. In the literature there are numerous examples of texts making a comprehensive treatment of specific aspects of agrochemicals (such as toxicity, analytical measurements of pesticides and their derivatives in the environment and in living organisms or practical uses of insecticides), but there are few attempts of bringing together the wide variety of topics covered in this book. With this publication, we intend to provide a useful guide for a broad range of readers, from students, to teachers or professionals in the agrochemical sector.

In each of the chapters, the following information is taken into account: a description of useful synthetic and biorational insecticides, which highlights environmentally-friendly processes; the mode of action, emphasizing selectivity towards targeted species and environmental effects (*e.g.* toxicity, bioaccumulation and metabolism and a comparison with classical insecticides); as well as human and environmental risk assessments. Furthermore, formulation, dispersal and persistence in the environment are covered as key aspects in developing greener agrochemicals.

There are numerous insect species considered to be pests which cause health problems in humans, cattle and crops, and even affect some human constructions. Many insects are known to act as vectors in spreading important diseases such as malaria, dengue or typhus, which continue to be responsible for a large number of deaths. Moreover, the crops devoted to human and cattle feeding

RSC Green Chemistry No. 11
Green Trends in Insect Control
Edited by Óscar López and José G. Fernández-Bolaños
© Royal Society of Chemistry 2011
Published by the Royal Society of Chemistry, www.rsc.org

suffer from important losses in productivity every season because of insect pests acting on them.

The control of such pests has been an arduous task that can be traced back to the establishment of the first known civilizations, with numerous techniques and approaches being employed over the centuries.

Historically, the first attempts to fight non-beneficial insects included the use of natural sources, either minerals or botanical extracts, which became the first available chemical substances against insects. After that, an arsenal of synthetic derivatives started to be commercially available in what seemed to be a chemical panacea for controlling pests. It became evident quite soon, however, that most of the compounds successfully applied to crops, cattle and even humans exerted severe effects on the environment.

Many of the original chemicals, both natural and synthetic, were initially used indiscriminately, and proved to be highly toxic against living organisms, some of them acting indiscriminately on beneficial insects and other animals. Many other insecticides were found to be highly persistent in the environment, as a result of their slow environmental degradation and were bioaccumulated through the trophic chain. Acute toxicological and long-term effects included teratogenic damage, impairment of the central nervous system and malfunctioning of numerous biological processes, among others.

Humanity slowly started to be conscious of the necessity of replacing the classical approaches for insect control with "greener" substitutes, with an increased selectivity and a reduced toxicity, as a way of minimising environmental impacts. Furthermore, a deeper knowledge of insect physiology has allowed the discovery of targets to which more selective insecticides can be aimed (*e.g.* sodium channel, chloride channel, acetylcholinesterase, nicotinic acetylcholine receptors, moulting processes, *etc.*).

The first chapter of this book (S. Manguin *et al.*) is devoted to a general introduction to entomology, with a special emphasis on those insects that act as vectors in diseases caused by microbial pathogens. This chapter covers the key biological aspects of insect vectors, such as their morphology, habitats, feeding and reproduction. The main insects acting as vector of diseases covered in this section are: mosquitoes, biting midges, flies, fleas and lice which are responsible for spreading severe diseases such as malaria, typhus, lymphatic filariasis, dengue, hemorrhagic fevers and certain encephalitis. The methods by which insects spread infection and the most efficient strategies for fighting against them are widely covered in Chapter 1. The information presented allows for a logical understanding of the mode of action of insecticides, for the design of novel, more efficient compounds.

The second chapter (Ó. López *et al.*) provides a historical review of classical insecticides, as examples of non-green agrochemicals that must be replaced and their use revised. The chapter is divided into two main sections In the first section, the approaches used for insect control before the development of synthetic insecticides are covered. The use of non-rational insecticides, developed from mainly mineral sources (*e.g.* sulfur, arsenicals, copper and boron derivatives) and, in most cases, lacking specificity is widely reviewed.

The second part of the chapter is focused on the three main families of synthetic insecticides that have dominated the agrochemical sector from the 1940s to today: organochlorine, organophosphorus and carbamate insecticides. The high efficiency and economic viability of such compounds allowed an indiscriminate use on crops, households and cattle. Throughout this chapter, however, it is strongly emphasized that many of these insecticides are bioaccumulated and can provoke severe toxicological effects in mammals, fish and beneficial insects, such as bees. The restrictions and bans exerted by highly recognized organizations (such as the US EPA, the European Union Committees and the World Health Organization) for the reduction and even the elimination of the use of such insecticides are also discussed.

The rest of this book is devoted to the different approaches for overcoming the adverse effects exerted by the classical insecticides. In this context, the third chapter (Schleier and Peterson) focuses on pyrethroids, a family of insecticides which emerged in the 1970s, inspired by natural pyrethrins (botanical extracts from the *Chrysanthemum* genus) but lacking their intrinsic photolability. Pyrethroids possess a series of features which have been shown to be greener than the precedent organochlorine, organophosphorous and carbamate insecticides; although pyrethroids are lipophilic molecules, they do not undergo biomagnification and exert no toxic effect on mammals. A critical overview of these compounds is covered, in which they are compared with the parent pyrethrins and classical synthetic insecticides, in terms of toxicity, environmental stability and future expectations.

Also inspired by botanical extracts, the neonicotinoids (Chapter 4, Nauen and Jeschke) are a relatively new family of insecticides developed in the 1990s. With a surprisingly fast growing market, the development of neonicotinoids has allowed for the control of some of the most destructive crop insect pests. In this section, the different neonicotinoid generations, their strong specificity against insect pests and their low mammal toxicity are discussed. The numerous advantages of this group of agrochemicals are reported, and they are proposed as being one of the most promising insecticides in the near future.

Chapters 5 (Dripps *et al.*) and 6 (Gomez *et al.*) describe work in which DowAgrochemicals has been involved over the last few years. The first of these chapters is devoted to spinosyns, macrocyclic lactones resulting from the metabolism of the soil-dwelling bacteria of the *Saccharopolyspora* species, which were discovered in the late 1980s and firstly marketed in the 1990s. The authors highlight the green character of this kind of insecticide, not only in terms of the production process (fermentation or semi-synthesis), but also in terms of its environmental behaviour with regard to bioaccumulation and selectivity. Spinosad and spinetoram insecticides received the Presidential Green Chemistry Challenge Award in 1999 and 2008, respectively, and when spinosad received its crop registration in 1997, it was among one of the first insecticides to be considered "reduced-risk" by the US EPA. The authors give a detailed description of the mode of action, and QSAR and biological studies in both targeted and not-targeted species.

An example of chemicals aimed at the moulting process of insects, and thus at insect growth and development, is the bisacylhydrazines, which are reviewed in Chapter 6. The bisacylhydrazines are a non-steroidal family of Insect Growth Regulators (IGRs), for whose initial development Rohm & Haas was awarded the US EPA's Presidential Green Chemistry Challenge Award in 1998. Structural studies, the design of new examples, environmental effects and formulations are covered in this chapter.

In Chapter 7 (Isman and Paluch) the use of botanical extracts as insecticides is described. Although, when compared to some other families of agrochemicals, the market of botanical insecticides is much more limited (approximately 1% of total market sales), compounds such as terpenes (from essential oils), or azadirachtin (from neem extracts), deserve attention because of their beneficial biological properties. Plants provide an arsenal of compounds which exert a variety of different insecticidal modes of action: from direct toxicity to insects, to inhibitory properties on growth, development, reproduction and feeding, as well as repellent activities. The chances of developing and marketing new botanical insecticides are also covered.

Besides conventional insecticides (based on the use of natural or synthetic compounds), the use of living systems (*e.g.* viruses, bacteria, fungi, insect predators and engineered-plants) as highly selective insect control mechanisms is included in Chapter 8 (Soberón *et al.*). The mode of action, specificity, environmental risk assessments and application indications are widely covered in this chapter, with a particular emphasis on the most widely used bioinsecticides, the endotoxins produced by the bacteria *Bacillus thuringiensis*. In order to obtain more efficient insecticides from microbial sources, the use of genetically-modified organisms is also reviewed.

Finally, the current philosophy in the agrochemical sector of the Integrated Pest Management programmes (IPM) is introduced in Chapter 9 (Castle and Prabhaker). IPMs attempt to reduce the use of synthetic pesticides while keeping pest populations to an acceptable level. The authors claim that a change in agricultural policies is needed for increasing crop production to the rate required while competing with predatory pests. They suggest that key aspects in pest control which are usually disregarded include the intensification of crops, the appropriate selection of plant species and the appropriate use of fertilizers. The main tactics of defence against pests from a green perspective are also reviewed: the combination of bioengineered crops (*e.g.* producing the bacterial endotoxins described in Chapter 8); an increase of the plant types in cropping systems; semiochemicals (*e.g.* repellents, antifeedants, sex pheromones, *etc.*, usually in baits); and insect growth regulators. The authors describe how the development of more ecological crops would help reduce insect populations and at the same time our dependence on synthetic insecticides.

Finally, we would like to express our gratitude to Dr. Manguin, Schleier, Nauen, Dripps, Gómez, Murray, Soberón and Castle, together with their co-workers, firstly for their valuable contributions and also for their constructive discussions and suggestions for this book.

Óscar López and José G. Fernández-Bolaños
Seville

Contents

RSC Green Chemistry No. 11
Green Trends in Insect Control
Edited by Óscar López and José G. Fernández-Bolaños
© Royal Society of Chemistry 2011
Published by the Royal Society of Chemistry, www.rsc.org

Chapter 4 Basic and Applied Aspects of Neonicotinoid Insecticides   132
R. Nauen and P. Jeschke

Chapter 5   The Spinosyn Insecticides   163

J. E. Dripps, R. E. Boucher, A. Chloridis, C. B. Cleveland,
C. V. DeAmicis, L. E. Gomez, D. L. Paroonagian,
L. A. Pavan, T. C. Sparks and G. B. Watson

Chapter 6   The Bisacylhydrazine Insecticides   213
Luis E. Gomez, Kerry Hastings, Harvey A. Yoshida, James E. Dripps, Jason Bailey, Sandra Rotondaro, Steve Knowles, Doris L. Paroonagian, Tarlochan Singh Dhadialla and Raymond Boucher

Chapter 7 Needles in the Haystack: Exploring Chemical Diversity of Botanical Insecticides   **248**

Murray B. Isman and Gretchen Paluch

Main Topics in Entomology: Insects as Disease Vectors

S. MANGUIN,[1][*] J. MOUCHET[2] AND P. CARNEVALE[3]

[1] IRD, Université Montpellier 1 (UM1), UMR-MD3, Laboratoire d'Immuno-Physiopathologie Moléculaire Comparée, Faculté de Pharmacie, 15 Avenue Charles Flahault, F-34093 Montpellier, France; [2] 15 Avenue de Lattre de Tassigny, F-43300 Langeac, France; [3] Immeuble "Le Majoral", Avenue Tramontane, F-34420 Portiragnes, France

1.1 Introduction

Insects of medical interest are numerous, and some have had a major impact on the course of human history due to epidemics of vector-borne diseases which have led to millions of deaths. Mosquitoes are at the origin of severe epidemics of malaria, dengue, yellow fever and chikungunya, all of which continue to kill millions of people throughout the world each year. Fleas are responsible for propagating the plague, which killed millions of people during three pandemics. Lice transmit typhus, which has been shown to weaken armies in periods of war. Among those insects of medical interest, the vast majority belongs to the insect order Diptera. This chapter deals with the most common insect vectors of diseases, whose classification is as follows.

RSC Green Chemistry No. 11
Green Trends in Insect Control
Edited by Óscar López and José G. Fernández-Bolaños
© Royal Society of Chemistry 2011
Published by the Royal Society of Chemistry, www.rsc.org

Classification

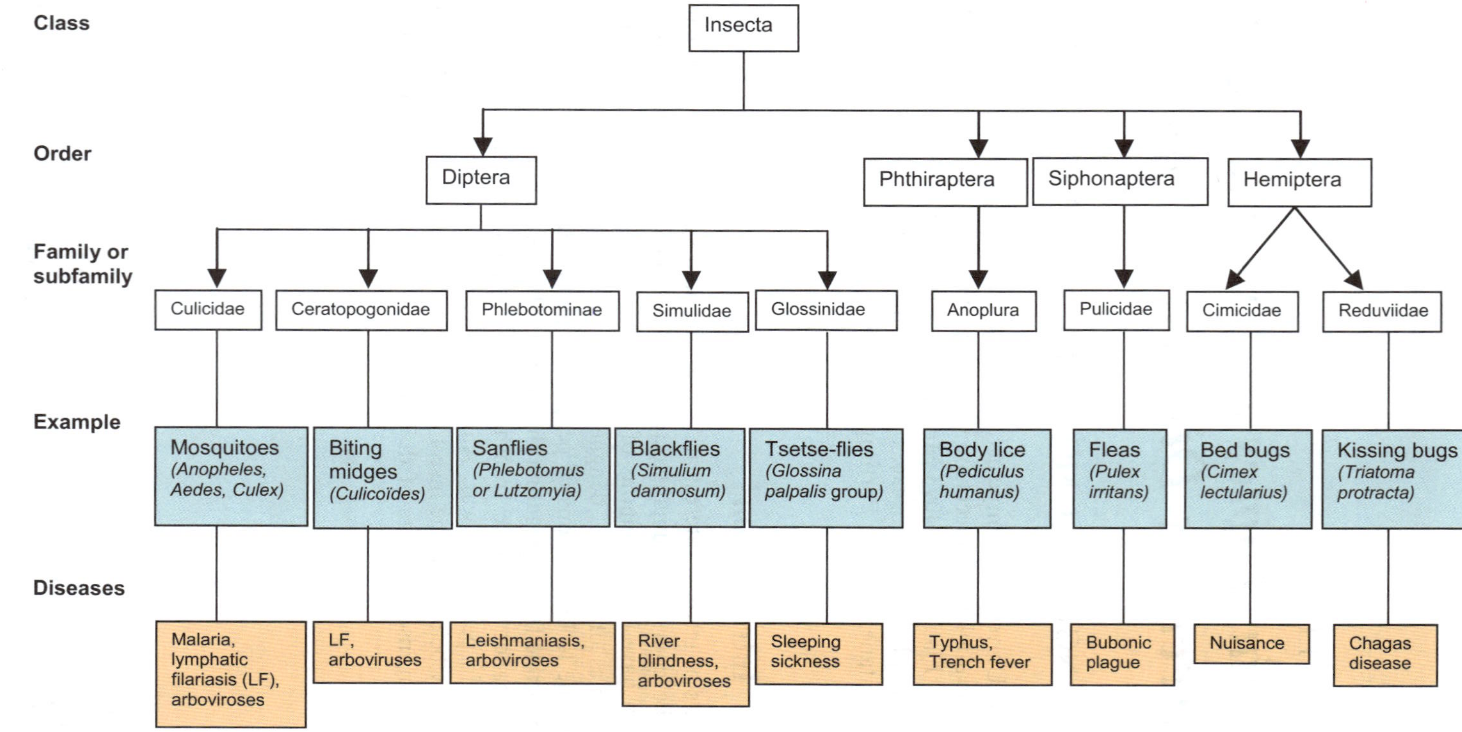

1.2 Mosquitoes

There are more than 3400 species of mosquitoes, which belong to 37 genera joined together in only a single family: Culicidae, which is divided into three sub-families: (1) Toxorhynchitinae, (2) Anophelinae and (3) Culicinae. Mosquitoes have a worldwide distribution. They occur in tropical and temperate zones, even on the level of the Arctic Circle; there are, however, no mosquitoes in the Antarctic. Mosquitoes are also found in mountainous regions (at 5500 m altitude), as well as in caves and mines (at –1250 m altitude).

The mosquitoes that bite humans and which are the most important vectors of disease, belong to the genera *Anopheles*, *Culex*, *Aedes*, *Mansonia*, *Haemagogus* and *Sabethes*.[1] *Anopheles* can transmit malaria parasites, but also lymphatic microfilariae (*e.g. Wuchereria bancrofti*, *Brugia malayi* and *Brugia timori*) and some arboviruses (*e.g.* O'nyong-nyong and equine encephalitis). Some *Culex* species also transmit the microfilariae responsible for Bancroftian lymphatic filariasis, as well as many arboviruses, such as the West Nile, Japanese encephalitis, Saint Louis encephalitis, Murray Valley encephalitis and Rift Valley arboviruses. The *Aedes* genus includes important vectors of the viruses responsible for yellow fever, dengue (and hemorrhagic dengue), chikungunya, eastern equine encephalitis (EEE), and many other arboviruses. Some *Aedes* species also transmit microfilarial parasites (Bancroftian lymphatic filariasis), such as *Ae. polynesiensis* in the Pacific islands.

1.2.1 General Morphology

The development cycle of mosquitoes includes two phases (see Figure 1.1): (1) an aquatic phase, with the succession of immature stages, such as eggs, larvae (four stages) and pupae, and (2) an air phase with the male and female adults.

The eggs are ovoid and measure approximately 0.5 mm. They are laid either on the surface of water (*e.g. Anopheles* and *Culex*) or near the surface of water (*e.g. Aedes*). The eggs may be laid separately (*e.g. Anopheles* and *Aedes*) or close together in the form of an "egg raft" at the time of oviposition (*e.g. Culex* and *Coquillettidia*). The eggs are able to float due to side floats (*e.g. Anopheles*) or apical floats (*e.g. Culex*). The variations in egg ornamentation have been used to dismember the complex *Anopheles maculipennis* in order to understand the phenomenon of "anophelism without malaria" in Western Europe.

This raises the concept of "species complex", in which sibling species are important disease vectors whilst others are not involved in pathogen transmission at all, despite the fact that these species cannot be differentiated morphologically. Identification of the individual species must be based on more sophisticated techniques, in particular molecular ones. Many of the main vectors of pathogenic diseases belong to a species complex, such as *Anopheles gambiae*, *An. dirus*, *An. farauti*, *Culex pipiens*, etc.

The larvae which emerge from the egg evolve in four stages (L1, L2, L3 and L4), intersected with three moults which allow the larvae to grow from 1 mm (L1) to 15 mm (L4) in a week (longer in temperate regions). The four larval stages have a comparable general morphology. The larva is composed of three parts: head,

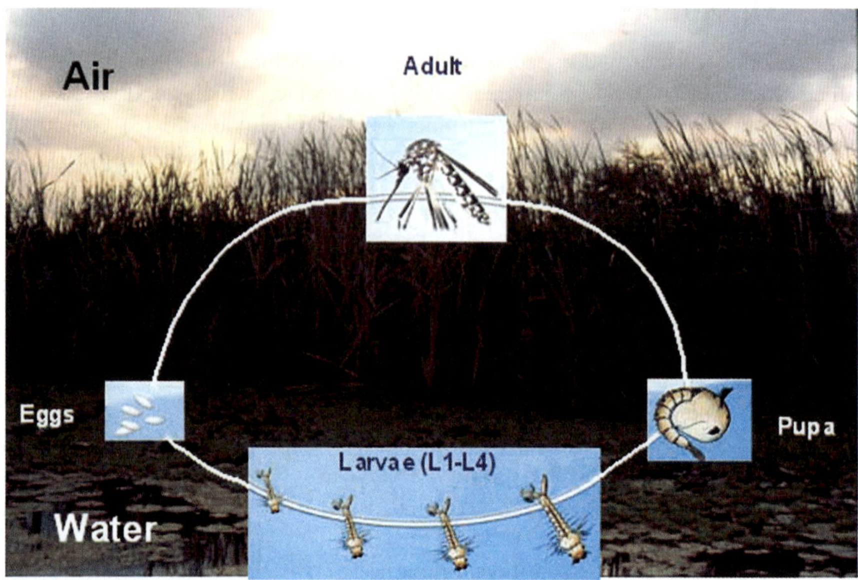

Figure 1.1 Mosquito cycle in its habitat.

thorax and abdomen. The 8th abdominal segment is modified (the "respiratory segment") with two important structures: (1) on the lateral side, the "comb" (pecten) which is made up of spines and scales of different forms, sizes and numbers, according to the genus and species of the mosquito, and (2) on the dorsal face, the spiracular apparatus which is located either directly on the tegument for *Anopheles*, or at the end of a respiratory siphon for Culicinae. This is a useful characteristic with which to differentiate the position of the larvae, as *Anopheles* larvae (without siphon) stay parallel to the water surface, whilst *Culex* or *Aedes* larvae (with a siphon) have an oblique angle of suspension to the water surface. For *Mansonia*, the end of the siphon is modified in a hard organ used to bore plants. *Mansonia* larvae do not breathe like other mosquitoes but attach themselves to the roots, leaves and stems of aquatic plants in order to obtain their air supply.

The pupa's morphology is completely different from that of the larva, consisting of two parts: (1) a prominent cephalothorax equipped with two respiratory trumpets (the pupa does not have an oral apparatus as it breathes but does not feed), and (2) an abdomen made up of eight visible segments (the ninth segment is barely visible), the eighth segment carries a pair of swimming paddles. The pupa is quite mobile and dives when disturbed. Its lifespan is short (one to two days).

The adults. Male and female mosquitoes can be easily differentiated by observing the head and the end of the abdomen. The head comprises of two compound eyes made up of hundreds of *ommatidia*, and two antennae with 15 articles in the male and 16 articles in the female. In the male, there is a great number of large setae which allows for the easy recognition of the "plumose" antennae, whereas the female has "pilose" antennae. The oral apparatus is of the "sucker" type for the male and the "biter" type for the female which includes: (1) a labium folded up in a gutter and finished by two labellums; in this gutter, there are six piercing stylets which will penetrate the skin and search for a capillary for the intake of blood; (2) the labrum (or upper lip), which serves as the "roof" of the food channel; (3) the hypopharynx which is connected to salivary glands by the salivary channel and forms the floor of the food channel; and (4) two mandibles and two maxillae. On both sides of this female piercing apparatus, there is a pair of maxillary palps, which are as long as the proboscis for *Anopheles* but shorter than the proboscis for Culicinae. This difference allows for easily differentiation of *Anopheles* from other mosquitoes.

The thorax includes three segments: the prothorax, the mesothorax, and the metathorax. Each segment comprises a pair of legs made up of a hip (or "coxa"), a trochanter, a femur, a tibia and a tarsus with five articles; the last article carries at its end two claws which help the mosquito to hold on to the support. The legs carry more or less coloured scales which are used for mosquito identification. The second segment, or "wing segment", is the largest and carries a pair of wings (two wings = "Diptera") with veins and scales whose form and colour ("wing ornamentation") are also used for species identification. The third segment carries halters which are used for balance during flight. The morphology of the lateral parts of the thorax called "pleurites" is very much used in systematics. There are two respiratory spiracles (on the second and third segments). The dorsal part of the second segment is called the "scutum" which is prolonged by a "scutellum". This is simple and rounded for *Anopheles* or trilobed for Culicinae. The abdomen is composed of ten segments of which eight are quite visible. Each segment comprises a dorsal chitinized part (tergite) and a ventral chitinized part (sternite), connected by a very extensible pleural membrane which allows for the swelling of the abdomen of the female after a blood meal or the maturation of the ovaries. Segments nine (genital segment) and ten (anal segment) are quite modified. The genital apparatus on the male is very complicated and its morphology is used in systematics (especially for *Culex*). Between 12 and 24 hours after the emergence of the adult, the male genital apparatus undergoes a 180° rotation and becomes ready for mating. The terminalia surround a complex penis (the "phallosome") which is located on the tenth segment. The abdomen of the female ends with two cercus.

1.2.2 Internal Anatomy

The internal mosquito anatomy is composed of: a digestive tract with the pharynx and its pump which aspirate the blood; the oesophagus; the stomach (midgut); and the posterior intestine (hindgut) which ends in the rectum and the anus. The salivary duct arrives at the lower face of the pharynx and the salivary pump allows for excretion of saliva during the bite. The salivary channel is connected to a pair of trilobed salivary glands. If parasites (*Plasmodium*) are in the mosquito's salivary glands, they are inoculated during the bite. In the female, the genital apparatus is composed of two ovaries with many ovarioles, and their oviducts meet to form an odd oviduct which arrives at the vagina. During maturation, the ovarioles evolve in five "Christopher's" stages.[2]

The spermatheca, which is a duct that opens in the vagina, is an organ where the spermatozoa, inoculated by the male during fecundation, are stored. There is one spermatheca in the *Anopheles* female and two in the Culicinae.

1.2.3 Biology

According to the species, mosquito larvae can develop in practically all possible types of habitats: freshwater to brackish water; clean water to heavy polluted one; stagnant water to running water; natural habitats to man-made breeding sites; small habitats (puddles, footprints, artificial containers, *etc.*) to large ones (rice plantations, lakes, *etc.*). Information on larval ecology is crucial in order to carry out appropriate vector control programs targeting specific mosquitoes. Three keys elements concern the larvae: (1) they feed, therefore it is possible to use insecticides of ingestion like *Bacillus thuringiensis* or *Bacillus sphaericus* (useless against pupae which do not feed); (2) they moult, therefore growth regulators like juvenoids or ecdysteroids can be used; and (3) they breathe at the water surface, therefore it is possible to use methods aiming at asphyxiating them, like monolayers, oils or polystyrene chips for example.

Only females bite to take a blood meal for egg maturation, but males and females feed on flower nectar from which they get their energy necessary for flight. Fecundation occurs two to three days after adult emergence, with generally only one fecundation, although several can take place. The female's life is conditioned by the succession of blood meals and the development of the ovaries, this is the gonotrophic cycle, which starts with the unfed female, then after blood-feeding, it becomes half-gravid, and gravid. This cycle must be known for each species or each situation considered in vector control programs, as its duration conditions, the frequency of the contacts host/mosquito, and the ingestion (from man to mosquito) or the transmission (from mosquito to man) of pathogens responsible of the disease considered. After egg laying the female seeks another blood meal, and the "gonotrophic cycle" repeats itself every two to three days. In tropical regions, the blood meal is accompanied by a maturation of the ovaries; this is the "trophogonic concordance". On the other hand, in temperate regions during a cold period there can be a "trophogonic dissociation" for which the blood meal is not followed by the development of the ovaries; the females can even enter into complete diapause, allowing hibernation. Mosquitoes can take their blood meal from humans (anthropophilic) or animals (zoophilic), or both. The trophic preferences of species are very important to know, the more anthropophilic a mosquito, the higher its vectorial role. The blood meal can be taken indoors (endophagic) or outside (exophagic). Some species bite essentially during the night (nocturnal) like *Anopheles*, others during the day (diurnal) like *Aedes*, and others during the morning or at sundown. After the blood meal, the mosquitoes have a phase of digestion which lasts approximately 48 hours, during which they rest either indoors (endophilic) or outdoors (exophilic). All these behaviours are very important to know in the definition of vector control strategies. It is clear that indoor spraying with remanent insecticides will be particularly efficient against anthropophilic, endophagic and endophilic mosquitoes, but effects against the exophagic and exophilic mosquitoes will be quite reduced, and the addition of insecticide could even increase exophilic behaviour and greatly reduce the impact. Some products have effects known as "deterrent" (the mosquito avoids entering the treated house), "excito-repulsive" or "irritant", where the mosquito avoids contact with a treated surface or remains in contact with the product for only a short period of time and so the insecticide does not have a lethal effect. The mosquitoes can thus survive outdoors and continue to bite the human population in spite of indoor treatment. Mosquitoes generally have a lifespan of about one month in tropical areas, although in temperate areas, mosquitoes can survive during the winter in diapause or semi-diapause. Their range of active flight is generally rather weak (active dispersion), three kilometres for *Anopheles* and *Aedes* hardly move away from their larval habitat, therefore vector control can target areas near breeding sites. Mosquitoes can, however, be transported by the wind (passive dispersion) and by modern means of transportation (*e.g.* "airport malaria" cases can occur as a result of *Anopheles* vectors travelling by plane from an endemic zone into a malaria-free area and inoculating *Plasmodium* parasites).

1.2.3.1 Culicinae

The subfamily of Culicinae includes 33 genera; the most important ones in medical entomology are the *Culex, Aedes, Mansonia, Sabethes* and *Haemagogus. Aedes, Culex* and *Mansonia* are found in the temperate and tropical regions; the genera *Sabethes* and *Haemagogus* are found only in Central and South America. Culicinae are easily distinguishable from *Anopheles* at the larval and adult stages (see Table 1.1).

1.2.3.1.1 Culex. *Culex* species are widespread in the whole World, except the most northern zones of temperate regions and the poles. There are thought to be some 800 species divided into 21 sub-genera.

The eggs, brown, long and cylindrical, are deposited on the surface of water and bound to form a "raft" composed of some 300 eggs which are laid in a large variety of aquatic habitats: small puddles, pools, permanent or temporary ponds, flooded marshes, borrow pits, ditches, rice plantations, as well as

Table 1.1 Morphological differences between Anophelinae and Culicinae in relation to the stages.

Stages	Anophelinae	Culicinae
Eggs	Visible lateral floaters	
Larvae	No respiratory siphon	Highly visible siphon
Adult females	Palps as long as proboscis	Palps smaller than proboscis
Adult males	Palps with rounded extremities	Palps with tapered extremities
Adult resting position	Oblique on the surface	Parallel to the surface

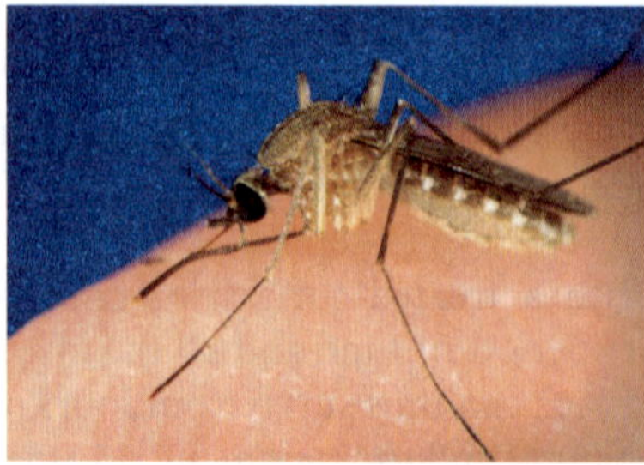

Figure 1.2   Adult *Culex quinquefasciatus* (Photo courtesy of CDC/James Gathany).

anthropogenic sites such as cans, cisterns, and even sewage drains with polluted water. The most important species, *Culex quinquefasciatus*, is strongly associated with anarchistic urbanization, with poor hygiene conditions and worn-out water drainage systems containing organic matter where the larvae can develop (*e.g.* polluted stagnant water, gutters, septic tanks, sewage drains, *etc.*). The density of mosquito populations can be very high under such conditions and constitutes a major cause of nuisance for the people affected.

Transmission of pathogens

Culex quinquefasciatus (Pipiens subgroup) has a nocturnal activity with an endophilic tendency (see Figure 1.2). It is the main vector of Bancroftian lymphatic filariasis in urban settings in tropical regions throughout the world,[3] while this parasite is transmitted by *Anopheles* in rural areas and *Aedes* in most Pacific Islands, a curious situation, as it is usually accepted that co-evolution occurs between the parasite and its vector.

Culex tritaeniorhynchus (Vishnui subgroup) is an important vector of Japanese encephalitis in India and South-East Asia and develops in clear water, in particular rice plantations, but also fish farm basins where manures are added. The Japanese encephalitis virus is also transmitted by *Cx. pseudovishnui, Cx. vishnui* and *Cx. gelidus*. Many other arboviruses, such as West Nile, are transmitted by species of the *Culex pipiens* complex in the USA, by *Cx. univittatus* and *Cx. theileri* in Africa, *Cx. modestus* and *Cx. molestus* in the Western and Eastern zones of the Mediterranean basin, and *Cx. vishnui* in India, *etc.* The virus of Saint Louis encephalitis is also transmitted by species of the *Cx. pipiens* complex and *Cx. nigripalpus* in the USA. The Murray Valley encephalitis virus is transmitted by *Cx. annulirostris* in the USA. In South America, the Bunyavirus of the groups "C", "Guana" and "Nyando" are also transmitted by various *Culex* species of which *Cx. portesi* and *Cx. vomerifer* are examples. The Rift Valley fever is a Phlebovirus transmitted by *Cx. pipiens* in Egypt and by other Culicidae elsewhere.

1.2.3.1.2   *Aedes.* *Aedes* mosquitoes are widespread throughout the World, even in Arctic zones where they represent an important nuisance for the human populations and cattle. There are some 870 species divided into 36 subspecies.

The eggs, generally black and ovoid, are laid separately on a wet substrate. The eggs are resistant to desiccation for several months, then, when they are covered with water, the eggs hatch quickly. Many *Aedes* larvae develop in small temporary habitats like tree holes (*e.g. Ae. pseudoscutellaris*), rock holes (*e.g. Ae. Togoi* and *Ae. vittatus*), bamboo stumps, coconut shells (*e.g. Ae. polynesiensis*) and crab holes. *Aedes* larvae are also well adapted to colonising anthropogenic habitats, such as worn and abandoned tyres (*e.g. Ae. albopictus*), funeral urns, cans, domestic containers to preserve water (*e.g. Ae. aegypti*), or any other small domestic containers. Their ability to develop in a large variety of habitats, with both freshwater and brackish water, and to resist desiccation, confers great advantages on *Aedes* mosquitoes in terms of adaptation and colonization of new sites. As a consequence, larval control of *Aedes* mosquitoes is quite problematic. In addition, as the larvae develop in water containers reserved for drinking, it is necessary to pay great attention in the choice of the larvicides employed. The majority of *Aedes* adults have a particular and quite visible ornamentation of white and black scales on the thorax and legs, which often allows for their rapid identification (see Figure 1.3). *Aedes albopictus* has characteristic black and white scale stripes, at the origin of his name "Tiger mosquito", as well as a central line of white scales on the thorax (see Figure 1.3B). *Aedes aegypti* has a characteristic shape of white scales drawing a "lyre" on the thorax (see Figure 1.3A). The normal duration of the cycle, egg to adult, is from one week to 10–12 days.

Aedes tends to be a rural insect which flees the urban areas and pullulates everywhere else, especially in natural sites; except *Aedes aegypti* which is an urban mosquito that reproduces in all domestic and peridomestic containers (*e.g.* cans, gutters, dustbins, wheelbarrows, flower pots, drums, dugouts, tyres, concrete pits, grease-boxes, *etc.*). It bites only during the day, with a peak of activity in the early morning and at sundown. Their behaviour is especially exophagic and exophilic, hence the use of insecticides in indoor spraying (inside residual spraying or IRS) for adult control is inefficient. Control is therefore often based on the elimination of larval habitats, with the participation of local communities. During epidemics, spatial and focal pulverizations must be repeated to eliminate transmission. Epidemics of dengue (in Cayenne) have

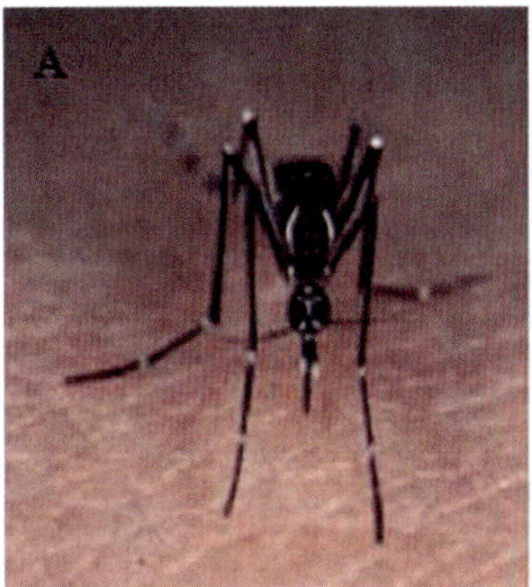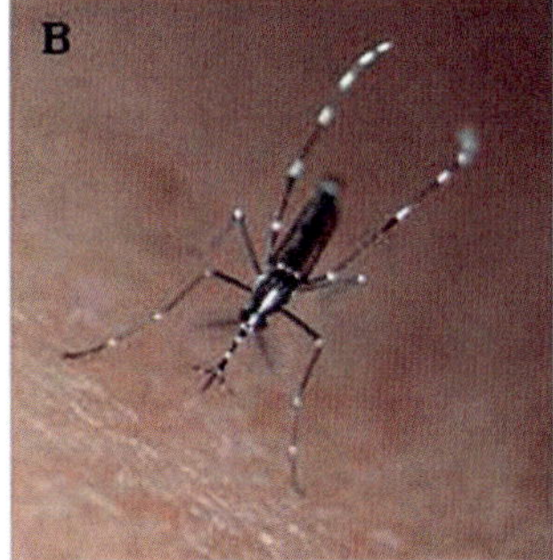

Figure 1.3   Adults of *Aedes aegypti* (A) (Photo © IRD/Jean-Pierre Hervy) and *Aedes albopictus* (B) (Photo © IRD/Michel Dukhan).

been analysed through geospatial studies, for monitoring the expansion of the disease and in order to adapt the vector control program according to the local context.[4]

Some species are involved in the transmission of pathogens, including: *Aedes aegypti,* the main vector of yellow fever of urban type; along with *Ae. africanus* and *Ae. simpsoni,* the vectors of the sylvatic yellow fever virus; and *Ae. albopictus,* a vector of dengue and chikungunya viruses which is spreading across the World (*e.g.* epidemics in Reunion Island in 2005–2006).[5] *Aedes polynesiensis* and *Ae. pseudoscutellaris* are important vectors of diurnal and subperiodic Bancroftian filariasis, as well as *Ae. togoi,* which is also a vector of *Brugia* filariasis.

Aedes aegypti has a worldwide distribution and has colonized the majority of the tropical countries. A species complex has been recognized with two subspecies (or forms), *Aedes aegypti aegypti* and *Aedes aegypti formosus,* which differ in their biology, behaviour, susceptibility to dengue viruses and present genetic variations.[6,7] The pale, anthropophilic, domestic form (*Ae. ae. aegypti*) develops mainly in anthropogenic habitats and is involved worldwide in dengue epidemics. The dark, non-anthropophilic, peri-domestic form (*Ae. ae. formosus*) occurs primarily in Africa, it preferentially colonizes natural sites and has only been reported in a forest cycle of dengue in Western Africa.[8]

Aedes albopictus is described as being the principal vector of dengue only when *Ae. aegypti* is absent or present at low density, generally in continental regions and suburban or rural zones. Its receptivity to the dengue virus is less than *Ae. ae. aegypti* (pale domestic form), but better than *Ae. ae. formosus* (dark peri-domestic form). *Aedes albopictus* is also thought to be responsible for the maintenance of infection because its rate of sexual and transovarian transmission is higher than for *Ae. aegypti.*[9]

1.2.3.1.3 *Mansonia.* Very aggressive during the day, *Mansonia* mosquitoes (25 species) occur in wet tropical regions, but some species have also been found in Sweden and Tasmania. Some *Mansonia* species are of medical importance. *Mansonia uniformis* is a vector of lymphatic filariasis agents,[3] such as *Brugia malayi* in India and Southeast Asia and *Wuchereria bancrofti* in Asia and New Guinea. *Mansonia dives* and *Ma. titillans* (see Figure 1.4)

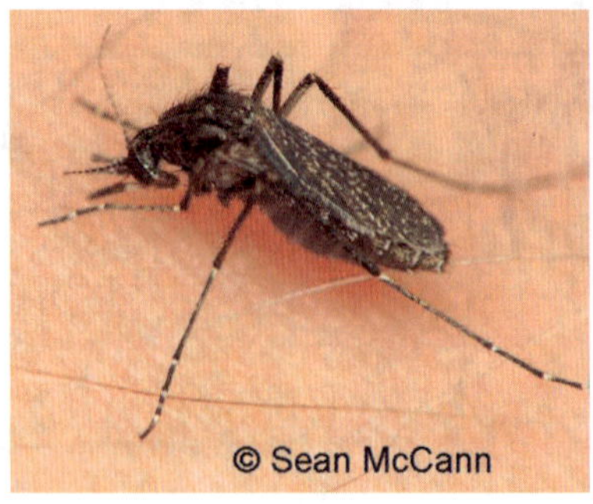

Figure 1.4   Adult *Mansonia titillans* (Photo © Sean McCann).

are also vectors of *W. bancrofti* in Asia and the tropical Americas, respectively; *Ma. annulifera, Ma. annulata* and *Ma. indiana* are vectors of *B. malayi* (nocturnal subperiodic) in Southeast Asia and sometimes also in India. In Africa, *Mansonia* (*Ma. africana* and *Ma. uniformis*) transmit arboviruses, such as "Spondweni" (Flavivirus) and Rift Valley fever (Phlebovirus). Its control is very difficult due to its peculiar larval ecology.

1.2.3.1.4 *Haemagogus.* *Haemagogus* mosquitoes (24 species) are only found in Central and South America (Neotropical region). They are first and foremost forest mosquitoes. The adults bite during the day, primarily in the canopy feeding on monkeys. Under certain environmental conditions, in particular at the edge of forests, during tree cutting or in the dry season, they can leave the canopy and bite humans. Several species of *Haemagogus* are involved in the transmission of the selvatic yellow fever (*e.g. Hg. spegazzinii, Hg. leucocelaenus, Hg. capricornii, Hg. janthinomys, etc.*).[10]

1.2.3.1.5 *Sabethes.* *Sabethes* mosquitoes, which include 39 species classified into five subgenera, are distributed in the Neotropical region. They are diurnal biters and forest (canopy) mosquitoes but, as with *Haemagogus*, they can bite humans when flying near the ground. *Sabethes chloropterus* has been involved in the transmission of selvatic yellow fever in monkeys, accidentally in man, as well as arboviruses.[10]

1.2.3.2 Anophelinae

The 484 described species of *Anopheles*[11] are distributed all around the world, except the polar zones and most of the Pacific islands. *Anopheles* species are distributed by geographic zone,[12] whereas some species of *Aedes* and *Culex* have a panmictic distribution. The adults are night biters (from sundown to sunrise), although some specimens can bite during the daytime in forests or when cloudy.[13] Their behaviour can be anthropophilic/zoophilic, endophagic/exophagic or endophilic/exophilic. The biting activity is important to know in order to establish appropriate vector control methods. Insecticide treated nets (ITN) are obviously particularly efficient against endophagic species biting especially during the second part of the night. Biting behaviour is more commonly not restricted but depends on the accessibility of hosts and resting places. Some species are more opportunistic than others, but even strongly anthropophilic species can blood feed on animals if their preferential hosts are unavailable. Some authors have proposed a vector control method called "zooprophylaxy" by deviating the anthropophilic *Anopheles* species towards animals.[14]

Anopheles species are principally known for their involvement as vectors of malaria pathogens (see Figure 1.5). The efficiency of vector control programs requires a precise identification of the species concerned, especially important and arduous for species complexes, and a thorough knowledge of the biology of the actual vector in order to set up and evaluate strategies adapted to the local

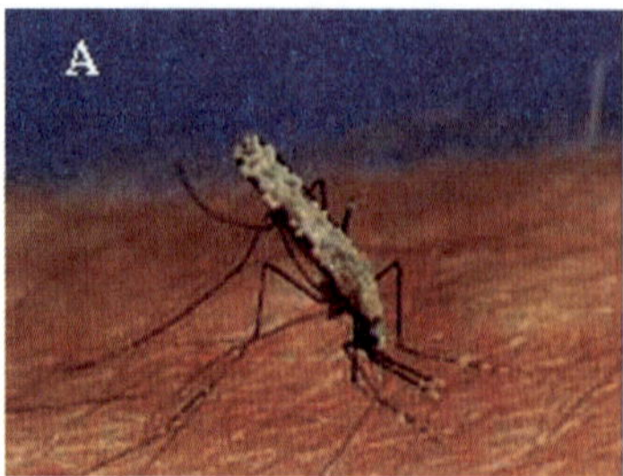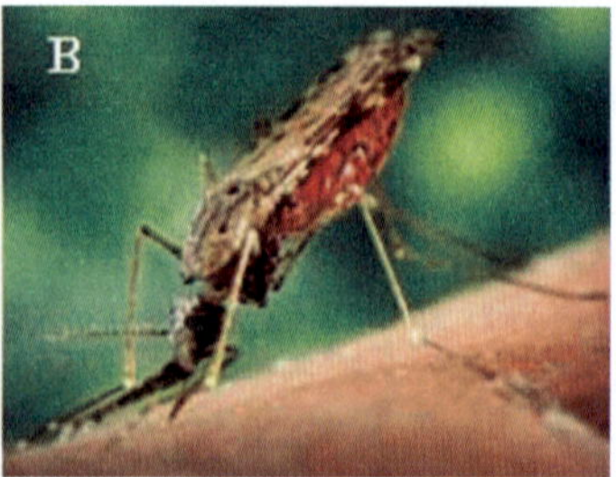

Figure 1.5   *Anopheles* mosquitoes: *Anopheles gambiae* (A) (Photo © IRD/Jean-Pierre Hervy) and *Anopheles albimanus* (B) (Photo courtesy CDC/James Gathany).

eco-epidemiological context and the behaviour of the targeted species. Vector control encounters many technical constraints, such as vector resistance to insecticides, or socio-cultural problems, for example the acceptability of control methods such as IRS or the regular use of ITN. However, vector control remains the best strategy for malaria prevention and control, while vaccines are unavailable and drug resistance in *P. falciparum* strains continues to develop worldwide.

1.2.4   Main Mosquito-Borne Diseases

1.2.4.1   Malaria

Only *Anopheles* species are vectors of malaria parasites which affect nearly 50% of the world's inhabitants living in the 109 endemic countries, especially the poor developing countries (see Figure 1.6). The WHO evaluated that there are approximately 250 million cases of malaria annually, with nearly 1 million deaths, 90% in Africa, south of the Sahara which pays a heavy debt to this disease.[15]

Malaria is due to Protozoan parasites of the genus *Plasmodium* of which four species are primarily involved: *Plasmodium falciparum*, *P. vivax*, *P. malariae*, and *P. ovale*.[17] Recent reports have suggested a possible fifth species, *Plasmodium knowlesi*, as an important and common emerging zoonotic pathogen responsible for human infections in Southeast Asia.[18] Globally, *P. falciparum* is the most common cause of malarial infection, responsible for approximately 80% of all cases and 90% of deaths. *Plasmodium* transmission from the *Anopheles* vector to humans is accomplished by direct injection of the parasite at the sporozoite stage, contained in salivary gland fluid during blood feeding. Of the 484 recognized species of *Anopheles*, only about 15% or less are generally involved in malaria parasite transmission. From a biological point of view, the *Anopheles* mosquito is the definitive host for the parasite, where sexual reproduction between male and female gametes occurs, whereas humans are the intermediate hosts only where asexual multiplication (schizogonic cycle)

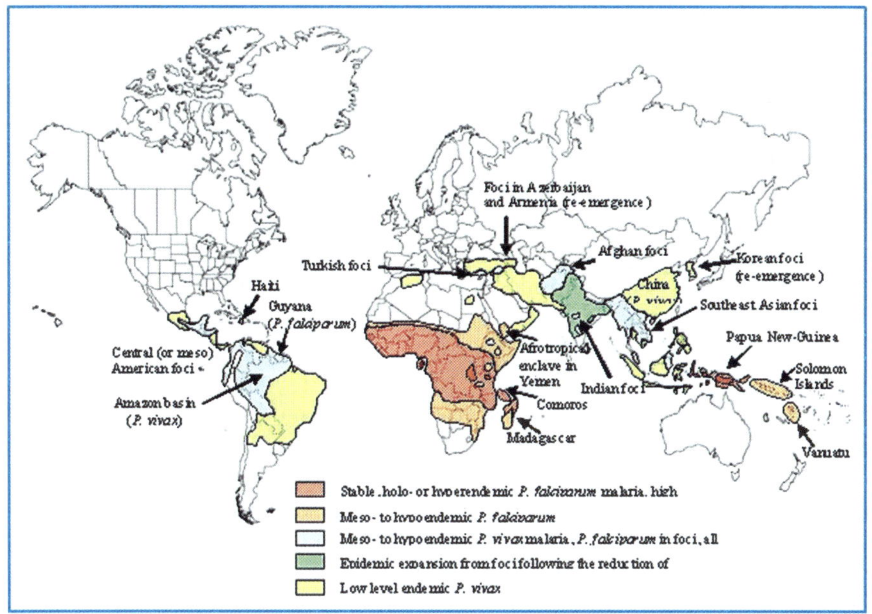

Figure 1.6 Malaria distribution around the world (Photo courtesy of John Libbey Eurotext).[16]

takes place. *Anopheles* females become infected by ingesting sexually mature gametocytes present in the peripheral blood of the host. In the mosquito midgut fertilization produces the ookinete which traverses the mosquito gut and forms an oocyst under the outermost layer of the gut wall. After repeated multiplication of sporoblasts, each oocyst eventually ruptures, releasing hundreds of sporozoites into the mosquito body cavity, a proportion of which will invade the salivary glands awaiting the opportunity to infect another human upon the next blood feeding by the mosquito. This "sporogonic phase" (from gametocytes ingestion to infective sporozoite) takes on average 8–14 days in tropical conditions depending on the ambient temperature for *Plasmodium falciparum* or *P. vivax*, and much longer for *P. malariae* or *P. ovale*. Sporogonic development does not occur at temperatures below 16 or 18 °C. Infective female mosquitoes will generally remain infectious during their entire life and can therefore inoculate sporozoites at each blood feeding.[17]

Control: The current recommended treatments for malaria, namely artemisinin-combination therapies (ACT), Intermittent Presumptive Treatment (IPT) of pregnant women, together with vector control using primarily insecticide-treated nets (ITN), long-lasting insecticidal nets (LLINs), or indoor residual spraying (IRS) remain as effective methods for controlling malaria when used properly. Larviciding can be used in special situations (well known breeding sites, mainly man-made) and biological or genetical methods are still not operational in spite of great advertising.[19]

1.2.4.2 *Lymphatic Filariasis (LF)*

Lymphatic filariasis is regarded as the second most common global arthropod-borne infectious disease, with an estimated burden of 128 million infected people distributed over 78 endemic countries (see Figure 1.7) and an estimation of 1.3 billion people at risk from developing new active LF infection annually.[20] Like malaria, the predominance of LF infections are found in humid tropical areas of Asia, Sub-Saharan Africa, the western Pacific and scattered areas of the Americas. Although not fatal, LF is considered a leading cause of infirmity, permanent disability and chronic morbidity, often resulting in a societal stigma of disfigured victims.

This disease is caused by macroscopic nematode pathogens, of which *Wuchereria bancrofti* is responsible for 90% of human LF infections. The remaining 10% are due to two species of the genus *Brugia* (*B. malayi* and *B. timori*) and occur only in Asia. There are three variants of *W. bancrofti* recognized on periodicity patterns of circulating microfilaria (mf) found in the peripheral blood of humans, namely, the nocturnally periodic (NP), the nocturnal subperiodic (NSP) and the diurnal subperiodic (DSP) forms. Periodicity is based on the prevailing circadian distribution of mf in the peripheral blood, *e.g.*, the nocturnally periodic (NP) form presents the majority of mf by night (peak periodicity 22.00–03.00 hrs). The primary vectors of the NP filariae are nocturnally active mosquitoes, such as *Anopheles* species in rural areas and

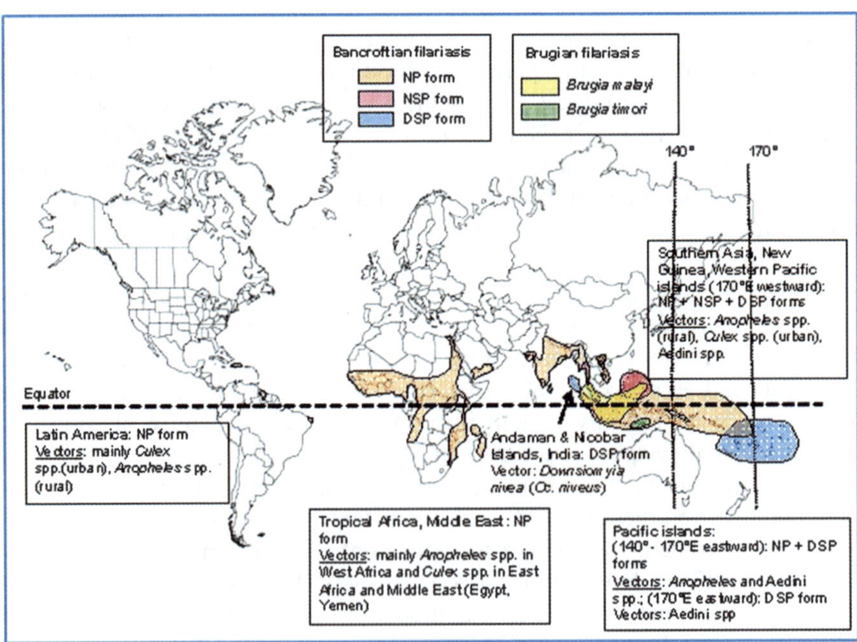

Figure 1.7 Lymphatic filariasis distribution around the world (Photo courtesy of Elsevier).[3]

Culex quinquefasciatus in urban settings. The NP variant is responsible for the vast majority of infections occurring worldwide in patchy foci distributed along the tropical and subtropical belt. The subperiodic microfilariae are strongly correlated to a transmission by vectors of the genera *Aedes* and *Ochlerotatus*, which are diurnally active species.

More than 70 species of mosquitoes within six different genera are known vectors of *W. bancrofti*, including *Anopheles* (43 spp.), *Aedes/Ochlerotatus/Downsiomyia* (approximately 20 spp.), *Culex* (6 spp.), and *Mansonia* (3 spp.).[3] Among the anophelines, 36 species are capable of both malaria and LF transmission, 26 of which are regarded as major LF vector species. Bancroftian filariasis and *Brugia malayi* are unique among the vector-borne parasitic diseases in that larval development can take place in several genera of mosquitoes. Three main zones of LF transmission are recognized herein: (1) West Africa, Southeast Asia (rural areas), New Guinea Island, Vanuatu and Solomon Islands, where *Anopheles* mosquitoes are the principal vectors; (2) East Africa, Middle East (Egypt, Yemen), Southeast Asia (urban zones), and the Latin American region (*e.g.* Haiti) where the infection is transmitted mainly by the *Cx. quinquefasciatus* and *Cx. pipiens* group; and (3) the south-western Pacific islands (including French Polynesia) and limited areas of Southeast Asia where Aedine (*Aedes, Ochlerotatus, Downsiomyia* spp.) vectors play a dominant role.

Brugian filariae are predominately found in rural locations and are vectored by *Anopheles* and *Mansonia* species for *B. malayi,* and by *An. barbirostris* for *B. timori. Brugia malayi* occurs from scattered areas of India (south and north-east) and Sri Lanka, to Southeast Asia and the Philippines. *Brugia timori* is restricted to a small group of islands of the Lesser Sunda Archipelago, primarily Timor and Flores.

The cycle starts with the absorption of mf by the female mosquito during the blood meal. They reach and cross the abdominal midgut (stomach) wall into the hemocoele to migrate to the insect's thoracic flight muscles to begin development. Microfilariae do not reproduce in the vector, but rather each worm completes two intermediate larval stages (L1 and L2) moults to become a third-stage (L3) infective parasite. The L3 eventually breaks free from the flight muscles into the hemocoele and ultimately ends up in the insect's head lodged in or near the labium of the proboscis. As for the malaria parasite, the filarial development within the mosquito takes approximately 10–14 days and is also temperature dependent. When the mosquito takes a blood meal, the 1.2–1.6 mm long L3 infective larvae will break free through the cuticle or emerge from the tip (labellum) of the labium onto the skin. In contrast to malaria parasites, filarial nematodes are not inoculated but deposited on the skin and they must actively enter the host body *via* an open portal (*e.g.*, the mosquito bite wound or a nearby break in the skin). High ambient humidity and skin moisture favour successful transmission. After entering the vertebrate host, the L3 is transported *via* the lymphatic vessels to the lymph nodes to begin development into mature adult worms (0.2 mm wide, up to 10 cm long). Contrary to *Plasmodium*, the mosquito acts as the intermediate host and humans serve as the definitive host for *Wuchereria* and *Brugia* species. It is thought that microfilariae survive and circulate freely in the blood of the human host for many months, possibly longer, while awaiting an opportunity of being picked up by mosquitoes.

Depending on transmission intensity, the LF infection is usually acquired early in childhood, although a period of ten to 20 years of exposure may be required before presenting the characteristic morbid manifestation called "elephantiasis", visible at adolescence and adulthood. Although the chronic physical phase of the disease afflicts only a small percentage of those infected, in its most apparent forms, LF morbidity can result in temporary or permanent infirmity which is often the painful and gross enlargement of the legs and arms, the genitals, vulva and mammary glands. Additionally, adult worms and mf can also cause internal damage and disease to other organs such as kidneys and lungs. The psychological and social stigmas associated with the disease are immense and it has a major social and economic impact in countries where 10–50% of men and up to 10% of women can be adversely affected due to permanent damage to the lymphatic system.[21]

Control: Since 1997, the Global Programme to Eliminate Lymphatic Filariasis (GPELF) has been directed to people living in at-risk communities by providing once annual oral treatment, using a two-drug combination, either albendazole + ivermectin, or albendazole + diethylcarbamazine (DEC) for the elimination of microfilariae (mf) in the blood and disruption of the adult female reproductive capacity.[20] It is expected that reducing the number of mf in humans will lower vector infection through their bite and therefore stop further transmissions. However, this doesn't occur with *Aedes* in Pacific Islands where *Aedes* become more infected when few mf are available ("facilitation phenomena") and strong vector control is thus also needed in such situations. It is generally assumed that LF elimination, in areas where *Anopheles* species are transmitting NP strains of *W. bancrofti*, will be relatively easy to achieve. Implementing synchronous and multifaceted strategies, with MDA and comprehensive vector control as central components, can stop both filarial and malaria transmission. As such, integrated control strategies targeting both diseases in areas sharing the same *Anopheles* vector species are highly recommended as the most cost-effective approach.

1.2.4.3   *Main Tropical Arboviruses*

Arthropod-borne-virus, more commonly called Arbovirus, is a virus transmitted to a vertebrate by hematophagous arthropods which constitute the biological vector. The arboviruses include different diseases for their symptomatology and especially their epidemiology (see Table 1.2). Their precise diagnosis is delicate and requires recourse to biological examinations in specialized laboratories. Approximately 110 viruses are pathogenic for humans, 40 of them are also the cause of identified animal diseases.

The tropism of the viruses explains the principal clinical symptoms observed. All the arboviruses present a certain neurotropism. Three general clinical pictures

Table 1.2 Classification of arboviruses.

Family	Genus
Togaviridae	*Alphavirus* (28 viruses, including chikungunya, O'nyong-nyong, Ross River, Sindbis, Mayaro, equine encephalitis)
Flaviviridae	*Flavivirus* (68 viruses, including yellow fever, dengue, Japanese encephalitis, West Nile, Kyasianur Forest disease, Omsk hemorrhagic fever)
Bunyaviridae	*Bunyavirus* (138 viruses, including Bunyamwera)
	Phlebovirus (43 viruses, including Rift Valley fever)
	Nairovirus (24 viruses, including Crimean-Congo hemorrhagic fever) + 41 unclassified viruses
Reoviridae	*Orbivirus* (69 viruses); *Coltivirus* (2 viruses); + 6 unclassified viruses
Rhabdoviridae	*Vesiculoviris* (18 viruses); *Lyssavirus* (16 viruses); + 36 unclassified viruses

can be observed: (1) Acute febrile syndromes ("dengue-like"): *alphavirus* (Chikungunya, O'nyong-nyong, Ross River, Sindbis and Mayaro); *flavivirus* (dengue and West Nile); *bunyavirus* (Bwamba, Bunyamwera and Tataguine); *phlebovirus* (Rift Valley fever); (2) Encephalitic syndromes: *alphavirus* (equine encephalitis); *flavivirus* (Japanese encephalitis and West Nile); (3) Hemorrhagic syndromes: *flavivirus* (dengue hemorrhagic fever, yellow fever, Kyasianur Forest, Omsk hemorrhagic fever); *phlebovirus* (Rift Valley fever); *nairovirus* (Crimean-Congo hemorrhagic fever).

The arboviruses affect vertebrates and arthropods; this is referred to as "horizontal transmission". In certain cases, a "vertical transmission" may occur, as in the case of transovarian or trans-stadial transmission, *i.e.* the viruses pass through the genital tractus and preimaginal stages and a new adult generation is infected. The arthropod remains infective all its life. Any climatic change is likely to involve important effects on the foci of the arbovirosis.

General arbovirosis prevention is difficult but actions can be taken such as: (1) monitoring the epidemiologic foci, *e.g.* human, vertebrate and vectors, to prevent any outbreak; (2) surveillance of the wild vertebrate hosts (illusory) or domestic ones (limited effect); (3) control of the wild vectors (impossible) or domestic or peri-domestic vectors (possible but difficult due to diversity of man-made breeding sites); (4) protection of the receptive human population (strongly recommended) based upon skin repellents or treated clothes, even mosquito nets, while vaccines are available only against yellow fever and Japanese encephalitis.

1.2.4.3.1 Alphavirus (Chikungunya, Equine Encephalitis)

1. *Chikungunya and O'nyong-nyong Viruses*
   - *Chikungunya* causes dengue-like syndromes which comprise fugacious rash and arthralgia or arthritis during five to seven days. Chikungunya virus prevails on the endemic mode, with great epidemics in Sub-Saharan Africa (*e.g.* Senegal in 1996 and 1997 and Democratic Republic of Congo in 1999–2000), in Southeast Asia, Indian sub-continent. Reservoir hosts are primates, domestic cattle and birds.

Vectors belong to the genus *Aedes*, particularly *Ae. aegypti*. Epidemics occurred in the Comoros Islands in early 2005 with extension to other islands of the archipelago: Mauritius, Mayotte and La Réunion, with *Ae. albopictus* as the vector.[5,22] Since 2005, massive outbreaks of chikungunya have also been reported in India.[23] After four to seven days of incubation, a brutal fever of 40 °C starts followed by intense joint pain of the member extremities, arthritis, cephalgias, and sometimes organ failure and heart attacks. There is a high frequency of asymptomatic infections. Serology may not be able to distinguish between chikungunya and O'nyong-nyong virus. The immunity is durable. Treatment is based on anti-inflammatory drugs. Vector control is difficult due to the resistance of the eggs to desiccation and the multiplicity of small man-made breeding sites suitable for *Aedes* development.

- *Virus O'nyong-nyong* occurs in Sub-Saharan Africa. The first recorded epidemics occurred in East Africa (1959–1963; 2 millions people infected), with more recent episodes in Uganda and Tanzania (1996–1997). Reservoir host: humans; vectors: *Anopheles gambiae* and *An. funestus*.

2. *Equine Encephalitis Viruses*
   - *Western Equine Encephalitis* (WEE) has symptoms which range from a mild flu-like illness to frank encephalitis, coma and death in man and equine (horses/donkeys). Occurs west of the Mississippi Basin. Closely related to eastern and Venezuelan equine encephalitis viruses. Reservoir hosts: birds; vectors: *Culex tarsalis*.
   - *Eastern Equine Encephalitis* (EEE) is a rare illness in humans, and only a few cases are reported in the United States each year. The severe cases involve encephalitis, inflammation of the brain and may progress into disorientation, seizures, or coma. Reservoir hosts: wild birds; vectors: *Culiseta melanura*, transmission to man and equine by *Aedes* mosquitoes.
   - *Venezuelan Equine Encephalitis* (VEE) occurs in tropical America, Haiti and Trinidad. After infection, equines may suddenly die or show progressive central nervous system disorders. Humans may present flu-like symptoms (high fevers, headaches). The young and elderly people can become severely ill or even die. Reservoir hosts: equines, rodents; vectors: *Culex, Aedes taeniorhynchus, Ae. serratus, Mansonia, Psorophora*.

3. *Mayaro Virus*
   The geographical distribution of the *Mayaro virus* is in the forest belts of the Caribbean, Central and South America (Amazon). Cases were reported from French Guiana (Kourou) in 2001. Syndromes include dengue-like syndromes, with arthralgia or fugacious arthritis. Reservoir host: howling monkey (*Alaouatta seniculus*); Vectors: *Haemagogus* mosquitoes.

4. *Epidemic Polyarthritis Virus*
   Epidemic polyarthritis virus is caused by two viruses: Ross River and Barmah Forest viruses. The former circulates on an epidemic mode in

Australia, New Guinea and in certain archipelagos of the southern Pacific, such as New Caledonia (epidemic in 1979–80). Barmah Forest virus occurs in Australia. Flu-like symptoms are common with fever and joint stiffness (arthralgia, arthritis or even acute polyarthritis). Vectors: *Aedes vigilax* and *Culex annulirostris.*

1.2.4.3.2 Flavivirus (Yellow Fever, Dengue, Japanese Encephalitis)

1. *Yellow Fever Virus* (YF)
   The YF remains a serious endemic disease and a constant threat in Sub-Saharan Africa and South America (including Trinidad). It does not prevail in Asia, Oceania, nor the Indian Ocean. The incidence of YF varies greatly from one year to another: in 1997, the WHO considered an annual incidence of YF of 200 000 cases, with 30 000 deaths per year. Many YF infections are mild, but the disease can cause severe, life-threatening illness. Symptoms can include high fever, chills, headache, muscle aches, vomiting, and backache. After a brief recovery period, the infection can lead to shock, bleeding, and kidney and liver failure with "black vomit". Liver failure causes jaundice (yellowing of the skin and the whites of the eyes), which gives YF its name. Symptoms start three to six days after being bitten by an infected mosquito. Yellow fever virus is diagnosed by a blood test. There is no specific treatment, but infected people should rest and drink plenty of fluids. Yellow fever virus can be prevented by vaccination. Epidemiology is complex as, in its natural environment, the YF virus circulates permanently within the monkey populations being transmitted by simiophilic mosquito vectors. Human cases occur only accidentally when in contact with a selvatic cycle. In 2001, YF epidemics occurred in Ivory Coast with the first cases reported in the western forests, then reaching the capital, Abidjan (epidemic with urban prevalence).[24] Reservoir hosts: monkeys in viremy phase and mosquito vectors which stay infective for life, some can even transmit the virus to their descendants.
   In Africa: In East Africa, there are three cycles: selvatic, rural and urban. There are three zones. (1) In forest zones, *Aedes africanus* is the vector of the selvatic YF which prevails between populations of monkeys. It is aggressive especially at twilight when the monkeys are already immobilized for the night. Under woods, it also can transmit the virus to man. (2) In rural zones, the principal vector is *Ae. simpsoni* which is active at daytime. They bite monkeys and are also aggressive for man. (3) In urban zones, *Ae. aegypti*, is the exclusive vector which ensures a strictly inter-human epidemic transmission.
   In West Africa, there are three different zones: endemic, emergence and epidemic zones. (1) The endemic zone is located in the forest belt (selvatic YF); the circulation of the virus is discrete; and the principal vector is *Ae. africanus.* (2) The emergence zone is located in the area of the mosaic

forest savannas; the selvatic circulation of the virus is variable in time and space and proceeds by cyclic paroxysms during which the human contaminations can be numerous (end of the rainy season). The principal vectors are *Aedes* of the Furcifer group, *Ae. luteocephalus*, *Ae.* subgroup *africanus*. (3) The epidemic zone is found in dry savannas and the Sahel; the virus circulates only episodically during epidemics according to the level of protection of the human population and the abundance of the main vector, *Ae. aegypti*, which also occurs in urban zones. Urban zones and dry savannas are both regarded as epidemic zones.

In South America: The American monkeys are sensitive to the YF virus and their rarefaction is at the origin of the epizootic waves. The vectors are *Haemagogus* mosquitoes. In the event of ecosystem changes (*e.g.* deforestation, building sites, *etc.*), *Haemagogus* can become active at the ground level, biting humans in their camp sites which is at the origin of sporadic YF cases. The urban transmission is ensured by *Ae. aegypti*. Three types of measurements are associated: (1) systematic vaccination of the exposed populations using vaccine 17 D, a vaccine consisting of live viruses (not recommended for pregnant woman); (2) vector control, particularly targeting *Ae. aegypti* and personal protection; and (3) insulation under mosquito net (*e.g.* patients, suspected cases).

2. *Dengue Virus*

Transmission of dengue virus occurs in all tropical and subtropical zones of the world. This is the main public health problem among arboviruses. There are four viral serotypes, dengue 1, 2, 3 and 4, which do not induce cross-protection. Currently, dengue occurs throughout the world, from India to Brazil, Venezuela to China, and on islands such as the West Indies (Martinique, Cuba, *etc.*) or French Polynesia, being reported in more than 100 countries. Worldwide, more than 50 million cases of dengue fever occur every year. Roughly 500 000 people with dengue haemorrhagic fever (DHF) require hospitalization each year, a very large proportion being children, and about 2.5% of those affected die. The increase in the number of dengue cases can be explained by uncontrolled demographic growth, wild urbanization, the absence of an adequate policy of water management, the virus propagation through travellers and commercial trades, and a general decrease in vector control programs.

Clinically, there are several dengue forms: asymptomatic dengue, classical dengue and the serious forms, in particular DHF and dengue shock syndrome (DSS), which can result in death, especially among children. The diagnosis is biological with the search for specific IgM, virus isolation and PCR. The treatment is symptomatic, but for DHF an intensive care unit is required. Man is the principal reservoir host and disseminator, although infected monkeys have been found in Asia and Africa. Vectors belong to the genus *Aedes*. Although *Ae. aegypti* is the major vector, *Ae. albopictus* plays an important part in rural and peri-urban areas and develops under temperate climates (*e.g.* dengue prevailed in Greece in 1927–1928).[25]

There are schematically two zones: (1) endemic zones, where the four serotypes circulate permanently, such as South-East Asia where the DHF is observed; and (2) epidemic zones, where a serotype spreads due to the movement of populations, such as Oceania, the Indian Ocean (all islands, except Madagascar), the eastern coast of Africa, tropical America and the Caribbean and Polynesia. West Africa represents a poorly explained situation with circulation of the virus among monkeys and rare human cases. Central Africa remains apparently dengue-free, although a lack of reliable diagnoses makes this uncertain. The Mediterranean Basin is again threatened by the recent introduction and spread of *Ae. albopictus*. Since 1989, several French overseas departments and territories have suffered from major epidemics with the emergence of serious forms of the dengue virus responsible for deaths, the most recent of which occurred in 2010 in Martinique with at least 16 deaths reported. In the Indian Ocean, recent epidemics have been reported in the Comoros Islands (1993), the Seychelles and La Réunion (2004). As there is no vaccine, a constant monitoring of the populations and vectors is required for implementing adapted vector control programs against *Aedes* using insecticides, larval habitat removal, and appropriate modes of water storage.

3. *Japanese Encephalitis* (JE)
Japanese encephalitis is an expanding disease distributed across the Asian continent, from India to the Philippines, north of China to Indonesia and New Guinea, especially in rural zones (*e.g.* in piggeries and rice plantations). The symptoms range from asymptomatic forms to encephalitis with coma and death or sometimes with very serious after-effects (*e.g.* children). Approximately 50 000 cases per year are reported, with 25 to 30% resulting in death. Reservoir hosts are wild or domestic birds (*e.g.* ducks) and pigs; vectors: *Culex*. A vaccine is available for local populations living in rural zones, travellers or expatriates staying in highly endemic rural zones. The main vectors belong to the Vishnui subgroup, especially *Culex tritaeniorhynchus*, along with *Cx. pseudovishnui* and *Cx. vishnui*.

4. *Murray Valley Encephalitis* and *Kunjin Virus*
Murray Valley encephalitis and Kunjin virus occur in New Guinea, Australia. Symptoms range from asymptomatic forms to encephalitis. Reservoir hosts: birds; vectors: *Culex*.

5. *West Nile Virus* (WNV)
West Nile virus has a wide distribution covering Sub-Saharan Africa, Madagascar, Middle East, India, North America and the Mediterranean Basin. The symptoms range from asymptomatic forms to fever, encephalitis with death or serious after-effects. Epidemics occurred in the United States (New York) in 1999, Israel in 2000, and again in the USA in 2002, 2003 and 2004 (California) with more than 7000 cases reported. In France (the Var department), there were equine and human cases in 2003 (meningo-encephalitis). Epidemics in horses occurred in the Camargue with human cases also reported (the vector was *Cx. modestus*).[26]

Reservoir hosts are mammals and birds, the major vectors are *Culex pipiens* and *Cx. tarsalis* in North America (south), and *Cx. modestus* in the Mediterranean Basin.

6. *St Louis Encephalitis Virus*
   St Louis encephalitis virus is found on the American continent (Canada to Argentina). Symptoms range from asymptomatic forms to deadly encephalitis (with a 30% death rate for adults over 75 years old). Reservoir hosts: birds; vectors: *Culex.*

7. *Wesselbron, Banzi* and *Zika viruses*
   Wesselbron, Banzi and Zika viruses are located in Sub-Saharan Africa. Syndromes are dengue-like and symptoms are mild and short-lasting (two to seven days). Reservoir hosts: wild and domestic mammals; vectors: *Aedes* mosquitoes.

1.2.4.3.3 Bunyavirus (Bunyamwera, La Crosse)

1. *Bunyavirus*
   Bunyaviruses, such as the *Bunyamwera, Ilesha* and *Bwamba* species, are found in Sub-Saharan Africa. Syndromes are dengue-like. Reservoir host is unknown; vectors: *Aedes* mosquitoes.

2. *Bunyavirus of the California group*
   Bunyaviruses of the California group are especially known in North America: encephalitis of California. Reservoir hosts: rodents, deer; vector: *Aedes.*

3. *Group C and Guama viruses*
   Group C and *Guama viruses* are distributed in Central and South America. Syndromes are dengue-like. Reservoir hosts: rodents; vector: *Culex* and *Aedes.*

4. *Nyando virus*
   Nyando virus is transmitted by *Anopheles funestus, Tataguine virus* agent of exanthematic fever in Central Africa; vector: *Anopheles gambiae.*

5. *La Crosse virus*
   La Crosse virus affects the central nervous system and can cause severe complications leading to death. Reservoir hosts: chipmunks and squirrels; vector: *Aedes triseriatus.*

1.2.4.3.4 Phlebovirus (Rift Valley Fever). *Rift Valley fever virus* (RVF) is at the origin of epidemics in Africa, Kenya (1930, 1997), Chad and Cameroon (1967), Egypt (1977), Mauritania (1987), Malagasy (1991), the Arabian Peninsula (2000). Its vector is *Culex pipiens*. In 2001, two strains of RVF virus were isolated on 50 French soldiers in Chad. Symptoms range from asymptomatic forms and mild forms with fever, muscle pain, joint pain and headache, to serious forms with ocular disorder, meningo-encephalitis or haemorrhagic fever. Reservoir hosts: livestock (RVF is a disease of herders, farmers, slaughterhouse workers and veterinarians); vectors: *Aedes, Mansonia*

and *Culex*. Transmission occurs either by mosquito bites or by contact with blood, secretions or the meat of sick animals. An inactivated vaccine has been developed for human use.

1.3 Ceratopogonidae (Biting Midges)

The Ceratopogonidae is a large and diverse family of small Nematocera flies, mostly crepuscular, commonly known as "biting midges". They occur in huge numbers and Kettle stated "One midge is an entomological curiosity, a thousand can be hell".[27] There is little information on their ecology and bionomics. Especially important are those vectors of filariae and viruses to domestic animals or humans (the *Culicoides* genus).

1.3.1 Systematics

The Ceratopogonidae family includes nearly 6000 known species, divided into at least 103 genera. The four genera which comprise species of medical and veterinary interest are: (1) *Leptoconops*: present in tropical and warm temperate regions, diurnal activity, important nuisance; (2) *Austroconops*: only one diurnal anthropophilic species in Australia, *A. macmillani*; (3) *Forcipomyia*: subgenus *Lasiohelea* comprises around 50 diurnal species biting vertebrates including man in tropical zones; and (4) *Culicoides*: cosmopolitan genus comprising more than 1300 species, they are crepuscular biters of mammals and birds, they constitute a great nuisance and are vectors of pathogens (mostly arboviruses).

1.3.2 Morphology

The eggs are elongated, 300 to 500 μm in length. Larvae (four stages) are aquatic, semi-aquatic (*e.g. Culicoides*) or terrestrial, slender (5–6 mm) and nematode-like. The head capsule is generally small, sclerotized, dark followed by three thoracic segments and nine abdominal segments. Pupae are long 2–4 mm, with a cephalothorax followed by an abdomen with a pair of respiratory trumpets. They are relatively unactive and do not feed. Adults are small flies of 0.6–5 mm (see Figure 1.8). Their legs are short and stout. The wings are deprived of scales but their pattern of pale and dark is of great taxonomic importance. Female mouthparts are short and small with armed proboscis mandibles which act as scissors for dilacerating skin to take blood. They have one to three spermathecae. Most species are brown or black.[1]

1.3.3 Biology

There is little information on the specific life cycle of Ceratopogonidae, especially in the Palaearctic region and their larval ecology is still unstudied (biting midges cannot be maintained in laboratory conditions).

The females lay their eggs on moist substrate or water. Breeding and larval habitats are not well known. The eggs are often joined to each other in a chain

Figure 1.8   Female *Culicoides sonorensis* blood feeding (Photo courtesy of P. Kirk
               Visscher).

of 60 to 250 eggs laid on mud, leaf litter, humus or plants near water, depending
on the species.[28] Incubation lasts two to 15 days according to temperature and
species, but certain species hibernate (diapause) at egg stage. Larvae are
important detritivores or predators in semi-aquatic and aquatic habitats of all
sizes, with either freshwater, brackish water or even seawater (*e.g.* estuaries,
mangrove or coastal ponds). The larvae are often localised in muddy habitats,
in wet sites rich with organic matter of plant origin (*e.g.* tree holes, crab holes,
trunks of banana trees, cacao, mushrooms, *etc.*). Some species develop well in
sandy areas near the sea, which constitutes a nuisance on the beach. The larval
development takes two to three weeks (in tropical regions), but can reach seven
months (in temperate regions). The larvae feed on plants in decomposition.
Pupae are generally sluggish and this stage lasts from two to ten days, then the
adult hatches. The adults feed on flower and plant nectar; only the females are
hematophagous, taking their blood meal from hot or cold blooded vertebrates.
Certain species feed on hemolymph of other insects (*e.g. Culicoides anophelis*
bites *Anopheles* mosquitoes in the Southeast Asia). Biting activity is variable
according to species, but it is generally done during crepuscular or twilight,
though some night biting also occurs. Biting behavior is usually exophagic.
Swarms of biting midges will come to people for biting the head or any other
exposed areas, their small oral parts do not allow them to bite into blood vessels
such as mosquitoes, but they can dilacerate skin to suck the blood (pool
feeding). This method of blood feeding allows the transmission of skin filariae.
Their active dispersal is usually short and limited ($<500\,$m), but passive dispersal through wind can spread populations over large areas. Little is known
about their natural longevity; *Culicoides obsoletus* live more than 50 days in
captivity, but one month seems to be their average longevity. Adults are cap-
tured using animal-baited traps or by black light traps. The nuisance or damage
of biting midges is due to their occurrence in large numbers, which affects
tourism, forestry or farming (*e.g.* as many as 10 000 *Culicoides nubeculosus* per
cow have been found in Denmark). The bites of ceratopogonidae are often
painful with intense local rash, oedema and pruriginous reactions being able to
persist up to three weeks.

Transmission of filariae

Biting midges are vectors of filariae considered low or non pathogenic for humans.

Mansonella perstans is a filaria widespread on the Atlantic side of tropical America, the West Indies and Sub-Saharan Africa (East, Central and West Africa to Zimbawe), also found under the term of *Dipetalonema perstans*. In Sub-Saharan Africa, the principal vector is *Culicoides milnei* (or *C. austeni*) in forested area with a nocturnal activity. Larvae develop in six to nine days in the vector. *Culicoides garnhamii* are found in zones of savanna and banana plantations.

Mansonella streptocerca is a dermic filaria from the dense forests of Central and West Africa transmitted by *C. garnhamii*. The larvae develop in seven to ten days into the vector encountered in Ghana, Burkina Faso, Nigeria, Cameroon and the Democratic Republic of Congo.

Mansonella ozzardi is an American filaria present in the forests of the Caribbean Islands, Central and South America. The larvae develop in seven to nine days in *Culicoides furens*, *C. paraensis* in Brazil, *C. phlebotomus* in Trinidad, and *C. guttatus* in Surinam. *Mansonella ozzardi* can also be transmitted by black flies in Brazil and Colombia.

Transmission of Arboviruses

Worldwide, more than 50 arboviruses have been isolated from Culicoides, most within the following families: Bunyaviridae (20 viruses), Reoviridae (19 viruses) and Rhabdoviridae (11 viruses). Many of these viruses have been isolated from other arthropod groups and their association with Culicoides species is probably incidental. The only significant human viral pathogen transmitted by Culicoides is Oropouche virus (OROV) and the most important ones infecting animals are African horse sickness virus (AHSV), Bluetongue virus (BTV), Epizootic hemorrhagic disease virus (EHDV) and Equine encephalitis virus (EEV).

Bluetongue virus is an orbivirus infecting ruminants, but severe disease usually occurs only in certain improved breeds of sheep and cattle. The vector species are *Culicoides sonorensis* in North America, *C. imicola* and *C. bolitinos* in Africa, species of the Obssoletus and Pulicaris groups for western and northern Europe. African horse sickness virus can be devastating with mortality rates frequently exceeding 90% in horses from Sub-Saharan Africa. The virus belonging to the *Orbivirus* genus and its vectors are *C. imicola* and *C. bolitinos*. Epizootic hemorrhagic disease virus is also an orbivirus which occurs in North, Central, and South America, Africa, Southeast Asia, Japan and Australia, and its vertebrate hosts include domestic and wild ruminant species. OROV is a member of the Simbu group of bunyaviruses and is the cause of one of the most important arboviral diseases in the Americas. In the field, OROV has been frequently isolated from *C. paraensis* which occurs in high density during epidemics and bites humans both indoors and outdoors. OROV is responsible for fever in the Brazilian Amazon (80 000 cases in 1979–1980).

1.3.4   Control

Attempts to control biting midges are based on insecticide spraying of their larval habitats or mechanical control such as land drainage or immersion of habitats, but the effects are of short duration. Large scale control of biting midges is nearly impossible to achieve as their breeding sites are often widely scattered or sometimes unknown, and adults are reintroduced by the wind into treated areas. In such instances, protection measures using repellent treated screens, netting, personal protection with skin repellents, treated cloths or treatment of animals with pyrethroids or ivermectin, are essential.

1.4   Phlebotominae (Sandflies)

1.4.1   Systematics

Phlebotomine sandflies belong to the family of Psychodidae (Diptera, Nematocera). These insects with complete metamorphosis are of a great medical importance as biological vectors of the leishmaniasis (kala-azar, oriental sore, espundia, *etc.*). There are more than 600 species of sandflies classified into five genera. The two genera of greatest medical importance are: (1) *Phlebotomus*: distributed in the Old World (*e.g.* the Oriental, Afrotropical and Palaearctic regions), they bite mammals and humans and are vectors of leishmaniasis; and (2) *Lutzomyia*: occurring in the New World, its anthropophilic species are vectors of visceral leishmaniasis in South America. There are approximately 17 species of the genus *Leishmania* distributed in 88 countries throughout the five continents, with 350 million people living in at-risk areas of leishmaniasis. It has been estimated that 12 million people are affected by leishmaniasis. There are four main forms of the disease: (1) the benign form causing permanent cutaneous lesions; (2) the intermediate form, which attacks the muco-cutaneous membranes, disfiguring patients for life; (3) the diffuse cutaneous form; and (4) the visceral form which is fatal if left untreated. Currently, the estimate of the world annual incidence of cases of leishmaniases ranges between 1.5 and 2 million, being distributed between 1 to 1.5 million cutaneous cases and about 500 000 visceral cases.

1.4.2   Morphology

Eggs are elongated, ovoid and slightly curved; their size is 300–400 µm. Larvae resemble that of a caterpillar, with a well-sclerotized head capsule with robust mandibles, followed by a thorax and an abdomen. Pupae measure 3 mm. The adult is a delicate insect of small size (1 to 4 mm), yellow grey or brownish, strongly hairy with long and slender legs (see Figure 1.9).

1.4.3   Biology

The eggs (around 50) are laid separately in various types of habitats, such as cracks in the ground, burrows of small mammals, caves, slits of walls and the

Figure 1.9 Female *Phlebotomus papatasi* sandfly (Photo courtesy of CDC/Frank Collins/James Gathany).

grounds of dwellings or cattle sheds. Habitats are generally quiet, dark, cool and damp places with the presence of organic detritus. The egg incubation lasts 4 to 17 days according to temperature and season. Larvae are sedentary, saprophagous and phytophagous. There are four larval stages and the last moult produces the pupa which remains fixed by its abdominal end in the larval habitat and does not feed. The immature development lasts 20 to 75 days.[1] The adults encounter each other in habitats that are calm, in the proximity of vertebrate hosts (blood meals), and near favourable larval habitats. Females are hematophagous (pool feeding lasting 10–30 min). They can be found in both domestic and natural (*e.g.* associated with reptiles and rodents) environments. The majority of the species are rather zoophilic (*e.g.* bat, rodents, carnivorous, marsupials, birds, reptiles, *etc.*), but the main vectors of leishmaniasis, *Lutzomyia intermedia* and *L. longipalpis* in America and *P. papatasi* in the Mediterranean basin, are anthropophilic and endophilic. Sandflies are generally active at sundown or at night, but they can be diurnal in the canopy (*e.g. L. trapidoi* in Panama). They have a short flight (< 1 km), but considerable dispersion with the wind is possible.

Transmission of *Leishmania*

Sandflies are the only vectors of *Leishmania*, a protozoan parasite.[29] More than 50 species of sandflies are involved in the transmission of several *Leishmania* species to man, such as species of the *Lutzomyia donovani* complex (*e.g. Le. donovani, Le. infantum* and *Le. chagasi*); the *Le. mexicana* complex (*e.g. Le. mexicana, Le. amazonensis* and *Le. venezuelensis*); *Le. tropica*; *Le. major*; *Le. aethiopica*; species of the sub-genus *Viannia* (*e.g. Le. braziliensis, Le.*

guyanensis, *Le. panamensis* and *Le. peruviana*). Morphologically indistinguishable, species within a complex require molecular techniques for reliable identification.

The human parasites at the amastigote stage are ingested during the bite, and they transform into promastigotes with flagellum and multiply within the sandfly gut. The promastigotes are highly motile and they migrate forward to the oesophagus and the pharynx, and gain the oral parts about 9–10 days after the blood-meal. *Leishmania* parasites are deposited passively on the wound of the bite. The entire development lasts up to two weeks, depending on the *Leishmania* species and temperature, and it stays infective throughout the life of the infected sandfly. *Phlebotomus ariasi* of southern France (see Table 1.3) infected by *Le. infantum* can normally survive 30 days carrying out three to four gonotrophic cycles (3–4 changes to transmit).

Visceral leishmaniasis or kala-azar: More than 90% of the world's incidences of visceral leishmaniasis occur in India, Bangladesh, Nepal, Sudan and Brazil, but cases are also found along the Mediterranean basin, the Middle East, Central Asia and northern China (see Table 1.3). The parasites are *Leishmania* of the *Donovani* complex, especially *Le. infantum* (Mediterranean basin), *Le. donovani* (India and Bangladesh), *Le. archibaldi* (East Africa) and *Le. chagasi* (Latin America). The hosts are man (children), dog, fox, jackal and rodent. The vectors are *Phlebotomus major* (Israel), *P. perniciosus* (Northern Africa), *P. longicuspis*, *P. ariasi* (Southern France), *P. chinensis* (Northern and Eastern China, Central Asia) and *P. sergenti* (Afghanistan).

Cutaneous (CL) and Muco-Cutaneous (MCL) Leishmaniases

Ninety percent of cases of CL occur in Afghanistan, Saudi Arabia, Brazil, Iran, Peru and Syria; 90% of the MCL cases occur in Bolivia, Brazil and Peru, with 1 to 1.5 million new cases in the world each year. The main vectors belong to the genus *Lutzomyia*, subgenus *Nissomyia*. According to the region, these leishmaniases differ quite markedly in their clinical and epidemiological aspects.

Table 1.3 Main regions and foci of occurrence of Leishmaniasis with their respective vertebrate hosts and some of their vectors species.

Regions	*Foci*	*Vertebrate hosts*	*Vectors*
Palaearctic	Mediterranean	Dog, human, wild carnivorous	*P. perniciosus* *P. ariasi*
Oriental	Central Asiatic	Wild carnivorous, dog, human	*P. longiductus* *P. smirnovi*
	Chinese (northern, central)	Dog, human	*P. chinensis*
	Indian	Human (only)	*P. argentipes*
Afrotropical	East-African	Human, wild carnivorous, wild rodents	*P. orientalis* *P. martini*
Neotropical	Latin American (South & Central)	Dog, human, wild carnivorous	*Lutzomyia longipalpis*

(1) Oriental sore of humid form, the parasite is *Le. major* (= *Le. tropica major*) and it occurs in semi-desert and rural regions of northern Sahara (*e.g.* Algeria, Tunisia), Arabia, Central Asia (*e.g.* Turkmenistan, Uzbekistan), Iran, Afghanistan and India. The vectors are *P. papatasi, P. caucasicus* (Central Asia, Iran), *P. mongolensis* (Iran) and *P. saheli* (India).

(2) Oriental sore of dry form, the parasite is *Le. tropica* (*Le. t. minor*), the reservoir host is the dog. This form of leishmaniases occurs in urban areas along the Mediterraean basin, Middle-East, Central Asia (*e.g.* Azerbaidjan, Uzbekistan, Turmenistan), Iran, Afghanistan and Pakistan. The vectors are *P. papatasi* (France, Tunisia, Turkey), *P. perfiliewi* (Italia), and *P. sergenti* (Portugal, Spain, Cyprus, Iran).

(3) Tegumentary diffuse leishmaniasis is present in the mountainous areas of Ethiopia, Kenya, Tanzania, and Namibia, the parasite is *Le. aethiopica* and the vectors are *P. longipes* in Ethiopia, and *P. pedifer* in Ethiopia and Kenya.

(4) Chiclero ulcer is due to the parasite *Le. mexicana mexicana*. This is a benign leishmaniasis found in Mexico (*e.g.* Yucatan, Oaxaca), Guatemala, Honduras and Belize among workers in rubber plantations. In the Yucatan, the vector is *Lutzomyia olmeca*. There is a similar infection due to *Le. m. amazonensis* present in sporadic or epidemic form in Brazilian Amazon, Bolivia and Trinidad; the vector is *Lutzomyia flaviscutellata*.

(5) Panama leishmaniasis affects wild mammals of Panama, Costa Rica, Honduras, especially sloth (*Choloepus*), man is infected when deforestation occurs; the parasite is *Le. braziliensis panamensis* and the vectors *Lutzomyia trapidoi, L. gomezi, L. ylephiletor, L. panamensis*.

(6) Pian Bois occurs in forests of Guyana and bordering Brazilian regions, the parasite is *Le. b. guyanensis,* the vector *Lutzomyia umbratilis*, and the reservoirs are sloths and marsupials.

(7) American tegumentary diffuse leishmaniasis resembles the equivalent Ethiopian leishmaniasis, its parasite is *Le. m. pifanoi*, the main vector is most likely *Lutzomyia complexa*; it occurs in Brazil and Venezuela. In the Andes (at altitudes of 800 to 1800 m), another form due to *Le. garhnami* is transmitted by *Lutzomyia towsendi*.

(8) Espundia is a serious muco-cutaneous leishmaniasis occurring in sporadic or epidemic form, it is found in low altitude areas of the Amazon region of Brazil, Colombia, Peru, Ecuador, Chile, Bolivia, Paraguay and Venezuela; its parasite is *Le. b. braziliensis;* rodents are the natural reservoir and the vectors are *Lutzomyia wellcomei, L. intermedia* and *L. pessoai.*

(9) Uta is a dry leishmaniasis due to *Le. b. peruviana* with a peridomestic transmission; it occurs in the Andes (at altitudes above 1000 m) of Peru and Bolivia, the reservoir are dogs; the vectors *Lutzomyia peruensis* and *L. verrucarum.*

Transmission of *Bartonella*

Bartonella bacilliformis is responsible for a human infection present in certain dry valleys on the Western slope of the Andes (at altitudes of 500 to 3000 m) of Peru, Ecuador and Colombia. There are two clinical forms: (1) Peruvian Verruga and (2) the fever of Oroya or Carrion disease. The principal vector is *Lutzomyia verrucarum*, the other vectors are *L. noguchii, L. peruensis* and *L. pescei*.

Transmission of arboviruses
At least 25 viruses are transmitted by sandflies and are responsible for sandfly fevers. These include the papataci fever, or "three-day fever", caused by distinct virus serotypes (Naples and Sicilian) and which results in acute febrile illness in man for approximately two to four days. These sandfly fevers occur along the Mediterranean basin (vector: *Phlebotomus papatasi*), Middle East (*e.g.* Iran, Afghanistan), Central Asia, Pakistan, India, Oriental Africa and tropical America.

1.4.4   Control

The principal methods used in sandfly control are based on the application of insecticides in the peridomestic environment, especially houses, impregnated bednets and curtains and DDT is still one of the most used. It has been noticed that in malarious areas when house sprayings were stopped (after a malaria eradication program), the leishmaniasis situation greatly worsened (North Africa). When sandflies are exophilic or bite away from human dwellings, insecticide control is inefficient. Some successful programs against sandflies have involved integrated controlled targeting of both vector and reservoir, such as rodents in arid regions of the Old World, or dogs in Old and New World foci.[30]

1.5   Simuliidae (Blackflies)

1.5.1   Systematics

Blackflies are Diptera Nematocera, family of Simuliidae with more than 1500 species distributed worldwide (Holarctic, Palaearctic, Nearctic, Neotropical, Afrotropical, Asiatic, and Australian regions). The anthropophilic species belong to the genera *Simulium*, *Prosimulium* and *Austrosimulium*. In the *Simulium* genus, there are more than 1200 species in the world. This genus includes the most important vectors of human onchocerciasis or "river blindness", such as species of the *Simulium damnosum* complex (*e.g.* *S. neavei* and *S. woodi* in Africa), *S. ochraceum*, *S. exiguum* and *S. metallicum* in America. The *Prosimulium* genus has 110 species distributed in the Holarctic and eastern and southern African regions. The *Austrosimulium* genus includes 25 species of the Australian region (Australia, New Zealand) with several species being anthropophilic, such as *A. pestilens*.

1.5.2   Morphology

Eggs have a sub-triangular form of 0.15 to 0.30 mm in length. Larvae have a cylindrical shape of grey or brown colour; they measure from 4 to 12 mm in

Figure 1.10   Adult blackfly.

length. They have a head capsule with well-developed, crushing mouth parts. The larva is fixed on its support by hooks present on its abdomen. The larva is mobile with characteristic movements. Pupae are immobile, enclosed in a silk cocoon which remains fixed on its support. Adults are bulky, dark and similar to a small fly of 1 to 6 mm in length (see Figure 1.10).

1.5.3   Biology

The major bio-ecological characteristic of the blackflies is the development of the immature stages in running freshwater habitats. The female lays about 150 to 500 eggs on a partially immersed support: stone, various plants or branches.[1] The female generally lays four to five batches of eggs during its lifetime. The egg incubation is short (two to seven days), except in temperate climate (diapause of several months). The larval phase comprises six to eight stages which last from one to two weeks in tropical zones, but there can be hibernation in a cold period/zone. The larvae live in great numbers anchored to immersed inorganic (*e.g.* stones) or vegetational substrates (*e.g.* plants, branches), or even the bodies of other arthropods (*e.g.* freshwater crabs, shrimps, dragonfly larvae). The larvae are mobile as they use their circle of hooks and the prothoracic pseudopodes for moving. The larvae are especially prone to live in strong running water (0.30 to 1.50 m s^{-1}), that is highly oxygenated, such as cascades, rapids, dams, *etc.* The pupa is motionless and does not feed (stage duration: 2–10 days). Emergence of the adult takes place at daytime in about one minute. Females bite (pool feeding) during the day, outdoors and they visually target their hosts. Blackflies are not exclusively anthropophilic but rather opportunistic. There are marked seasonal variations with important pullulations representing a considerable nuisance. Their longevity is about two to three weeks. Blackflies develop in zones of savannas, forests and mountainous areas, in the torrents at heights of up to 4520 m in Kenya and 4700 m in the Andes, but the anthropophilic blackflies do not seem to go above 1500 m. Blackflies are excellent fliers with an active flight of several kilometers (15 to 35 km, and up to 80 km).

Transmission of *Onchocerca volvulus*

The human onchocerciasis is due to the filarial nematode worm, *Onchocerca volvulus*, responsible for a serious filariasis where man provides the reservoir of parasites. Onchocercal blindness is endemic in savanna and rural areas. The microfilariae (mf) of *Onchocerca volvulus* occur in the human skin and are ingested by the blackfly when sucking blood. The vector can absorb several hundreds of mf (315 to 360 µm in length) which will go into the blackfly midgut. They will then pass into the hemocoele and gain the thoracic wing muscles, where they will transform into stage 1 and 2 larvae (400 to 500 µm). After a second moult, the mobile larva stage 3 (450 to 800 µm), which is the infective stage for man, migrates towards the proboscis (head) of the blackfly. Larvae 3 are released on the skin of the host during the next blood meal. The development of *O. volvulus* in blackflies lasts from 10 to 14 days. The longevity of blackflies is reduced when they are extensively infested by microfilariae.

Onchocercal blindness is a non-fatal dermal and ocular disease caused by *O. volvulus* which prevails primarily in foci along streams (vector habitats), over vast areas of tropical western and equatorial Africa (at a latitude of 15° N to 13° S), Yemen and tropical America (*e.g.* Mexico to Brazil). Estimations report 20 to 30 million infected people in the world, including 95% in Sub-Saharan Africa. In some villages of West Africa, 10% of the population can be blind and up to 30% of men of reproductive age. Onchocerciasis has been considered as the major cause of depopulation of many fertile valleys in West Africa, with serious economic consequences and was one of the reasons the extensive onchocerciasis control program (OCP) took place.

The vectorial efficacy of blackflies is high and the critical density necessary in order to maintain transmission is low. In Africa, transmission is especially pronounced due to *S. damnosum* s.l. in the western regions, the *S. neavei* group in the eastern regions and the Democratic Republic of Congo, as well as *S. albivirgulatum*. In West Africa, *S. damnosum* includes two species in the savanna areas (such as *S. damnosum* s.s. and *S. sirbanum*) and four species in the forest areas (*S. damnosum, S. yahense, S. sanctipauli* and *S. soubrense*). Their dispersion and longevity would explain the epidemiology of onchocerciasis according to their biotope (see Table 1.4). *Simulium yahense* remains close to their habitat, in cool and shaded, small rivers; *S. sanctipauli* and *S. soubrense* can fly long distances (several hundreds of meters) from their larval habitats which are large and sunny rivers; *S. damnosum* and *S. sirbanum* move

Table 1.4   The three environmental and related vectorial systems of the human onchocerciasis.

Foci	*Vectors*
Savanna	*S. damnosum s.s.; S. sirbanum*
Forests: large rivers	*S. sanctipauli* (forest); *S. soubrense* (forest + forest fringe)
Forests: streams, small rivers	*S. yahense, S. squamosum* (forest and humid savanna)

only a few kilometres along their rivers in the gallery forest of savanna zones, but the wind can push them on great migratory flights which will participate in the recolonisation of the treated sites.

Longevity is higher for the savanna species than forest species, with a maximum longevity of one month. Blackflies bite at daytime and *S. damnosum* s.l. is zoo-anthropophilic. In Eastern and Central Africa, the vectors belong to the *S. neavei* group whose larvae are associated to crabs, but in Sudan and Ethiopia it seems that the transmission is still due to S. *damnosum* s.l. The dispersion of *S. neavei* is less than that of *S. damnosum*. The species of the *S. neavei* group are closely linked to the forest biotope. In Yemen, there are foci which extend to Saudi Arabia and transmission is ensured by *S. damnosum* s.l. In America (onchocerciasis or Robles disease), several foci are known in the mountainous areas of Mexico (Oaxaca, Chiapas), Guatemala (three foci near the Pacific side), Venezuela (three foci in the north), Colombia and Ecuador (one focus each) and Brazilian Amazon. In Mexico and Guatemala the foci are at altitude (450–1500 m) in forest belt or coffee plantations; the main vector is *S. ochraceum* which develops in small, shaded streams. It is an anthropophilic species, diurnal, exophagic, but which can also bite at night, indoors. Longevity and dispersion are important. Two other vectors are also incriminated: *S. metallicum* and *S. callidum*. In Venezuela, onchocerciasis was observed at altitude, as well as on plains, in forests and on open land. The vectors are *S. metallicum, S. exiguum* and *S. ochraceum. Simulium metallicum* is quite anthropophilic, it develops in medium size rivers (<5 m broad) whereas *S. exiguum* develops in the large rivers and has a greater longevity. In Colombia, the disease is restricted to the wet southern forest, the vector is *S. exiguum* with strong anthropophily. In the Brazilian Amazon, the disease occurs in hilly areas covered with dense forest near the Venezuelan border, the vectors are *S. oyapockense* in lowlands and *S. guianense* in highlands. In Ecuador, there is a focus in the north with *S. exiguum* as the main anthropophilic and endophagic vector, as well as *S. quadrivittatum*.

Transmission of other pathogens

Blackflies are suspected to play a role in the transmission of Venezuelan equine encephalomyelitis (VEE) arbovirus. They are also involved in the transmission of pathogens to different animals: *S. ornatum* vector of *Onchocerca gutturosa* parasite of cattle in England; *Austrosimulium* sp. vector of *Onchocerca gibsoni* among Australian cattle; and *S. amazonicum* is the vector of another filarial, *Mansonella ozzardi*. Blackflies can also transmit the virus responsible of myxomatosis in rabbits.

Nuisances

In Canada, there have been reports of violent reactions to blackfly aggressions with fever, cephalgias, nauseas and local oedema lasting 48 hours. In Brazil, the "hemorrhagic fever of Altamira" due to the saliva of blackflies in cases of pullulation can be fatal to children. In Guatemala, the pullulation of blackflies can stop the work in coffee plantations. Pullulation is also a nuisance in temperate countries, such as Russia (Siberia), France, Canada and the United States.

1.5.4 Control

Three methods of control can be used: (1) physical control such as the destruction of larval habitats by environmental modifications; (2) chemical control based on insecticide spraying; and (3) biological control by using bio-larvicides. Physical control is sometimes possible by removing substrate or by altering river discharge or configuration. The construction of dams has a great impact on the populations of blackflies (*e.g.* damming the Niger River decreased the population of the *S. damnosum* complex). Chemical control is based on regular spraying of insecticides/larvicides by terrestrial or aerial application, according to the nature of the habitats and their size. The residual effect of these insecticides, however, is low (about 10 days), application can require considerable skill and costs can be very high. Sometimes several insecticides are used in rotation to avoid problems of resistance; the basic larvicide used is temephos (Abate®), growth regulators are also used. Biological control is essentially based on *Bacillus thuringiensis* H14, which was widely used in the OCP program.

The first Onchocerciasis Control Program (OCP) was launched by the World Health Organization in 1974 with the sponsorship of the World Bank, as well as the Program of the United Nations. Initially founded on vector control, this program was directed towards chemotherapeutic control, especially with the discovery of Mectizan®. The Onchocerciasis Control Program was extended across 11 countries in West Africa and completed in 2002. It was a true success because it practically reduced to zero the transmission of the onchocerciasis parasite in West Africa.[31,32] More recently, two other control programs against the "river blindness" were launched in 19 African and six American countries.

1.6 Glossinidae (Tsetse Flies)

1.6.1 Systematics

There are about 30 species of tsetse fly belonging to the genus *Glossina* (Diptera, Brachycera) subdivided into three subgenera: *Austenina* (or *G. fusca* group), *Nemorhina* (or *G. palpalis* group) and *Glossina* (or *G. morsitans* group), of which the latter two are of medical importance. The *Nemorhina* subgenus includes tsetse fly species found in forest galleries, vegetation close to rivers, lakes, and mangroves of Western and Central Africa. The main vectors are *G. palpalis palpalis*, *G. p. gambiensis*, *G. tachinoides* and *G. fuscipes quanzensis*. The *Glossina* subgenus includes xerophilous or savanna species occurring in Central, West and East Africa, associated to woodlands, deforested savannas, but seldom associated to cultivated lands.

Tsetse flies are confined to tropical Africa (at a latitude of 15° N to 20° S), but absent from Madagascar. Some species have a broad distribution like *Glossina morsitans* present in Eastern, Central and West Africa, whereas *G. palpalis* is found only in West Africa. Tsetse flies are vectors of human (sleeping sickness) and animal trypanosomiasis. The most important species as vectors are *G. palpalis*, *G. tachinoides*, *G. fuscipes*, *G. pallidipes*, *G. swynnertoni* and *G. morsitans*.

1.6.2 Morphology

The first and second instar larvae are found directly in the uterus of the female. The female lays a third-instar larva which measures 1 cm, with the shape of a maggot. This free-living third-instar larva has a brief existence. The pupa is protected by a dark and hard "shell". The adult tsetse flies have a long piercing proboscis and their wings are closed over the abdomen as scissors (see Figure 1.11). The length of the adults ranges from 8 to 15 mm.

1.6.3 Biology

The immature development of the tsetse fly is peculiar as only one egg develops at a time. The first instar larva hatches *in utero* after three to four days, the second and third instar larvae follow one another in eight to twelve days. The larva stage 3 (8–9 mm in length) is deposited by the female; this reproduction is referred to as "adenotrophic viviparity".[1] A female will lay six to ten larvae during its lifetime. The first larva is deposited approximately 16–20 days after adult emergence. The larva is deposited on the ground in shaded places and in soft soil where it hides, and 15 minutes later it becomes a pupa. This stage lasts about 30 days according to the species, sex and ecological conditions. The imago emerges from the pupa and it will take its first blood meal one to two days after the adult

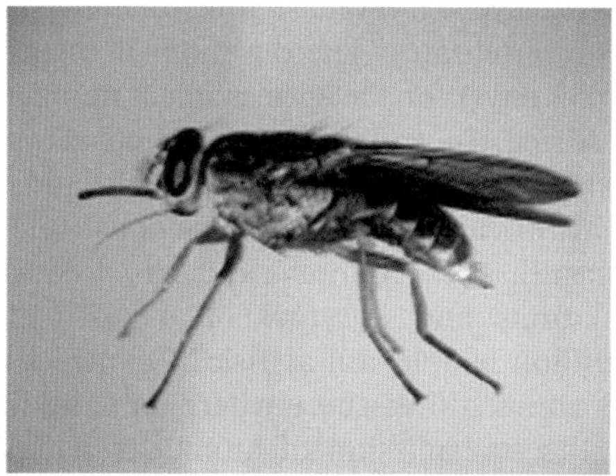

Figure 1.11 Adult tsetse-fly (Photo © IRD, Michel Dukhan).

emergence. Both females and males are hematophagous and bite men, domestic and wild mammals, as well as reptiles and birds. Tsetse flies do not limit themselves to a single host, but the majority of species have food preferences. *Glossina palpalis* bites human and reptiles, *G. tachinoides* bites human and cattle, and *G. longipennis* bites the elephants, rhinoceroses and hippopotamuses. They take a blood meal every two to four days (shorter periods during the dry season and longer periods during the rainy season extending to 10 days). They bite during the daytime. Mating takes place only once. The adult lifespan is about two to three months (up to five to six months). Tsetse flies are capable of flying at 25 kilometres per hour but they cannot sustain flights for more than a few minutes. The vision of tsetse flies is important as they are attracted by dark colours and this is used to catch them with black and blue traps.

Transmission of *Trypanosoma*

Eight species of *Trypanosoma*, a protozoan blood parasite, are transmitted by tsetse flies. Two species are pathogenic for humans and six for domestic and wild animals. These parasites are responsible for human or animal trypanosomiasis, also called "sleeping sickness". The human pathogenic species are *Trypanosoma brucei gambiense* and *T. brucei rhodesiense*. The animal pathogenic species are *T. vivax* (cattle, sheep, goats of tropical Africa); *T. uniform* (cattle, goats, sheep, buffaloes, antelopes and giraffes from south of the equator); *T. congolense* (domestic and wild mammals of Eastern Africa); *T. simiae* (pigs and dromedaries of limited distribution); *T. suis* (domestic and wild pigs); and *T. brucei brucei* (domestic and wild mammals, very pathogenic for equidae and dogs). The main vectors are *Glossina palpalis*, *G. fuscipes*, *G. tachinoides*, *G. morsitans*, *G. pallidipes* and *G. swynnertoni*.

Tsetse flies contract the *Trypanosoma* parasite (trypomastigote form) while feeding on an infected subject (human or animal). It is considered that the more infective meals are those taken by the young flies (1 to 2 days). The tsetse flies digest the blood-meal and the trypomastigotes multiply. They leave the midgut and change into the epimastigote form which multiplies in the salivary glands where they evolve in metacyclic trypomastigote ("metatrypanosomes") which is the infective form for vertebrates. The duration of the cycle in the tsetse fly takes 30 days (up to 53 days) for the species infecting man.

In intertropical Africa, in spite of nearly one century of control, the sleeping sickness still prevails in endemic and epidemic forms in the majority of the old historical foci. This disease constitutes a crucial problem in terms of public health and economy, and has a considerable impact on both human populations and livestock. It is estimated that there are approximately 50 million people still at risk of contracting this disease, but there are very many zones still insecure where no valid medical information exists. In addition, the transmission of animal trypanosomiasis represents an important obstacle to economic development. In humans, sleeping sickness is characterized in its early stages by an absence of clinical symptoms, there is a long clinical latency and there can be some "trypanotolerance", but clinical expression differs according to the particular trypanosome species. *Trypanosoma gambiense* seems to infect only man. The principal reservoir of parasite remains men with "asymptomatic carriers". The disease known as "Gambian type" is thus acquired primarily in the vicinity of humans: along backwaters near villages which provide an ideal site for tsetse flies, in villages bordered by orchards (cacao, coffee, *etc.*), or trails going from village to plantation and passing by rivers and forest galleries. The main vectors are *G. palpalis* and *G. tachinoides* in West Africa and *G. fuscipes* in Central Africa to Eastern Africa. It is interesting to note that even in the foci of the disease, a very low number of tsetse flies is infected (1 in 1000). The disease is relatively chronic and the death occurs only after several years. It prevails in the form of foci, in fact historical foci burst in the forms of epidemics which are spread along transportation routes (*e.g.* roads, tracks, rivers, *etc.*).

Trypanosoma rhodesiense (East Africa) is a typical parasite of wild fauna (*e.g.* antelopes, giraffes, hippopotamuses, warthogs), as well as carnivores (*e.g.* lions, hyenas, *etc.*). The antelopes are a reservoir of parasites. The main vectors are *G. morsitans* (Mozambique, Zimbabwe, Malawi, Bostwana), *G. swynnertoni* (Kenya, Tanzania), *G. pallidipes* (Uganda, Kenya, Tanzania, Zambia, Malawi, Bostwana), and *G. fuscipes* (Uganda, Kenya). The trypanosomiasis due to *T. rhodesiense* generally appears in the form of isolated cases. It is a zoonosis which prevails in the wild and domestic animals (*e.g.* cattle) in savannas. Men, such as fishermen, hunters and honey collectors, are infected while interfering accidentally with the natural cycle of tsetse flies/wild animals. It is a virulent and acute disease in man, which can be fatal. It prevails in Eastern and Southern Africa, especially in the area of the Great Lakes. Trypanosomiasis due to *T. gambiense* is therefore a chronic anthroponosis and the trypanosomiasis due to *T. rhodesiense* is a zoonosis or acute anthropozoonosis. They both co-exist around Lake Victoria.

1.6.4   Control

Control techniques based on the removal of the vegetation on which tsetse flies depend for their habitat (*e.g.* selective clearing), and killing the wild animals on which they depend for their food have been undertaken.[33,34] These techniques can be effective but it is labour-intensive and unecological. Nowadays, control is mainly based on the use of traps to catch and kill tsetse flies, which can greatly reduce the vector population and significantly affect the disease transmission. Various models of traps have been designed, such as biconical, pyramidal ones placed in "strategic" sites. The insecticide impregnation of either biconical traps (see Figure 1.12) or blue cloth screens hung from vegetation in the vector habitat has allowed a control of 90% of the vectors of the *G. palpalis* group.

Figure 1.12   Biconical trap of Challier-Laveissière used for *Glossina* (*G. fuscipes fuscipes*) by the Mbororos herders in Central African Republic (Photo courtesy of Dominique Cuisance).

1.7   Anoplura (Lice)

1.7.1   Systematics

Lice belong to the order of Anoplura (Phthiraptera, Heterometabola) which are small insects (0.5–8 mm in length). There are 42 genera, but only two are of medical importance, *Pediculus* genus with *P. capitis*, the head louse and *P. humanus*, the body or clothing louse; and *Phthirus* genus with *P. pubis*, the crab-louse. Lice are ectoparasites of man and other mammals.

1.7.2   Morphology

***Pediculus capitis* (head lice), *Pediculus humanus* (body lice) and *Phthirus pubis* (crab lice)**
Eggs (nits) are oval and measure 0.8 mm by 0.3 mm. They are laid by the female louse who glues single eggs with a drying cement at the base of the hair-shaft nearest the scalp or onto cloth fibre. Larvae (or nymphs) have the same morphology as the adults; they differ only by their smaller size. Adults are small

Figure 1.13   Adult head lice.

(3 mm in length), apterous (wingless), with the body flattened dorso-ventrally (see Figure 1.13). The thorax carries three pairs of strong and bulky legs ending in a big claw which represents a powerful grip. Adult crab lice are smaller (1.5 to 2 mm in length), they have a broad thorax and legs with enormous claws and grips.

1.7.3   Biology

1. ***Pediculus capitis* (head lice) and *Pediculus humanus* (body lice)**
   The *P. humanus* louse is specific to man and cosmopolitan.[1] The lice are permanent and obligatory ectoparasites and hematophagous at all stages and in both sexes.[35] The egg (nit) hatches to release a nymph after six days at 35 °C and nine days at 29 °C. The nit shell remains attached to the hair shaft. Nymphs mature after three moults and become adults seven to 12 days after hatching. Females can lay six to nine nits per day and around 200 to 300 nits during its life. Adult lice can live up to 30 days on its host and blood feed 2–3 times daily. Temperature is a key factor in their biology, during variations in the host (fever or death), the louse leaves the body leading to a dissemination of the lice. An unfed louse survives only 2 to 5 days away from its host, but well engorged ones can survive 8 to 10 days. Meals can be taken during day and night. The head louse is found behind the ears and on the nape of the neck; the body louse is found on clothing. There are generally about 100 to 200 lice per host, but massive infestations may occur with more than 20 000 nits and adults on a shirt. The dissemination takes place by physical contact with an individual carrying lice or contaminated clothing, which is supported by poor hygiene conditions and promiscuity due to poverty, misery or war.
   Nuisance
   Lice are biting pests which are at the origin of dermic diseases such as pediculosis, which is a real public health problem in developing countries, as well as developed and industrialized ones. Prevalence of head lice

among school-aged children is a growing concern due to their resistance to common insecticides. Itching is intense and scratching of bites can cause impetigo and even carvical adenopathies. For body lice, itching occurs principally on the shoulders, the back, the armpits and the belt, which can also result in impetigo with scratching of the bites. In developed countries, it is usually found on homeless vagrants. The skin can present dark and hard areas (melanoderma or "vagrant's disease") due to biting with the injection of saliva by the lice followed by continuous scratching.

Transmission of pathogenic bacteria

– Epidemic typhus is caused by *Rickettsia prowazeki*, a Gram-negative intracellular bacteria. It is an endemo-epidemic disease transmitted by body lice and specific to man who constitutes the sole reservoir of the disease. Epidemics of typhus are due to the pullulation of lice during defective hygiene situations, such as wars (*e.g.* prisoners), refugee camps and displaced populations. The greatest epidemics took place during the First and Second World Wars with around three million deaths from typhus in each. Currently, typhus persists in many foci located at altitude with a cold climate, such as the mountainous areas of Africa (Ethiopia, Rwanda, Burundi), South America (Bolivia, Peru, Ecuador) and Central Asia. There are an estimated 4000 to 20 000 cases annually with low death rates, but figures are often of doubtful validity.

The cycle starts with *P. humanus* acquiring the *Rickettsia* from the blood of an infected human during the first ten days of illness. The bacteria invade and multiply in the epithelial cells of the louse midgut. The bursting of these cells, after four to five days, releases enormous numbers of pathogens which are found in the intestinal cavity of the insect and pass into the louse faecces. *Rickettsia* are deposited on the skin of the host who infects himself when scratching. Moreover desiccated dejections become pulverulent and disseminate in the environment, clothing and bed linen. *Rickettsia* can survive 60 days in the desiccated dejections. The louse will stay infective all its life but it seems that the bacterium has a negative influence on the vector (vector death occurs 20 days after being infected).

– *Pediculus humanus* is also the vector of *Rickettsia quintana*, responsible of the "five-day fever" or "trench fever", named after its devastating impact on soldiers of the First World War. The mode of contamination is comparable with that of epidemic typhus; dejections of lice contain the bacteria and remain infectious even several months after desiccation. Man is the natural reservoir of this cosmopolitan disease.

– Louse-borne relapsing fever is caused by spirochetes (bacteria) of the genus *Borrelia*, such as *B. recurrentis*. *Borrelia* are absorbed by *P. humanus* with blood, they penetrate into the intestinal epithelium where they multiply, then the spirochetes gain the general cavity of the insect. *Borrelia* escape from the insect only when the hemolymph of the contaminated louse spreads onto the skin of a host (when crushing the insect), and contamination occurs during scratching or when touching the mucous membrane (ocular). The hemolymph of the louse is contaminanted all its life. Man is the only reservoir of *B. recurrentis*. This disease is currently present in the Ethiopian massif. Epidemics burst out in the same circumstances as those of typhus, as was particularly true during the Second World War, with five million cases and a death rate of 5 to 10%. The two epidemics of typhus and louse-borne relapsing fever can occur on the same population since the conditions are the same (promiscuity, defective hygiene, *etc.*), with the borreliosis bursting out after typhus.

2. ***Phthirus pubis* (crab lice)**

The biological cycle of crab lice is comparable with that of the other lice. The nit is attached to a hair; incubation requires 7–8 days. The larval phase includes three stages and lasts 13–17 days. The female lives about a month, during which she lays about 30 eggs. Crab lice are localised in the pubic and perianal areas, but in the event of massive infestation they can also invade the beard, the lashes, eyebrows, *etc.* They are sedentary and remain fixed at the level of the same hair. Dissemination occurs by sexual contact. It is a cosmopolitan human parasite.

Nuisance

Crab lice are not vectors but cause a phthiriasis with intense itching involving scratching of bites, impetigo, pyodermitis with adenopathy. The formation of bluish spots observed on the site of the crab lice is due to the saliva injected at the time of the bite.

1.7.4   Control

1. ***Pediculus capitis* (head lice) and *Pediculus humanus* (body lice)**

There are many treatments against lice, including shampoos and powders which can be very effective but resistance occurs.[36] For the treatment of head lice, it is necessary to use a product against the adults and the nits, to scrupulously follow the instructions under penalty of failure and to repeat with a second treatment 7–10 days later. Most products contain pyrethroids (*e.g.* lotion, spray, shampoo) and insect growth regulators (IGR's) to kill the nits. In control operations, it is necessary to treat all clothing but also all the entourage (*e.g.* class, family members) or the community (*e.g.* school, prison, camps, *etc.*). In the event of epidemics, insecticides must be applied at two levels: individuals and clothing, bed linens, objects. To treat body lice, permethrin (0.5% dust) is currently used to pulverize clothing, but it is important to know the sensitivity of the lice to insecticides. Clothing must be washed with hot water (for at least one hour at 60 °C) to kill the nits which resist cold washings.

2. ***Phthirus pubis* (crab lice)**

The treatment of crab lice is carried out using water-based insecticide lotions. All hairy parts of the body should be treated, including head hair; eyelashes should be treated with petroleum jelly twice a day for ten days. It is also necessary to treat any partners.

1.8 Siphonaptera (Fleas)

1.8.1 Systematics

Fleas belong to the Siphanoptera order with approximately 2500 described species and subspecies grouped into two super-families: (1) Pulicoidea with two families – Pulicidae including 150 species comprised of common fleas such as the human flea (*Pulex irritans*), the dog flea (*Ctenophalides canis*) and the cat flea (*Ctenophalides felis*), and Tungidae with its tropical species; and (2) Ceratophylloidea which is composed of 15 families.

1.8.2 Morphology

Flea eggs are ovoid or round, 0.3 to 0.5 mm in length. Larvae are elongate (3–5 mm in length), without legs, acephalic and blind. There are three larval stages; the last stage forms a silken cocoon in which the pupa will develop. The adults are small (see Figure 1.14), flattened laterally, and have piercing-sucking oral parts. Both sexes are apterous with the third pair of legs adapted for jumping. The aptitude for the jump is related to a particular proteinic substance, the resiline which can release the physical energy necessary to jump 7 to 10 cm in height and more than 15 cm in length.

1.8.3 Biology

The eggs are laid separately or in small series of two to six, generally in the litter of the host or the dust of the dwellings.[1] A female flea deposits several hundred eggs during her life: 200 to 300 eggs for *Xenopsylla cheopis*, 800 for *Ctenocephalides felis* and several thousand for *Tunga penetrans*. Eggs will hatch after one or two weeks according to the species, temperature and humidity. Each of

Figure 1.14   Adult male *Oropsylla montana* flea (Photo courtesy of CDC/DVBID, BZB, Entomology and Ecology Activity, Vector Ecology & Control Laboratory, Fort Collins, CO/John Montenieri).

the three larval stages lasts approximately two to six days. The larvae have a negative phototropism and are very hygrophilic. They are active and feed on a variety of organic debris, the remnants of digested blood produced by adults, small arthropods present in the nest and even dead adult fleas. The larval lifespan is 10–21 days. The pupa is motionless living inside a cocoon and does not feed; this stage lasts one to two weeks. The adult leaves the cocoon and shortly afterwards, the male and female copulate and seek a host for the first blood meal. Both sexes are hematophagous. The saliva of fleas is irritating and certain people have a particular sensitivity, bites can result in sleep loss, nervous disorders, *etc*. Secondary infections can also occur when bites are scratched. A blood meal occurs every two to four days for species living in burrows, and every day (or more) for those species living on their host. The flea quickly leaves the host if death occurs and will rapidly seek out another host. This behaviour is of great epidemiologic importance in the propagation of diseases. The longevity of the adults is about ten months, although this varies according to species and conditions. The optimum range of temperature and humidity is generally narrow: 22 °C to 24 °C and greater than 80% humidity for *X. cheopis*. Three models of behaviour have been identified: (1) the species which live permanently on their host like the "fur fleas", *e.g. X. cheopis*, *P. irritans* (human flea), and jump from one host to another; (2) the "nest fleas" which live in the nest and visit the host to take their blood meals, and have only a weak aptitude for jumping; and (3) the sedentary species which live fixed by their oral parts on the host, like *Echidnophaga gallinacea* on poultry or *Spilopsyllus cuniculi* on rabbits, or embedded in their host like *Tunga penetrans*.

Transmission of plague
Plague is due to a Gram negative enterobacterium, *Yersinia pestis*, and it is a disease of rodents transmitted by fleas in the form of bubonic plague. This disease is still endemic in several regions of the world due to the fact that it is a zoonosis primarily infecting rodents (*e.g.* rats and other wild rodents). Its emergence in humans depends both on the frequency of infection among rodents and the promiscuity of human with these animals. From 1989 to 2003, 38 310 human cases and 2845 deaths were reported from 25 countries in the world, but 82% of all cases occur in Africa.[37] Twelve African countries had notified cases of plague: Algeria, Botswana, the Democratic Republic of Congo, Kenya, Madagascar, Malawi, Mozambique, Namibia, Tanzania, Uganda, Zambia and Zimbabwe, but there may also have been occurrences in other countries. Madagascar is particularly affected by plague, especially on the central and Northern Highlands, typically above 800 m altitude. There are three foci: (1) rural foci in the Highlands, the reservoir of pathogens is *Rattus rattus*; (2) urban foci in the Highlands and the capital, Antananarivo, *Rattus norvegicus*; and (3) a coastal urban focus at Mahajunga, *Rattus rattus* and *R. norvegicus*. The vectors are *Xenopsylla cheopis* (indoors: 95%) and *Synopsyllus fonquerniei* (outdoors: 86% to 95%). 14% of the world's cases occur in eight countries of Asia: China, India, Indonesia, Kazakhstan, Laos, Mongolia, Republic of the Union of Myanmar and Vietnam. 4% of the world's cases have been reported in five

American countries: The United States of America, Peru, Brazil, Bolivia and Ecuador. No cases of plague have been reported recently in Oceania or Europe. In France, the last cases occurred in 1945 in Corsica.

One of the characteristics of plague epidemics is their capacity to disappear during several years, before reappearing brutally in an epidemic form. In 1994, an epidemic of bubonic plague (128 cases) was recorded in Mozambique after more than 15 years of silence, which was then propagated to Zimbabwe and Malawi. At about the same time, an epidemic occurred in Peru (1031 cases in 1993–1994). In spite of their close temporal appearance, there is probably no epidemiologic link between these epidemics. More recently (June 2003), the plague reappeared in Algeria (Oran) after a period of 50 years of inter-epidemic silence. A pulmonary epidemic of plague was recently declared in the Democratic Republic of Congo (December 2004) and an epidemic of bubonic plague has been reported (June 2009) in Libya near the Egyptian border.

In man, the disease takes two principal forms: (1) bubonic (contracted by the flea bite) and (2) pulmonary (airborne transmission). The bubonic plague (most frequent) is characterized by a very severe infectious syndrome with strong fever, accompanied by a hypertrophy of the lymphatic ganglion (bubo). In 20% to 40% of cases, the patient recovers after a rather long time of convalescence. Otherwise, the disease develops towards an acute and fatal septicaemia. In certain cases, the bacteria reach the lungs and the disease develops towards a pulmonary plague. The inter-human contagion occurs *via* the infected expectorations. In the absence of an early and suitable treatment, pulmonary plague is systematically fatal within three days.

The plague has played an important role in human history in some spectacular deadly epidemics. Three great pandemics can be noted: (1) during the 6th century "the plague of Justinien" occurred in the countries of the Mediterranean basin and Northern Europe. There were one hundred million deaths with a 60-year epidemic, or over two centuries with regular reoccurences every 10–12 years; (2) during the 14th century (1348–1350), the Black Death came from foci of Central Asia and killed a quarter of the European population, with a long series of epidemic reocurrences through Europe, India and China; and (3) in 1855, a third pandemic started from the ancestral Chinese foci of Yunnan and spread across the whole world (Bombay in 1896; Calcutta in 1898, Alexandria 1899, the Mediterranean basin in 1901, northern Europe in 1908, and eventually to reach the New World and South Africa). There were 12 million deaths.

Plague involves flea–rodent–host transmission with three cycles: (1) wild plague prevails in wild rodents (squirrels, marmots, gerbils, prairie dogs, *etc.*) where the attack of humans is exceptional; (2) rural plague passes from wild rodents to the rats of villages; (3) urban plague is imported downtown from a rural focus, and is quickly propagated from rat to rat. The black rat (*Rattus rattus*) has the greatest sensitivity to the bacillus of plague with a phase of septicaemia favourable to the infection of fleas, and because of its contact with man, it can "transport" the infection from one continent to another, like from one house to another. The grey rat (*Rattus norvegicus*), a rodent of the cellars and sewage systems, has a stronger resistance to the infection.

There are more than one hundred species of fleas associated with the plague. The main vectors are: *Pulex irritans*, best adapted to man in temperate climates and cosmopolitan, involved in epidemics in Morocco, England, Algeria, Nepal, Brazil, *etc.*; *Xenopsylla cheopis*, cosmopolitan, related to black rat and bites in humans; *X. brasiliensis*, vector of plague in rural areas of South America, Africa and India; *X. astia*, vector in Asia; and *Nosopsyllus fasciatus*, a cosmopolitan species.

Transmission of other pathogens
Rickettsia: Fleas transmit *Rickettsia mooseri* (or *R. typhii*) responsible for the murine typhus which affects rats, mice and sometimes humans. This typhus is especially present in tropical and sub-tropical areas, especially in ports. The natural reservoir is the rat; the vectors are *X. cheopis*, *N. fasciatus* and *Leptopsylla segnis*. Contamination occurs through the faeces of fleas, where the concentration and pathogenicity of Rickettsiae is high.

Other bacillus and viruses: Fleas can accidentally transmit the bacillus of tularaemia (*Francisella tularensis*), contamination occurring as a result of contact with the infectious faeces of fleas present in the fur of contaminated animals handled by man. The flea, *Xenopsylla cuniculi*, transmits the virus of the myxomatosis (*Leporipox virus*) to rabbits and hares in Australia, Europe and the United States.

Animal pathogens: Fleas can transmit *Trypanosoma* parasites to animals (*T. lewisi* of rats transmitted by *N. fasciatus* and *X. cheopis*; *T. duttoni* of mice transmitted by species of *Ctenophtalmus* or *Nosopsyllus*; *T. nabiasi* of rabbits by *Spilopsyllus cuniculi*). The fleas are also intermediate hosts of helminths in animals and occasionally in man, *Dipylidium canidum*, a cestode of cats and dogs, accidentally human; vectors: *Ctenocephalides canis* and *Pulex irritans*. The contamination is made when the larvae of fleas ingest the cestode eggs; the infection of man is done through ingestion of the fleas carrying cysticercoids.

Jiggers and tungiasis
Jiggers are extremely small fleas, such as *Tunga penetrans*, which develop in the dry sandy soils of hot regions of America, Africa, India and China.[38] The fertilized female buries herself into the skin of the feet of large mammals, humans and pigs. This causes a strong cutaneous irritation and ulcerations. The young female is small (1 mm), but the fertilized female can measure 1 cm, due to thousands eggs accumulated in the slack abdomen. Great care must be taken when extirpating the *Tunga* to avoid breaking their bodies and releasing several hundreds of eggs into the wound.

1.8.4 Control

Within the framework of plague control, control must start with the fleas rather than the rodents, because if one kills only the rodents, the fleas on man will increase in number.[39] The control of fleas is carried out by spraying insecticides on the walls which follow the route of rats, but fleas have developed some resistance (in particular to organochlorines). Where malaria control programmes based on DDT spraying have been enforced, fleas have been found to be resistant to this insecticide. Thus, before beginning a campaign against fleas, it is necessary to know their sensitivity to particular insecticides. Malathion is used in the event of resistance to organochlorines. The ground level of houses and materials which rats carry into their burrows are treated with powders or aerosols. Tests have been carried out using systemic insecticides absorbed by rats (in soft foods) that fleas might swallow when biting their hosts, but certain products degrade within 24 hours, and fleas bite only every 48 or 72 hours. It is also possible to eliminate the fleas present on domestic animals, or to provide the domestic animals with insecticidal collars and treat their habitats.

1.9 Bloodsucking Hemiptera (Bedbugs and Kissing Bugs)

1.9.1 Systematics

Bloodsucking bugs belong to the order of Hemiptera, suborder Heteroptera, characterized by the presence of two pairs of wings with the anterior pair being of harder consistency, and a rostrum brought back under the body with an oral apparatus of the piercing and sucking type. Heteroptera have 75 families, of which only two are of medical interest: Cimicidae (23 genera and 91 species) and Reduviidae (over 6000 species), among which only two tribes, Triatomini and Rhodniini, are of medical interest and include the vectors of the American trypanosomiasis – *Triatoma* and *Panstrongylus* (Triatomini) and *Rhodnius* (Rhodniini).

1.9.2 Cimicidae (Bedbugs)

1.9.2.1 *Morphology*

Bedbug eggs are ovoid. The nymphs have five instars which are smaller but similar to the adults in morphology, behaviour, diet and habitat. Adults bedbugs are small (4–6 mm in length, 3 mm broad), oval flattened dorso-ventrally, without functional wings (see Figure 1.15).

Figure 1.15   Adult *Cimex lectularius* bedbug (Photo courtesy of CDC/Harvard University, Dr. Gary Alpert, Dr. Harold Harlan, Richard Pollack/Piotr Naskrecki). http://phil.cdc.gov/PHIL_Images/9822/9822_lores.jpg

1.9.2.2   Biology

Ten to 50 eggs are deposited by batch and during its entire life a female bedbug lays 200 to 500 eggs. They are laid in cracks in the ground, walls, beds, *etc.*[1] They are stuck to a support and hatch after ten days at 20 °C. Larvae moult every eight days under favourable conditions and their size varies from 1.5 mm (stage 1) to 3.5 mm (stage 5). The larval stages are hematophagous and nocturnal. The adult stage is reached after six to nine weeks. The adult meal lasts 10 to 15 minutes and both sexes are hematophagous. They feed on vertebrates, such as birds, bats and men, during the night. They feed once a week during the hot season, and twice a month during spring and autumn, but they can survive several months in the absence of a host. The mean lifespan is about 9 to 18 months at 18–20 °C with the availability of hosts. On the 91 known species of Cimicidae, three are parasitic to man: (1) *Cimex lectularius*, the familiar bedbug with a cosmopolitan distribution in temperate subtropical regions of the world; (2) *Cimex hemipterus* (= *Cimex rotundatus*) distributed in the tropical zones of the world; and (3) *Leptocimex boueti* mainly associated with bats in West Africa, but which can also bite man. Among the other species biting man, there are *Oecacius vicarius* in Canada and the United States, *Haematosiphon inodorus* in Mexico and Southern United States, and *Cimex hirundinis* in England.

Medical importance
Bedbugs do not transmit any parasite to man although they may be a potential vector of hepatitis B, exanthematic typhus, plague, leprosy, recurring fevers, leishmaniasis.[40] The bites from bedbugs can involve important pruriginous reactions, and even nervous or digestive disorders. The reactions can also be serious in children (signs of great exhaustion). The reactions can begin from two to three minutes after the end of the blood meal.

Bedbugs are becoming resistant to the insecticides available and they are a real nuisance worldwide, with a current alarming invasion spreading to even the most expensive hotels of North America.

1.9.2.3   Control

Prevention of bedbugs is carried out by physical or chemical methods. Physical methods consist of eliminating all cracks, and heating or cooling the rooms. Chemical methods are based upon the use of insecticides, but the selection of insecticide is of concern due to resistance to organochorines. Insecticidal control programs based upon indoor spraying of pyrethroids has been very successful in regions of Latin America (Venezuela, Brazil, Argentina, Chile and Uruguay).

1.9.3   Reduviidae – Triatominae (Kissing Bugs)

1.9.3.1   Morphology

The eggs of kissing bugs are oval (1.5 to 2.5 mm in length). Nymphs resemble the adult but are wingless (see Figure 1.16). They measure 2 to 3 mm and there

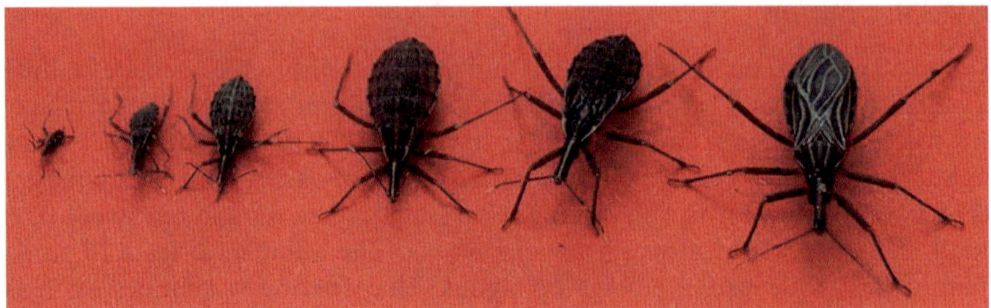

Figure 1.16 *Rhodnius prolixus* kissing bug (from nymphal to adult stage).

are five instars. Adult kissing bugs are large, brown or black in colour with often a red, yellow or orange ornamentation on the abdomen. They are elongated because of their long head and abdomen. They occur only in tropical zones, especially in America, there are also 11 species in Asia, Africa and Australia, as well as a pantropical species: *Triatoma rubrofasciata*. Adult kissing bugs have two pairs of wings folded up as scissors and long legs (insect runners) ending with two claws.

1.9.3.2 Biology

Eggs are laid in batches of 10–20 for *Rhodnius*, or individually for *Triatoma*.[1] The embryogenesis lasts around ten days (up to two months). Each larval stage takes at least one meal of blood. Copulation occurs shortly after emergence. Ten to 14 days later the blood fed female deposits her first batch of eggs. The total number of eggs laid by a female during her entire life is usually 200 to 300. The duration of the cycle from egg to adult can last three months (up to a maximum of six to ten months). The lifespan is several months, and even up to one to two years. Both sexes are hematophagous and they take their blood meal at night. In the absence of hosts, larvae and adults can survive for four to six months. The majority of the triatomines are parasites of wild animals (*e.g.* armadillos, bats, rodents), but also domestic animals, mammals, birds and men. The bites are not painful and are especially frequent on the face and arms. Triatominae live in cracks in the ground or in the walls of poor housing or in the burrows of wild animals. They are present on the American continent at a latitude of 42° N to 43° S, and at an altitude of up to 3600 m.

Transmission of *Trypanosoma*
In America, some kissing bugs are vectors of *Trypanosoma cruzi*, responsible for the American trypanosomiasis or Chagas disease. The disease prevails from northern Mexico to southern Argentina, and on both South American coasts. It is one of the major public health problems in Latin America; 15 to 25% of infected subjects present clinical symptoms.

Trypanosoma parasites are absorbed during the blood meal taken on an infected subject. In the triatome digestive tract, the parasite evolves in two to three weeks to the epimastigote form, then the metacyclic trypomastigote forms, which is the infective stage transmitted by the faeces of the kissing bugs. The infectivity of *Rhodnius prolixus* and *Triatoma infestans* has been well studied: 15 to 30 days after the infective meal, their faeces contain *Trypanosoma* which remain infective for a long period. When scratching the bite, the trypanosomes are brought into contact with the mucous membranes, in particular eyes (Romana's sign) or small cutaneous excoriations.

Schematically, there are three epidemiologic cycles from *T. cruzi*: (1) a wild cycle, the circulation of *Trypanosoma* occurs between wild animals (*e.g.* marsupial, rodent, bats, *etc.*) and the kissing bugs which live in contact with them. Man can be accidentally infected if he enters such environments and is bitten by triatomes, though the possibility of infection is low; (2) a peridomestic cycle, the circulation of the *Trypanosoma* occurs between the kissing bugs which live in hen houses, rabbit burrows, cattle sheds *etc.*, the domestic animals which are kept there and the wild animals which take refuge in this accommodation (*e.g.* rats, marsupials, cats, dogs, *etc.*). Risks of contamination are great; and (3) a domestic cycle, the circulation of the *Trypanosoma* is carried out between humans and triatomes living inside human dwellings in generally poor conditions. The reservoir of *Trypanosoma* is then man, while his domestic animals (cats, dogs) are definitely less infected. The risks for man are very high.

The main vectors are: *Rhodnius prolixus* in Guyana, Guatemala, Nicaragua, Panama, Venezuela and Surinam; *Panstrongylus megistus* in Brazil; *Triatoma dimidiata* in Belize, Colombia, Costa Rica, Ecuador, Guatemala, Honduras, Mexico, Nicaragua, Panama and Venezuela; *Triatoma infestans* in Argentina, Bolivia, Brazil, Chile, Paraguay, Uruguay and Peru. Among the other vectors with local distribution, there are *Triatoma maculata* in Brazil, Venezuela, Guyana and Colombia; *Triatoma brasiliensis* in Brazil; and *T. sordida* in Argentina, Bolivia, Brazil, Chile, Paraguay and Uruguay.

Transmission of *Trypanosoma rangeli* and other *Trypanosoma*
Some kissing bugs, in particular *Rhodnius prolixus*, can transmit *Trypanosoma rangeli*, usually a parasite of animals, to man. Care should be taken to differentiate between *T. cruzi* and *T. rangeli*. Although both are passed on by the same vertebrates, have the same vectors and are found in the same zones, *T. rangeli* is responsible for a relatively benign disease whilst *T. cruzi* can cause the potentially fatal American trypanosomiasis. In Asia, there exists other *Trypanosoma*, parasites of man transmitted by the pantropical species *Triatoma rubrofasciata*.

1.9.3.3 Control

The control of Chagas disease must integrate vector control because there is neither a vaccine nor a preventive drug. Control is done by mechanical (destruction or remodelling of houses) and chemical methods (insecticides). There are great inter-regional programs of vector control (program of the "Southern Cone") with construction of houses and insecticide spraying, along with the use of traditional methods, but it is important to understand the sensitivity of the local vectors to particular insecticides. Interesting results have recently been obtained by the use of insecticide paints (mixture of insecticide and insect growth regulators); insecticide-treated nets have also been promoted.

1.10   Conclusions

Many human and animal diseases are due to the pathogens transmitted by insects. Among the deadliest infectious diseases worldwide, malaria and dengue hemorrhagic fever are vector-borne diseases. Many insect vectors belong to the Diptera order, such as mosquitoes, biting midges, sandflies, blackflies and tsetse flies, but insects of other orders can also transmit important pathogens like lice, fleas and bloodsucking bugs. Pathogenic agents can be parasites (*e.g.* protozoa, filariae, cestods, *etc.*), bacteria (*e.g.* rickettsiae, borellia, *etc.*), and viruses (*e.g.* arboviruses) that undergo a development inside the insect vector to reach their infective stage. Transmission can be done through the insect's bite with direct inoculation, or with pathogens deposited by the insect saliva or faeces onto the skin and entering their host either through scratching or by active action of the pathogen into the host wound.

For these insect-borne diseases, vector control is one of the first preventive measures, if not the only one in most cases due to a lack of vaccine (except for yellow fever and Japanese encephalitis) and resistance issues in chemoprophylaxis. Therefore, vector control is an integral component of vector-borne disease control programmes. However, in order to reach their full efficiency, vector control operations and programmes must be based on a sound knowledge of the targeted vectors (*e.g.* species identification, biology, distribution, *etc.*), of the local and regional ecological context, and of the socio-cultural conditions of human populations.

References

1. R. P. Lane and R. W. Crosskey, *Medical Insects and Arachnids*, British Museum, Chapman & Hall, London, UK, 1993.
2. S. R. Christophers, *The Fauna of British India, including Ceylan and Burma. Diptera, Family Culicidae, Tribe Anophelini*, Taylor and Francis, London, UK, 1933, vol. 4.
3. S. Manguin, M. J. Bangs, J. Pothikasikorn and T. Chareonviriyaphap, *Infect. Genet. Evol.*, 2010, **10**, 159.
4. A. Tran, X. Deparis, P. Dussart, J. Morvan, P. Rabarison, F. Remy, L. Polidori and J. Gardon, *Emergerging Infect. Dis.*, 2004, **10**, 615.
5. P. Renault, J. L. Solet, D. Sissoko, E. Balleydier and S. Larrieu *et al.*, *Am. J. Trop. Med. Hyg.*, 2007, **77**, 727.
6. A. B. Failloux, M. Vazeille and F. Rodhain, *J. Mol. Evol.*, 2002, **55**, 653.
7. W. J. Tabachnick and J. R. Powell, *Genet. Res.*, 1979, **34**, 215.

8. E. Wang, H. Ni, R. Xu, A. D. Barrett and S. J. Watowich *et al.*, *J. Virol.*, 2000, **74**, 3227.

9. L. Rosen, D. A. Shroyer, R. B. Tesh, J. E. Freier and J. C. Lien, *Am. J. Trop. Med. Hyg.*, 1983, **32**, 1108.

10. S. C. Rawlins, B. Hull, D. D. Chadee, R. Martinez and A. LeMaitre *et al.*, *Trans. R. Soc. Trop. Med. Hyg.*, 1990, **84**, 142.

11. R. E. Harbach, *Bull. Entomol. Res.*, 2004, **94**, 537.

12. S. I. Hay, M. E. Sinka, R. M. Okara, C. W. Kabaria and P. M. Mbithi *et al.*, *PLoS Med.*, 2010, **7**, e1000209.

13. L. J. Bruce-Chwatt, *Essential Malariology*, Med. Books Ltd., William Heinemann, London, 1980, p. 354.

14. J. A. Rozendaal, *Vector Control: Methods for Use by Individuals and Communities*, World Health Organization, Geneva, 1997.

15. WHO, World Malaria Report, in *WHO/HTM/GMP/2008.1*, Geneva, 2008, p. 215; http://www.who.int/malaria/wmr2008/malaria2008.pdf: (last accessed 12th November, 2010).

16. S. Manguin, P. Carnevale, J. Mouchet, M. Coosemans and J. Julvez *et al.*, in *Biodiversity of Malaria in the World*, ed. J. L. Eurotext, John Libbey Eurotext, Paris, France, 2008.

17. W. H. Wernsdorfer and I. McGregor, *Malaria: Principles and Pratice of Malariology*, Churchill Livingstone, Edinburgh, UK, 1988.

18. B. Singh, L. Kim Sung, A. Matusop, A. Radhakrishnan and S. S. Shamsul *et al.*, *Lancet*, 2004, **363**, 1017.

19. J. Mouchet, P. Carnevale, M. Coosemans, J. Julvez and S. Manguin *et al.*, in *Biodiversité du Paludisme dans le Monde.*, ed. J. L. Eurotext, John Libbey Eurotext, Paris, France, 2004.

20. WHO, The Global Programme to Eliminate Lymphatic Filariasis (GPELF), 2008; http://www.who.int/lymphatic_filariasis/disease/en/ (last accessed 12th November, 2010).

21. T. B. Nutman, in *Lymphatic Filariasis. Tropical Medicine: Science and Practice*, ed. G. Pasvol and S. L. Hoffman, Imperial College Press, London, UK, 2000, vol. 1.

22. D. Sissoko, D. Malvy, C. Giry, G. Delmas and C. Paquet *et al.*, *Trans. R. Soc. Trop. Med. Hyg.*, 2008, **102**, 780.

23. S. P. Kalantri, R. Joshi and L. W. Riley, *Natl. Med. J. India*, 2006, **19**, 315.

24. C. Akoua-Koffi, K. D. Ekra, A. B. Kone, N. S. Dagnan and V. Akran *et al.*, *Med. Trop.*, 2002, **62**, 305.

25. L. Rosen, *Am. J. Trop. Med. Hyg.*, 1986, **35**, 642.

26. J. Mouchet, J. Rageau, C. Laumond, C. Hannoun and D. Beytout *et al.*, *Ann. Inst. Pasteur*, 1970, **118**, 839.

27. D. S. Kettle, *Ann. Rev. Entomol.*, 1962, **7**, 401.

28. D. S. Kettle, *Ann. Rev. Entomol.*, 1977, **22**, 33.

29. D. J. Lewis, *Ann. Rev. Entomol.*, 1974, **19**, 363.

30. S. S. Amora, C. M. Bevilaqua, F. M. Feijo, N. D. Alves and M. do V. Maciel, *Neotrop. Entomol.*, 2009, **38**, 303.

31. J. Ciment, *Br. Med. J.*, 1999, **319**, 1090.

32. D. H. Molyneux, *Trends Parasitol.*, 2005, **21**, 525.
33. J. Ford, *Bull. World Health Org.*, 1963, **28**, 653.
34. D. J. Rogers, G. Hendrickx and J. H. Slingenbergh, *Rev. Sci. Tech.*, 1994, **13**, 1075.
35. P. Kimmig, *Z. Allgemeinmed.*, 1983, **59**, 1427.
36. K. Y. Mumcuoglu, S. C. Barker, I. E. Burgess, C. Combescot-Lang and R. C. Dalgleish *et al.*, *J. Drugs Dermatol.*, 2007, **6**, 409.
37. WHO, *Wkly Epidemiol. Rec.*, 2004, **79**, 301.
38. J. Heukelbach, F. A. de Oliveira, G. Hesse and H. Feldmeier, *Trop. Med. Int. Health*, 2001, **6**, 267.
39. R. Marsella, *Vet. Clin. North Am. Small Anim. Pract.*, 1999, **29**, 1407.
40. O. P. Forattini, *Rev. Saude Publicat.*, 1990, **24** (Suppl. 1).

Classical Insecticides: Past, Present and Future

Ó. LÓPEZ,*[1] J. G. FERNÁNDEZ-BOLAÑOS[1] AND M. V. GIL[2]

[1] Seville University, Department of Organic Chemistry, Profesor García González 1, 41012, Seville, Spain; [2] Extremadura University, Department of Organic and Inorganic Chemistry, Avenida de Elvas, 06071, Badajoz, Spain

2.1 Introduction

Humanity's efforts to combat insect pests affecting humans' and cattle's health, crop productivity, and even constructions can be traced back to the establishment of the first civilizations.[1] For several centuries an arsenal of chemicals, natural extracts and living organisms have been employed to ameliorate the depredation caused by insects.[2] Such chemicals were, for many centuries, of natural origin, and then these were superseded by synthetic insecticides in the 1940s.

Although the most common way of combating insects is with the use of natural and synthetic chemicals (either inorganic or organic), attention has also been paid to the use of living organisms (*e.g.* microorganisms, other insects, birds, *etc.*), the so-called "biocontrol".[3] The idea of establishing the biocontrol of insects already arose in ancient Chinese civilizations by using predatory ants (*Oecophylla smaragdina*) to prevent citrus fruit tree damage caused by caterpillar larvae infection.[4] Much later, in 1752, Carl Linnaeus reintroduced the possibility of using other insects as insect predators: "Every insect has its predator which follows and destroys it. Such predatory insects should be caught and used for disinfesting crop-plants".[5]

RSC Green Chemistry No. 11
Green Trends in Insect Control
Edited by Óscar López and José G. Fernández-Bolaños

Published by the Royal Society of Chemistry, www.rsc.org

2.2 Insecticides Prior to the *Chemical Era*

To our knowledge, the first documented compendium of insecticide substances is the Egyptian *Ebers Papyrus* (*ca.* 1600 BC).[6] Another example of historical reports concerning the control of insect pests can be found in ancient Chinese civilizations, where fire was used to destroy plagues of migratory locust (*Locusta migratoria manilensis*).[7] Pre-Roman civilizations already reported the burning of brimstone (sulfur)[6] as an insecticide and purifying agent. Such an application was also described by Homer in *The Odyssey* (*ca.* 1000 BC). Currently, sulfur (applied as dust, granulated or as a colloidal formulation) is not only used as an insecticide in crops and in indoors applications against mites and some caterpillars, but also as a fungicide and fertilizer.[8] Sulfur was not found to cause adverse effects in the environment, and also lacked any remarkable damage against humans.[9]

Pliny the Elder recorded in his *Natural History* various types of insecticides from the preceding 300 years coming from folk culture,[10] comprising pepper, turpentine or fish oil among others. He mentioned the repellent activity towards insects of both, fresh and burning leaves from *Mentha pulegium* (from Latin "pulex", meaning flea).[11]

Inorganic chemicals, known to be agents for struggling insect pests since antiquity, were used extensively from the 19th century until the first third of the 20th century.[12] The use of arsenic sulfides in China against garden pests was already reported by the year 900 AD, although the employment of such inorganic chemicals in the Western civilization did not take place until the 17th century.[13] Copper acetoarsenite ($Cu(CH_3COO)_2 \cdot 3Cu(AsO_2)_2$, *Paris Green* pigment) is considered to be the first broad spectrum insecticide, first used in 1865 to protect potato plants against the Colorado potato beetles.[14] Lead and calcium arsenates were introduced in 1892 and 1907, respectively, for improving insect control in ornamentals, fruits, tobacco or cotton.[15] Lead arsenate was particularly effective against the gypsy moth, and soon replaced *Paris Green* because of its strong phytotoxicity when used in large amounts.[16] White arsenic (arsenious oxide) was used in large quantities against grasshoppers as poisoned baits and against cattle ticks; the basic character of arsenious oxide precluded its direct use on foliage.[17]

Arsenicals were quite efficient for controlling chewing insects. Unfortunately, both arsenic and arsenic-containing compounds were found to be highly toxic.[18] Arsenic induces skin and bladder cancer, and also affects vascular, neurological and cognitive functions.[19] The US Environmental Protection Agency (US EPA) banned the use of most arsenic-based pesticides in the 1990s, and only allowed those organo-arsenicals that were not suspected carcinogens.[20] Nevertheless, the lack of data on the stability and transformations of these compounds in soil does not guarantee that they can be considered as completely harmless.[21]

Due to the environmental concerns of arsenic salts, cryolite (sodium fluoro-aluminate, Na_3AlF_6) surged as an alternative in the beginning of the 20th century for the preservation of fruits and vegetables. Although the use of such mineral insecticide decreased upon the introduction of synthetic insecticides, it is still allowed due to its environmental safety, low toxicity and limited capacity to be dispersed by water.[22] Cryolite, unlike other mineral salts containing fluorine, does not release the fluoride ion upon decomposition,[22] which ensures a low toxicity against mammals. As a consequence, cryolite is currently considered as a suitable insecticide for organic crop culture.[23]

Another popular mineral insecticide was the *Bordeaux mixture*, a combination of copper (II) sulfate and hydrated lime (calcium hydroxide) introduced by French viticulturists as a fungicide, and still used as insecticide and fungicide.[24]

Boron-containing insecticides, comprising boric acid and its salts (*e.g.* borax and disodium octaborate), are also in current use. Their formulations were registered as pesticides in the USA in 1948.[25] Borate insecticides have been used successfully against fleas, beetles, ants, cockroaches and some species of termites,[25] thus acting as a wood-preserving agent. Besides being inexpensive chemicals, they are relatively non-toxic compounds, lacking mutagenic and carcinogenic properties.[26] Boric acid is frequently used as a bait formulation containing an insect attractant within IPM (Integrated Pest Management) programs[27] (see Chapter 9), and it is usually formulated as tablets and as powder.

Traditional insecticide preparations often included some other heavy and toxic elements as well (such as antimony, thallium, selenium or mercury) or hazardous substances (like hydrogen cyanide),[1,17] thus strongly contributing to environmental pollution and increasing the risk of human poisoning.

The ionic nature of inorganic insecticides allows a better absorption through the insect gut wall than through the integument,[28] a major difference with the highly lipophilic organic insecticides covered in the next section. Inorganic insecticides are therefore considered as *stomach poisons*, which act by ingestion.[28] These chemicals are quite stable in the environment, easily dispersed, so they can easily contribute to soil and water pollution. In many cases they also affect non-targeted species, such as beneficial insect families, aquatic organisms and mammals.[28]

Among the organic substances used in insect control, whale-oil soap (1842) and petroleum components, such as kerosene (1865) are remarkable.[29] Insecticidal soaps are considered to be selective insecticides, as they are almost innocuous for humans, because they mainly affect soft-bodied insects, such as mites and sucking insects.[30] The amphiphilic nature of these compounds allows cuticle penetration by contact and disruption of the cell integrity, which leads to dehydration and death of the insect.[30] Potassium salts of fatty acids can be used either alone or in combination with other insecticides.[31]

As indicated previously, another source of organic chemicals that has been historically used as insecticides are botanical extracts. As an example, pyrethrum (*Chrysanthemum cinerariaefolium*) is a plant commercially grown for the extraction of pyrethrins from the flowers.[32] Such extracts were already used by Chinese civilizations around the year 1000 BC. Nevertheless, the strong photolability of these compounds has limited them to indoor use.[1]

In the 17th century, it was observed that nicotine present in water extracts from tobacco leaves killed plum beetles. However, it was never used as a marketed insecticide, due to its low potency and high toxicity to mammals.[33]

Other botanicals are rotenone[34] (from cubé resin), ryanodine and related compounds[35] (from *Ryania speciosa*), veratridine[36] (from sabadilla), *d*-limonene (from the citrus extract), or the triterpenoid azadirachtin[37] (from neem tree, *Azadirachta indica* A. Juss). A more detailed compilation of botanical insecticides can be found in Chapter 7.

2.3 Classical Chemical Insecticides

2.3.1 Organochlorine Insecticides

Although the first commercially available synthetic insecticide was potassium 3, 5-dinitro-*o*-cresylate[38] (sprayed for the control of insects affecting fruit trees), it is generally accepted that the chemical era of insecticides started with the development of organochlorine insecticides, in particular with 1,1,1-trichoro-2,2-bis (4-chlorophenyl)ethane, more commonly known as DDT **1** (from its trivial name: **D**ichloro**D**iphenyl**T**richloroethane). DDT and its derivatives, together with many other chlorinated organic hydrocarbons, comprise the so-called "organochlorine insecticides", the first groups of agrochemicals of synthetic interest as pest control agents.

Organochlorine insecticides can be classified according to their chemical structure (see Figure 2.1), and they fall into three major groups: diphenylethanes

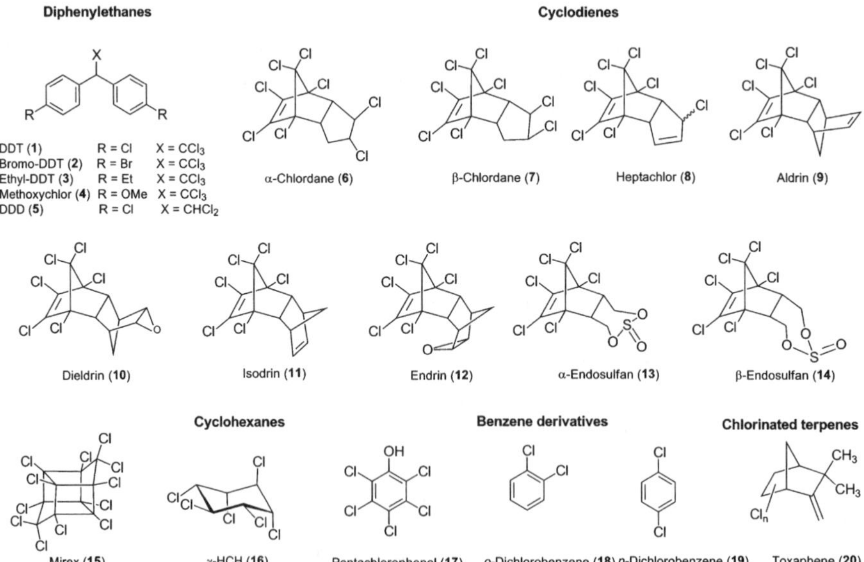

Figure 2.1　Main families of organochlorine insecticides.

(*e.g.* **1–5**), whose more noticeable example is DDT **1**, cyclodienes (*e.g.* **6–15**) and cyclohexanes (*e.g.* **16**), as well as two less numerous families: chlorinated benzenes (*e.g.* **17–19**) and norbornane derivatives (*e.g.* **20**).

2.3.1.1 Diphenylethanes

The development of DDT marked a milestone in the history of agrochemicals, as it inaugurated a new era where insects started to be efficiently and more selectively controlled by synthetic chemicals. It was also DDT that led to a time where humanity started to be increasingly conscious of the need for safer agrochemicals.

The discovery of DDT proved to be rather haphazard. The first synthesis was reported by Othmar Zeidler in 1874, a PhD student working under the supervision of Adolf von Baeyer at Strasbourg University.[39] The insecticidal properties of DDT were not discovered until 1939, however, when the Swiss chemist, Paul Hermann Müller, in his search for contact poisons, proved DDT's remarkable effectiveness against flies, mosquitoes and beetles.[39] Large production of DDT began in 1948.[40] The first uses of DDT involved soldier protection against several diseases during World War II, using a relatively low amount of insecticide.[41] In 1944, for the first time in history, people dusting with DDT managed to rapidly eradicate a typhus epidemic in Naples, Italy.[2] Müller was awarded the Nobel Prize for Physiology and Medicine in 1948 for his important contribution to pest control.[39]

DDT was originally prepared by Baeyer condensation of chloral with chlorobenzene in the presence of sulfuric acid (H_2SO_4) as catalyst (see Scheme 2.1). Oleum and chlorosulfonic acid were also proposed as catalysts.[42] Although the major product of the reaction was the *p,p'*-isomer **1** (up to 80%), the commercial formulation also included some other isomers (**21–23**) obtained as side-products.[43] Furthermore, traces of bis(4-chlorophenyl)sulfone **24** were also detected. The marketed formulations (Anofex®, Cearex®, Genitox®, Zerdane® and some others trademarks) were comprised of solutions (xylene, petroleum ether), emulsions, water-wettable powders, or aerosols, among others.[41]

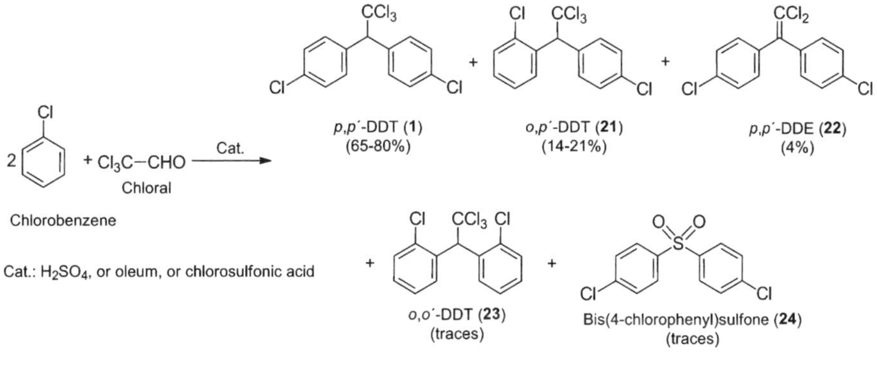

Scheme 2.1   Industrial preparation of DDT.

DDT was cheap to produce, exerted reduced acute toxicity against mammals (oral median lethal dose (LD$_{50}$) of 100–800 mg kg^{-1} in rats),[44] had a high toxicity against a broad spectrum of insects and showed prolonged stability in the environment as a result of a very low biodegradation, which allowed for a more prolonged action.[45] However, this last property also proved to be one of the main disadvantages of DDT. DDT soon became the most important chemical in insect control, and was considered a panacea for public health, reducing vector-borne diseases (see Chapter 1), mainly malaria and typhus,[46] and for controlling agricultural pests affecting cotton, fruits, potatoes and corn.[47] DDT was also used against head lice without observed human toxicity.[48] It is estimated that around 400 000 tons of DDT were used annually in the 1960s, and roughly 80% comprised agricultural use.[40]

The mode of action of DDT has been debated for a long time and was found to be similar[45] to that of natural pyrethrins and synthetic pyrethroids type I (see Chapter 3). It is generally accepted that DDT acts as a contact axionic nerve poison,[49] by binding the insect voltage-gated sodium channel proteins and affecting their peripheral nerves and brain; a high lipophilicity is required for the contact poison to penetrate the insect cuticle and reach the nerves.[50] Binding of DDT inhibits the closure of the sodium channel, which remains in its open-state, and as a result the inwards conductance of Na$^+$ is prolonged in time,[51] provoking a membrane depolarization and appearance of a residual slow-acting current, called "tail current".[52,53] The effects in the insect are hyperexcitation, tremors, paralysis ("knockdown") and death.[45] Despite the similarities between insect and vertebrate sodium channels, DDT binds selectively to the insect sodium channels. This is due not only to a higher affinity, but also to the body temperature difference between insects (15–20 °C) and vertebrates (37 °C for humans).[53] It has been demonstrated that the insecticidal potency of DDT decreased with temperature; the lower temperature of insects in comparison with that of vertebrates allows for a faster action. Furthermore, the lower temperature and reduced size of insects' bodies reduce the chance of deactivation by metabolization.[52] Furthermore, DDT penetrates the insect cuticle, but not the mammal's skin. Latter studies also suggested that DDT provoked disruption of some other cellular functions connected to membranes, such as oxidative phosphorylation and the Hill reaction.[54]

DDT was particularly effective in controlling the populations of *Anopheles* mosquitoes, the vectors transmitting malaria from the protozoal parasites of the genus *Plasmodium*,[55] especially in tropical countries comprising sub-Saharan Africa, Southeast Asia, the Pacific Islands and South America, where malaria was an endemic disease.[55] DDT was sprayed both indoors and outdoors as part of the *Global Malaria Eradication Campaign*, and by the end of the 1960s, malaria was eradicated in developed countries and in most of sub-tropical Asian and Latin American regions.[47] For instance, it was estimated that in 1947, 75 million malaria cases existed in India (roughly 23% of the population), whereas in 1964 only 100 000 cases were detected.[56] Conversely, the disease re-emerged again in subsequent years, reaching 1.1 million reported cases in 2000.[57]

A major environmental concern of DDT is its long persistence in the environment, and the first adverse effects started to be studied in the 1950s. The intrinsic low reactivity of DDT provokes a bioaccumulation *via* the food chain in adipose tissues, due to it pronounced lipophilicity.[40] As a result, a biomagnification takes place,[46] where those living organisms located on the high trophic levels of the ecological pyramid show DDT concentrations above the amount originally used, even when located quite far away from the initial application point. One of the most important observed effects of the bioaccumulation was the thinning of avian eggshells, probably by hormonal changes or disruption of the calcium metabolism.[46] The low water solubility led to accumulation as deposits in the water currents. In contrast, methoxychlor **4** is eliminated much faster than DDT, mainly by oxidation and dechlorination processes.[58]

The toxicological effects of DDT and its analogues have been extensively reviewed,[40,41,46,47] covering a vast number of epidemiological studies. Many of these studies are not easily carried out, however, and in some cases no definite conclusions can be reached. High concentrations of DDT were found in workers handling the chemical, and in the rest of the population, DDT was found to accumulate mainly through food.[47] In people having a strong occupational exposure, decline of the neurobehavioral function[46] was a clear symptom of DDT-induced neurological damage, probably by affecting human sodium channels.[41]

Slow environmental degradation of DDT furnishes, among other derivatives, 1,1-dichloro-2,2-bis(*p*-chlorophenyl)ethylene **22** (DDE), which is even more persistent than DDT itself. DDE is also one of the metabolites resulting from DDT biotransformation upon ingestion.[41] Exposure to DDT and DDE has been found to induce an increase in liver weight and hepatic necrosis.[47] Although there are some reports on DDT and DDE being carcinogenetic in mice, rats and non-human primates, DDT is only considered as a possible carcinogen in humans, as no definite data are available.[47] DDE was reported as an androgen receptor antagonist, whereas DDT showed estrogenic effects in rats.[40]

Another problem associated with DDT is that insects can develop resistance, so its effectiveness decreases.[45,51] It is for this reason that an increase is observed in the malaria cases in DDT-mediated *Anopheles* mosquito control in India. Resistance can be provoked either by mutations affecting the voltage-gated sodium channel, making it less sensitive to DDT binding (knockdown resistance, *kdr*) or by detoxification mechanisms that speed up metabolization. Insect resistance has been reported in important pests, usually accompanied by what is called *super-kdr*, a stronger resistance to pyrethroids,[45] which as indicated above shares the same insect target with DDT. Moreover, DDT possesses an irritant and repellent nature, so it is possible that mosquitoes may leave before ingesting the lethal dose.[47]

A large number of DDT derivatives were synthesized, as an attempt to obtain agrochemicals with enhanced activity, and improved environmental behavior.[59,60] SAR[54,61] and QSAR[62] studies were carried out to address the steric and electronic factors[61] that are important in active DDT analogues. Such studies revealed that *p*-disubstituted phenyl rings are essential for the derivatives to show insecticidal activity. Furthermore, the substituents on the ring must be relatively non-polar and no larger than a butoxy group (*e.g.* –alkyl, –OR, –SR or –X).[54]

DDT analogues bearing alkyl or alkoxy substituents (*e.g.* ethyl-DDT **3** or methoxychlor **4**) turned out to undergo biodegradation by enzymatic-mediated processes involving side-chain oxidation or *O*-dealkylation, respectively,[63] the former being the more effective process. The presence of alkylthio groups also allows side-chain oxidation.[64] Substitution in the phenyl rings, however, decreases the ratio of biodegradation by hydroxylation of the aromatic rings.

Methoxychlor **4** (oral LD_{50} of 5000–6000 mg kg^{-1} in rats)[41] is one of the most important commercially-available DDT analogues. It is less effective as an insecticide than DDT, however, and much more expensive to produce industrially (as anisole is more costly than chlorobenzene as starting material).[48]

The substituent on the methyl residue has also some restrictions concerning size and polarity. Optimal substituents are the trichloromethyl moiety, $-CH(NO_2)CH_3$ or $-CH(NO_2)CH_2CH_3$, and some DDT analogues of this type have been marketed.[41] Some other groups, such as hydroxyl or amino groups, were found to be too polar to confer insecticidal activity.[50]

Metcalf's group accomplished[63] the preparation of a series of symmetrical and asymmetrical α-trichloromethylbenzylanilines as DDT analogues, significantly more biodegradable than DDT starting by condensation of substituted aldehydes and anilines in refluxing ethanol to give Schiff's bases, followed by treatment with trichloroacetic acid.[63] α-(Trichloromethyl)benzyl phenyl ethers were also prepared by acid-promoted condensation reaction between α-(trichloromethyl)benzyl alcohols and phenol derivatives. Besides nitrogen, sulfur and oxygen have also been used in numerous derivatives as bridging atoms[39,63] between the two phenyl rings, and some of the corresponding derivatives also showed insecticidal activity close to that of DDT.

Fukuto and co-workers reported[59] the preparation of 2,2-bis(*p*-ethoxyphenyl)-1,1-dichloropropane as a DDT analogue with a higher toxicity against house flies. Such a derivative possesses two key structural aspects: the *p*-ethoxy substituent in the phenyl rings (which allows a biodegradable scaffold in the molecule), and the replacement of the benzylic proton by a methyl group.[59] Although it might confer more steric hindrance because of the larger size of the methyl compared to hydrogen, the trichloromethyl moiety of DDT was also replaced by a dichloromethyl group, thus compensating for the steric hindrance.[59]

Replacement of the methylene bridge or of the trichloromethyl group by silicon led to no insecticidal activity.[65] Hybrid structures of DDT-pyrethroids with insecticidal activity have also been reported.[66]

Due to ecological concerns, DDT was firstly banned in Sweden in 1970, followed by the former USSR (1970) and in most countries worldwide in 1972.[40] Nevertheless, in 1985, 300 tons of DDT were still exported from the USA. Underdeveloped countries nowadays claim for a controlled use of DDT against insect vectors transmitting malaria, still a major disease in many tropical underdeveloped countries. DDT is being produced in three countries: India, China and North Korea, and is currently used in 14 countries, although some others are still debating about its re-introduction.[67]

2.3.1.2 Cyclodiene Insecticides

Cyclodiene insecticides are comprised of compounds obtained by a Diels–Alder reaction, and were first marketed after World War II. The key synthetic intermediate is hexachlorocyclopentadiene **26** that can be obtained by chlorination of cyclopentadiene **25**, but also from pentane, cyclopentane or neopentane (see Scheme 2.2).[68] A Diels–Alder cycloaddition between cyclopentadiene and **26** gave an intermediate called chlordene **27**, whose chlorination afforded mainly a mixture of *cis-* and *trans-* isomers on 1 and 2 positions. These isomers are called α- and β-chlordane (**6** and **7**, respectively), which are the oldest cyclodiene insecticides (1944),[68] and whose technical product is a mixture of up to 14 compounds.[54] Chlordane was used against sucking and blowing insects and soil pests, and also as an acaricide as a contact, respiratory and stomach poison.[43]

Chlordane is roughly 300 times more active than its precursor chlordene **27**. Furthermore, β-chlordane is much more active than its α-isomer, whereas the toxicity against mammals is reversed.[43] These data suggest that for insecticidal activity, not only the number of chlorine atoms, but the stereochemistry of the compound is important.

A related compound that is also present in chlordane formulations is heptachlor **8**, whose synthesis also involves **27**, followed by chlorination with sulfuryl chloride or chlorine in the presence of a peroxide (see Scheme 2.2). Heptachlor has been used against termites and soil insects, and apparently undergoes epoxidation upon metabolization by the insect. The resulting epoxide has been proved to exhibit greater insecticidal activity than heptachlor itself.[54]

Another very popular organochlorine insecticide of the cyclodiene family is aldrin **9**, which can be obtained by Diels–Alder reaction between

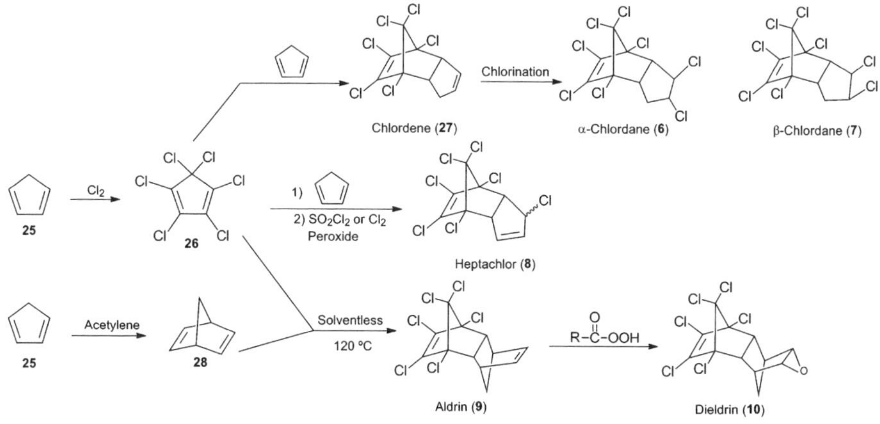

Scheme 2.2 Synthesis of chlordane, heptachlor, aldrin and dieldrin.

hexachlorocyclopentadiene **26** and norbornadiene[68] (available by another Diels–Alder cycloaddition between cyclopentadiene and acetylene) under solventless conditions (see Scheme 2.2), and the *endo-exo* isomer was obtained. This insecticide was especially useful for controlling soil-dwelling insects, due to its chemical stability and high vapor pressure,[43] with better results than DDT. Epoxidation of aldrin with a peracid (*e.g.* peracetic acid, perbenzoic acid), or alternatively with hydrogen peroxide and tungstic oxide, yielded dieldrin **10** (*exo*-epoxide), quite commonly used in the past on crops, soil and seed dressing applications,[68] and also for wood preservation in buildings.[69]

In a similar fashion, a Diels–Alder reaction between hexachlorocyclopentadiene as the diene and acetylene as the dienophile (see Scheme 2.3),[68] followed again by a cycloaddition reaction of the intermediate with cyclopentadiene, afforded isodrin **11**, an isomer of aldrin. Alternatively, isodrin was also prepared by a Diels–Alder reaction of **26** with vinyl chloride, followed by heating in the presence of cyclopentadiene.[43] Epoxidation of **11** with peracetic acid yielded endrin **12**,[68] of interest against insects affecting cotton. Endrin undergoes much faster biodegradation than isomeric dieldrin **10**, mainly by a hydroxylation reaction followed by biosynthesis of glucuronides or sulfates which are excreted *via* urine in mammals.[68]

Another derivative of commercial importance obtained *via* a Diels–Alder reaction is endosulfan, prepared as depicted in Scheme 2.3 by a cycloaddition reaction between hexachlorocyclopentadiene **26** and *cis*-pent-2-en-1,4-diol, followed by treatment with thionyl chloride.[70] Endosulfan exists as a mixture of α- and β-isomers **13** and **14**, respectively, resulting from chirality in the sulfur atom, although it has been demonstrated that on storage an irreversible conversion into the α-isomer can take place.[71] Endosulfan has been used in a 7 : 3 ratio for commercial purposes[71] for pests affecting cereals, fruits, vegetable, cotton and tobacco (*e.g.* chewing insects and mite pests), although it was found

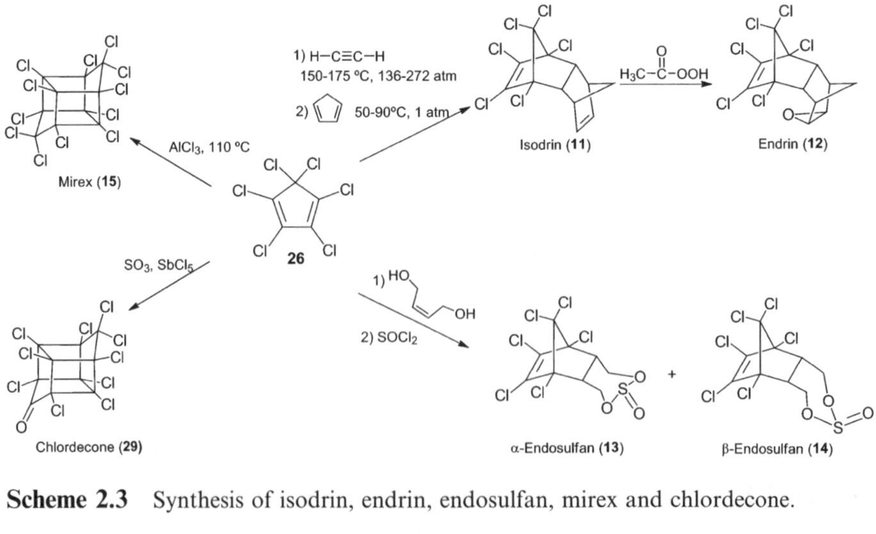

Scheme 2.3 Synthesis of isodrin, endrin, endosulfan, mirex and chlordecone.

to be quite toxic to mammals ($LD_{50} = 2\,mg\,kg^{-1}$ in cats) and highly toxic to aquatic organisms, especially fishes ($LD_{50} = 0.005$–$0.0010\,mg\,l^{-1}$).[43] Endosulfan is quickly excreted by initial hydrolysis of the sulfite moiety or by its oxidation to the corresponding sulfate.

Alternatively, another synthetic route used for the preparation of cyclodiene insecticides is dimerization of hexachlorocyclopentadiene (see Scheme 2.3). If such dimerization is carried out in the presence of $AlCl_3$, mirex **15** is obtained.[68] Mirex was widely used for the control of fire ants, but also as a flame retardant. Remarkably, mirex exhibits a worse toxicological profile than other related organochlorine insecticides, due to its high environmental stability and biomagnification, and toxic effects have been reported for occupational exposure.[68] Treatment of **26** with SO_3 in the presence of $SbCl_5$ furnished chlordecone (Kepone®) **29**.

Cyclodiene insecticides, like DDT, are also nerve poisons, acting in ganglia. In this case, they cause disruption of the normal functioning of the chloride channel, activated by neurotransmitter GABA (γ-aminobutyric acid). Cyclodiene insecticides are non-competitive inhibitors (antagonists) of post-synaptic binding of GABA to its receptor,[72] causing increased nerve activity and high-frequency discharges. As a result, the insect undergoes hyperactivity, hyperexcitability and convulsions.[73] Mutations leading to a modified GABA receptor have been reported to be responsible for cyclodiene insecticide resistance.[73]

In most cases, cyclodiene insecticides are more toxic than DDT, exhibiting lower LD_{50} values against mammals. In some cases, a cyclodiene insecticide dose of only $10\,mg\,kg^{-1}$ is sufficient to cause toxic effects in humans.[58]

In general, cyclodiene derivatives are quite persistent in the environment. Although S_N1 or E1 reactions on the chlorine atoms located on the bridgehead positions would lead to stabilized tertiary carbocations, such reactions are not possible due to strain restrictions for the planar conformation of the carbocation.[48] Instead, cyclodiene insecticides are slowly degraded in the environment.

2.3.1.3 Cyclohexane-Derived Insecticides

The most significant example of insecticides within chlorinated cyclohexane derivatives is γ-HCH **16**, one of the eight possible stereoisomers of 1,2,3,4,5,6-hexachlorocyclohexane (HCH), sometimes misnamed benzene hexachloride, and the only one with important insecticidal properties.[48] The insecticidal activities of γ-HCH were discovered in 1943, almost 120 years after its initial synthesis by Michael Faraday (1825). The structural difference between the hexachlorinated isomers is the number and arrangement of equatorial (e) and axial (a) chlorine atoms in the cyclohexane scaffold.

The preparation of hexachlorinated cyclohexane involves the photochemical chlorination of benzene, in a free radical-mediated process and in the absence of a catalyst that could promote substitution on the benzene ring. The reaction yields a mixture of the five major and more stable stereoisomers: α-HCH (aaeeee, 65–70%), β-HCH (eeeeee, 7–10%), γ-HCH (aaaeee, 14–15%), δ-HCH (aeeeee, 7%) and ϵ-HCH (aeeaee, 1–2%), together with three other minor ones: (1–2%): ζ-HCH (aaeaee), η-HCH (aeaaee) and θ-HCH (aeaeee).[74,75] The composition of the mixture can vary, however, depending on technical differences in the production process.[76] The resulting reaction mixture is called technical-grade HCH; when such a mixture is extracted with methanol, followed by fractional crystallization, it affords γ-HCH in a 90 to >99% purity; this formulation is called lindane.

HCH (either as a mixture or as lindane) emerged as a substituent for DDT, and has been extensively used since the 1940s for both agricultural (seed and soil treatment, *e.g.* against wood-inhabiting beetles, grasshoppers, rice insects) and non-agricultural uses (*e.g.* in pharmaceutical products against ectoparasites and control of vector-transmitted diseases in humans, poultry and cattle).[74,77] HCH was available in a large variety of formulations, such as wettable powders, emulsions, solutions in organic solvents, dust or aerosols. The mode of action of HCH against insects, either by contact or by ingestion, is the same as described above for cyclodiene insecticides, that is, acting as an antagonist of the post-synaptic binding of GABA to its receptor.[72]

Nevertheless, the disposal of large amounts of chlorinated cyclohexanes for the preparation of lindane formulations, as well as the indiscriminate use of this family of organochlorine insecticides was soon an environmental concern.[74] β-HCH isomer is one of the least volatile and least degradable derivatives (its blood half-life is 7.2 years), so it can be found in soil, blood or milk, as a result of local contamination with technical-grade HCH.[77] By contrast, more volatile derivatives, such as α- and γ-HCH isomers, can be found even in Arctic regions and seawater, indicating a long migration capacity. Some HCH isomers exert neuronal, reproductive or endocrine damage.[77]

When some stereochemical factors are fulfilled, certain isomers can undergo pH-dependent elimination reactions in soils, more favorable at basic pH values.[78] Another source of environmental degradation is reductive dechlorination reactions in the presence of certain catalysts. For such degradation processes, the rate has been found to increase with the number of axially-arranged chlorine atoms. Aerobic and anaerobic degradation of HCH has also been reported to be exerted by certain microorganisms (*e.g. Pseudomonas* sp. or *Clostridium* sp., respectively).[78]

Due to general high environmental persistence, bioaccumulation and toxicity effects, many developed and developing countries started banning or restricting the use of HCH-based insecticides in the 1970s–1980s. In this context, lindane has been considered by the US EPA[79] to be quite toxic, leading to liver, blood, nervous, cardiovascular and immune damage upon long-term exposure. Furthermore, lindane is also considered as a potential human carcinogen, although it is still allowed in the "second-line treatments" of human head lice and scabies.

2.3.1.4 *Minor Insecticides: Polychlorinated Benzenes and Terpenes*

Besides the three major families of organochlorine insecticides covered in sections 2.3.1.1–2.3.1.3, there are two other families of insecticides that have been produced in much lower quantities: polychlorinated benzenes and terpenes. The former family is prepared by catalytic chlorination of benzene or derivatives,[80] or, for example in the case of pentachlorophenol **17** (a termiticide used in wood preservation), by basic hydrolysis of non-active isomers of HCH, followed by subsequent aromatic chlorination. Such a process involves very toxic and persistent side-products, such as dibenzodioxins and dibenzofurans.

On the other hand, the most important polychlorinated terpene is toxaphene **20**, prepared by chlorination of camphene, which is obtained from turpentine.[2] Toxaphene is actually the most complex mixture commercially available, as it is comprised of more than 150 products (polychlorinated bornanes, camphenes and bornenes), and became one of the most popular insecticides in the USA in the mid-1970s. Toxaphene was used for protecting a plethora of crops such as cotton, soybeans or tobacco, as well as for the removal of exoparasites on livestock.[81] Nevertheless, it became a major pollutant in the Arctic, Antarctic and Scandinavian regions, due to the fact that some of the components remain unmetabolized and accumulated in the environment.

2.3.1.5 Current Status

All the organochorine products introduced in this section have been extensively revisited, and subsequently most of them have been banned and their production cancelled in most countries. In 2008, the *Stockholm Convention on Persistent Organic Pollutants* (*POPs*), a global treaty promoted by the UN in 2001 to eliminate or reduce the use of highly persistent chemicals,[82] managed to get ratification by almost 170 parties.

The Pesticide Action Network (PAN) dedicated their main focus on a group of polychlorinated chemicals considered to be especially persistent and environmentally hazardous compounds. These derivatives were the so-called "dirty dozen":[82] aldrin, chlordane, DDT, dieldrin, endrin, heptachlor, hexachlorobenzene, mirex, toxaphene, polychlorinated biphenyls (PCBs), polychlorinated dibenzo-*p*-dioxins and polychlorinated dibenzofurans (the two latter groups being obtained as side-products in some industrial processes). Initial exemptions were included for limited use of most of these compounds, although final revisions of the Convention led to a complete elimination of some compounds such as α- and β-HCH or chlordecone.

2.3.2 Organophosporous Insecticides

2.3.2.1 General Aspects

The organophosphorus (OP) insecticides are esters of phosphorus, mainly phosphates, phosphorothionates, phosphorothiolates, phosphorodithioates, phosphonates, and phosphoramidates.[83] These compounds were developed in the late 1930s by Gerhard Schrader, a German chemist working for IG Farben. Some of the synthesised organophosphorus compounds were extremely toxic,

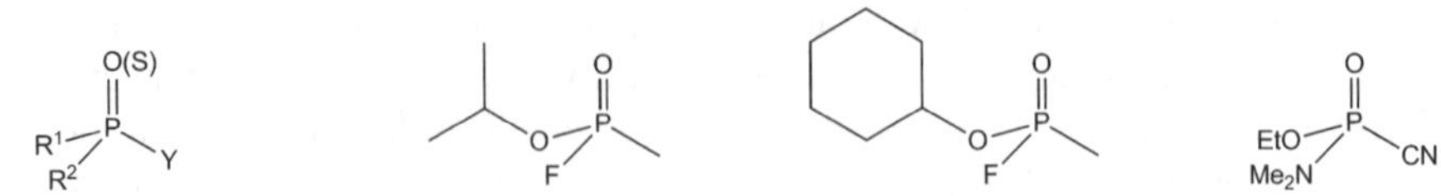

Figure 2.2   General structures for OP insecticides and main nerve gases.

and led to the development of nerve gases such as sarin **30**, cyclosarin **31** and tabun **32**.[84–86] These compounds were classified as "weapons of mass destruction" by the Chemical Weapons Convention, which was signed in 1993 to ban the manufacture and use of chemical weapons.[87] Almost all countries have now joined the Convention. The use of organophosphorus compounds as pesticides began in the 1930s and took off in the 1960s, when they were considered as an alternative to organochlorines.

Active organophosphorus insecticides have the general formulae shown in Figure 2.2, where R^1 and R^2 are short-chain alkoxy, alkyl or amino groups, mainly either methoxy or ethoxy groups, and Y is a displaceable group, including alkoxy, thioalkoxy, vinyloxy, aryloxy and heteroaryloxy groups.

Nowadays, among the various families of pesticides that are being used all over the world, organophosphorus pesticides are the most widely used group, accounting for more than half of the total world market.[88,89] Over 100 000 OP compounds have been screened as insecticides, of which about 100 are marketed in estimated quantities of *ca.* 2×10^5 ty^{-1}.[90] As an example, in the USA market, the use of organophosphates, as a percent of total used insecticides, increased from 58% in 1980 to 70% in 2001.[91] The main families of organophosphorus insecticides are depicted in Figures 2.3–2.5.

Organophosphorus insecticides are available with a wide range of properties. Some have a prolonged persistence, such as parathion **46** and azinphos-methyl **52**, whereas malathion **42** has a short residual action.[92] Compounds such as monocrotophos **35**, malathion **42**, parathion **46**, and chlorpyrifos **49** have a broad spectrum of action. By contrast, trichlorfon **57** is a selective insecticide.[93] Some can be used as plant systemic insecticides such as demeton-S **36** and dimethoate **38**.[94] Others, such as famphur **43** and phosmet **54**, act as animal systemic insecticides.[95,96]

2.3.2.2   *Preparation of Organophosphorus Insecticides*

The synthesis of organophosphorus insecticides involves numerous different reactions. In most cases, the preparation starts by treating phosphorous with sulfur or chlorine to give diphosphorus pentasulfide or phosphorus trichloride, respectively. Subsequently, they are transformed into the main intermediates **58–64** depicted in Scheme 2.4.[97] As an example, phenyl phosphorothionates **65** can be made by treatment of dialkyl phosphorochloridothionates **64** with phenols in basic medium (see Scheme 2.4). This is the approach used for the

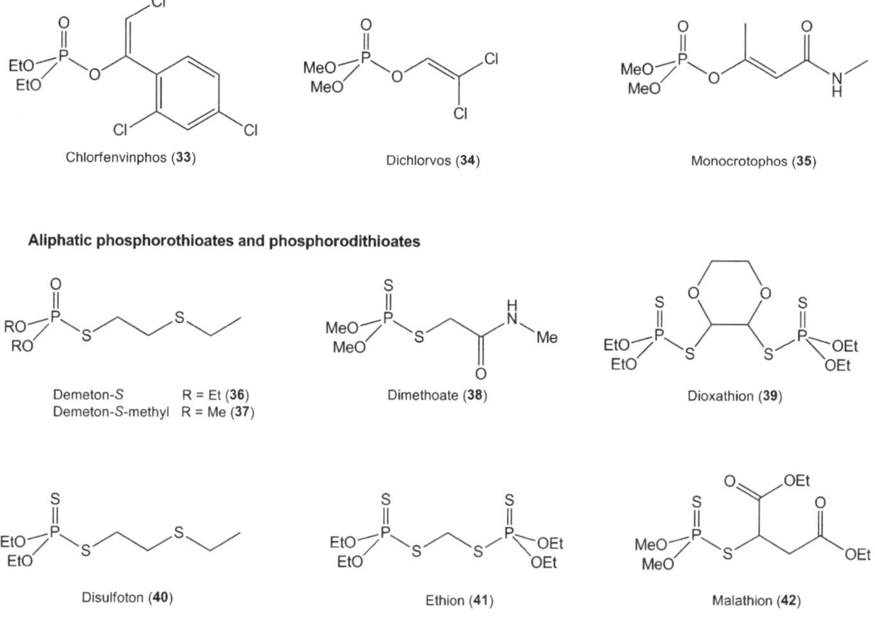

Figure 2.3 Main commercially available OP insecticides of vinyl phosphate and aliphatic phosphorothioate/phosphorodithioate families.

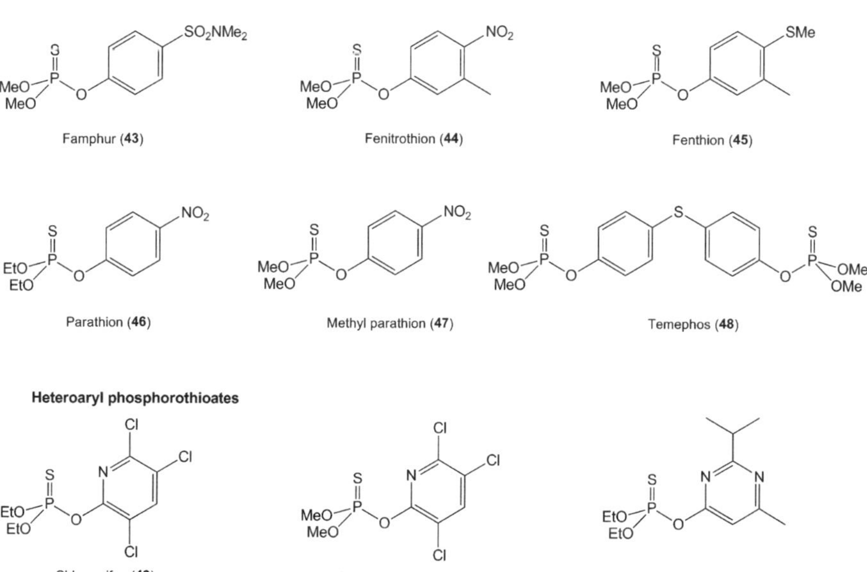

Figure 2.4 Main commercially available OP insecticides of aryl phosphorothioate and heteroaryl phosphorothioate families.

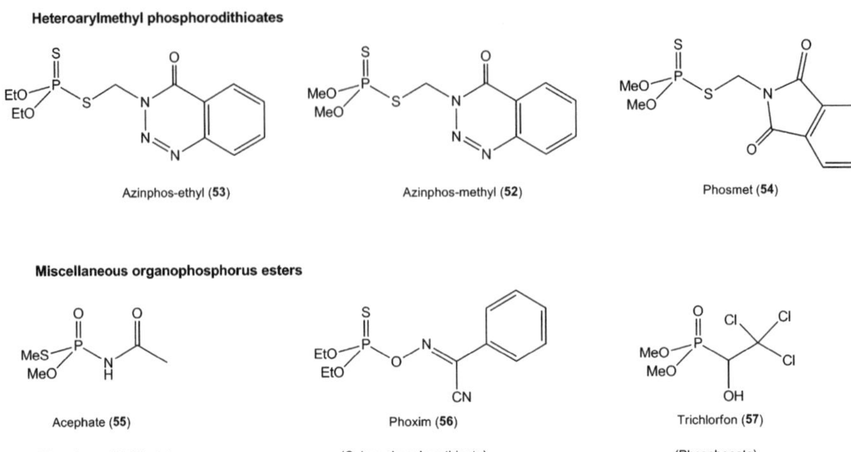

Figure 2.5 Main commercially available OP insecticides of heteroarylmethyl phosphorodithioate family. Miscellaneous structures.

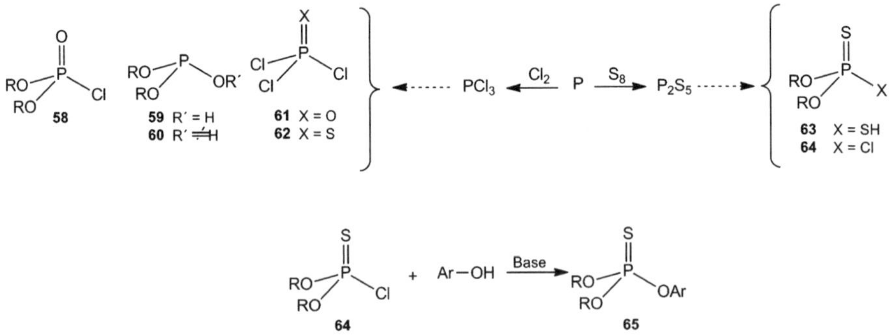

Scheme 2.4 General intermediates for the synthesis of OP insecticides

synthesis of parathion **46**. Furthermore, *O,O*-dialkyl phosphorodithioic acids **63** are used to yield phosphorodithioate esters. For example, malathion **42** is prepared by reaction with diethyl maleate.[97]

2.3.2.3 Classification of Organophosphorus Insecticides

2.3.2.3.1 Vinyl Phosphates. Substituted vinyl phosphates ($R^1CH{=}CR^2$) $P({=}O)(OR^3)(OR^4)$ have shown excellent insecticidal activity and desirable stability properties. Due to their volatility and biological instability, certain substituted vinyl phosphates are suitable for use on food crops near harvest time, as it is advantageous to have an insecticide with little persistence to alleviate the danger of toxic residues.[98,99] The phosphate esters are more reactive in organisms than the corresponding phosphorothioate esters, as a phosphorus atom doubly bounded to oxygen (P=O) is more electrophilic that its P=S counterpart. As significant examples of insecticides belonging to this group, chlorfenvinphos **33**, dichlorvos **34** and monocrotophos **35** should be mentioned (see Figure 2.3). The three examples considered in this family were classified in the group Ib (highly hazardous) by the WHO (World Health Organization).[100–102]

Chlorfenvinphos **33** is a hydrolysable insecticide, applied in soil to control root-flies, rootworms and other soil pests, although it was also used to control household pests such as flies, fleas and mites. It was banned by the US EPA in 1991,[103] and in the EU in 2006.[104]

Dichlorvos **34** is a rapid-acting insecticide, used in aerosols against flies and mosquitoes. It has also been incorporated in pet collars against ectoparasites. It was reregistered by EPA in 2006.[105] In the EU, **34** was re-evaluated and it was not included in Annex I to Directive 91/414/EEC in 2006,[106] and in 2010 virtually all use was prohibited.[107]

Monocrotophos **35** is a systemic broad spectrum insecticide, with stomach and contact action. It is used to control sucking, chewing pests and common mites, ticks and spiders, typically on cotton, citrus and olives. In the USA, this substance was voluntarily withdrawn by the registrant in 1989, and banned in the EU in 2003.[104] No remaining uses are currently allowed.[108] Monocrotophos is one of the most toxic pesticides to birds (LD_{50} 0.9–6.7 mg kg^{-1}) and is highly toxic to bees (33–84 µg bee^{-1}), as well as to humans.[108] Monocrotophos was also identified as the cause of paralysis in children in a cotton-growing area.[109]

2.3.2.3.2 Aliphatic Phosphorothioates and Phosphorodithioates.

Phosphorothioate esters have two general formulae: $R^1OP(=S)(OR^2)(OR^3)$ for phosphorothionates and $R^1SP(=O)(OR^2)(OR^3)$ for phosphorothiolates. Phosphorodithioate esters have the two following general formulae: $R^1SP(=S)(OR^2)(OR^3)$ for phosphorothiolthionates and $R^1SP(=O)(SR^2)(OR^3)$ for phosphorodithiolates (see Figure 2.3).

Demeton-*S* **36** is used as a mixture with its thiono isomer as a systemic insecticide to control aphids and other sucking insects, sawflies and spider mites on a range of crops. This mixture is classified by WHO as Ib, highly hazardous,[110] and as a result, all products containing this insecticide were banned in the USA by 1989.[111]

Demeton-S-methyl **37** is the *O,O*-dimethyl ester analogous to demeton-*S*, used to control aphids and other sucking insects, sawflies and spider mites on a range of crops as a systemic insecticide with contact and stomach action, manufactured by Bayer Cropscience. It is classified by WHO as Ib, highly hazardous,[110] with a LD_{50} of 30 mg kg^{-1} (oral, rat). In the EU, its registration was withdrawn in 2002.[112]

Dimethoate **38** is a systemic broad spectrum insecticide with contact and stomach action, used against sucking insects such as aphids, and against fruit flies. It can also be used for a postharvest dip treatment. The US EPA revoked certain tolerance of its use in 2008.[113] It is classified II by WHO, moderately hazardous, LD_{50} of $150\,mg\,kg^{-1}$ (oral, rat).[114] In the EU, it was re-evaluated and included in Annex I to Directive 91/414/EEC in 2006, allowing its use.[115]

Dioxathion **39** was used as a livestock insecticide and acaricide for the control of insects and mites on grapes, walnuts, ornamentals, apples, pears and quince. It is now considered obsolete and its use as a pesticide has been discontinued by WHO,[116] and is no longer allowed to be sold in the USA.

Disulfoton **40** was approved as a systemic insecticide, registered for use to control aphids, thrips, and other sucking insects, as well as spider mites on a variety of crops (frequently asparagus, beans, broccoli, Brussels sprouts, cabbage, cauliflower, lettuce, coffee, cotton and Christmas trees). It is also registered in the USA[117] for residential uses including flowers, shrubs and ornamentals, although it was classified by WHO in the group Ia, extremely hazardous.[116] Dioxathion and disulfoton registration in the EU were discontinued in 2002.[112] This insecticide is translocated through plant tissue, and activated by microsomal oxidation. This oxidation also takes place in the mammalian liver, making this insecticide extremely toxic to humans.

Ethion **41** is an acaricide and insecticide used on fruit trees, fiber and ornamental crops, including greenhouse crops, lawns and turf. It may also be used on cattle. It is efficient against aphids, mites, scales, thrips, leafhoppers, maggots and foliar feeding larvae. It is mixed with oil and sprayed on dormant trees to kill eggs and scales. WHO classification is II, moderately hazardous, exhibiting a LD_{50} of $208\,mg\,kg^{-1}$ (oral, rat). It was withdrawn from the EU in 2003.[118]

Malathion **42**, the most commonly used organophosphate insecticide[119] in the USA, is a persistent general-purpose insecticide. It is used in household, home garden, vegetable and fruit insect control, as well as in the control of insects affecting public health (*e.g.* flies, mosquitoes[120] and lice). Malathion is also sprayed, aerially, over cities, suburbs and farmland to control mosquitoes and Mediterranean fruit flies.[121] As a result of the emergence of pyrethroid-resistant head lice, malathion (0.5% formulation) has been approved by the FDA for the treatment of pediculosis.[122,123] However, oral ivermectin has been shown to be more efficient than malathion lotion for difficult-to-treat head lice.[124] Although malathion itself is of low toxicity, absorption or ingestion into the human body readily results in its metabolization to malaoxon, *via* oxidation of the P=S moiety to P=O, which is substantially (61 times) more toxic.[125] High levels of malathion metabolites in children's urine seem to be related to an increased risk of attention deficit hyperactivity disorder (ADHD).[126]

2.3.2.3.3 Aryl Phosphorothioates. The aryl OPs (see Figure 2.4) are generally more stable than the aliphatic analogues, therefore their residues are longer lasting. Famphur **43** is used as an insecticide for the lice and grubs of reindeer and cattle. WHO classification is Ib, highly hazardous, with a LD_{50} of $48\,mg\,kg^{-1}$ (oral, rat),[116] and it is quite toxic to birds.[127] In the US EPA its registration was cancelled in 1989.[128]

Fenitrothion **44** is a contact insecticide effective against penetrating, chewing and sucking insect pests on cereals, cotton, fruits, vegetables and forests, as well as against household insects (flies, mosquitoes and cockroaches). WHO classification is II, moderately hazardous, with a LD_{50} of 503 mg kg^{-1} (oral, rat).[116] In the USA, fenitrothion is registered and formulated for use only in indoor and outdoor bait stations, the target species being ants and roaches.[129] In the EU fenitrothion is severely restricted for pesticide use, and virtually all use is prohibited.[130]

Fenthion **45** is a persistent contact and stomach insecticide used against sucking and biting pests,[131] for veterinary hygiene purposes and as a mosquito larvicide. It has also been used for bird pest control, especially weaver birds. WHO classification is II, moderately hazardous, with a LD_{50} of 586 mg kg^{-1} (oral, rat). Fenthion exposure to the general population is quite limited based on its bioavailability. The EPA classified fenthion as a Restricted Use Pesticide because is very highly toxic to birds and estuarine/marine invertebrates.[132] In the EU general uses of fenthion were withdrawn in 2005, although, in the absence of efficient alternatives for certain uses in citrus, peaches and olives, it can be used with severe restrictions.[135]

Parathion **46** is highly toxic to non-target organisms and humans by inhalation, oral and dermal exposures. It is among the most highly toxic chemicals registered by the US EPA. WHO classification is Ia, extremely hazardous, with a LD_{50} of 13 mg kg^{-1},[116] and it is in the PAN "dirty dozen" list.[133] The US EPA and the manufacturer, Cheminova, signed an agreement in 2000 to cancel all remaining uses of parathion.[134] From 2006, it was no longer included in formulations in the EU.[135]

Methyl parathion **47** is a non-systemic pesticide that kills pests by acting as a stomach poison. It has similar properties to parathion, but hydrolyzes more readily. It has similar acute toxicity to parathion, with a LD_{50} of 14 mg kg^{-1}.[116] WHO classification is Ia, extremely hazardous, and it is in the PAN "dirty dozen" list.[136] In 2010, the US EPA cancelled all methyl parathion product registration, and all use of stocks shall be prohibited at the end of 2013.[137]

Temephos **48** is used as a larvicide to treat water infested with mosquito and black fly larvae. It is widely used against the mosquito, *Aedes aegypti*, the most important vector of both yellow fever and dengue viruses (see Chapter 1). Up to date, reduction in the population density of the vector mosquitoes is the only option for controlling the transmission of dengue virus, which infects 50–100 million people annually, with 500 000 people being admitted to hospital.[138] Temephos is applied in areas of standing water, shallow ponds, swamps and marshes, and water storage containers in tropical and subtropical urban and semi-urban areas.[139] WHO classification is III, slightly hazardous, with a LD_{50} of 4000 mg kg^{-1} (oral, rat).[116] It is currently registered by the US EPA, confirming temephos as the only organophosphate with any appreciable mosquito larvicidal use.[140]

2.3.2.3.4 Heteroaryl Phosphorothioates. The most significant examples of heteroaryl phosphorothioates are depicted in Figure 2.4. Chlorpyrifos **49** is used widely as a broad spectrum soil insecticide. Today it is the best-selling organophosphorus insecticide and one of the most widely-used pest control products in the world for both agricultural (more than 50 different crops) and non-agricultural pests. For home applications, chlorpyrifos is used to control cockroaches, fleas and termites. It is also used in some pet flea and tick collars. In farms, it is used to control ticks on cattle. In 2005, agricultural uses of chlorpyrifos received the EU approval (Annex I of the European Commission's Plant Protection Products Directive 91/414), allowing the EU members to renew their registrations of chlorpyrifos products.[141] In 2006, the US EPA completed the reregistration eligibility for agricultural use of chlorpyrifos, although its use in homes and other places where children could be exposed is not allowed, and its use on crops is severely restricted.[142] WHO classification is II, moderately hazardous, with a LD_{50} of $135\,mg\,kg^{-1}$ (oral, rat).[116] Chronic exposure has been linked to neurological effects and developmental disorders, and prenatal chlorpyrifos exposure can lead to an increased risk of delays in mental and motor development at the age of three.[143]

Chlorpyrifos-methyl **50** is much safer than chloryphos, and is used to control a range of insect pests in different crops and ornamentals. It is also used to control household pests and stored grain pests. WHO classification is U, unlikely to be hazardous, with a $LD_{50} > 3000\,mg\,kg^{-1}$ (oral, rat). The US EPA registered chlorpyrifos-methyl for use only as an indoor treatment for small grains (wheat, oats and barley) in storage bins and warehouses to control a variety of insects (including beetles, weevils, moths and grain borers).[144] In 2005, agricultural uses of chlorpyrifos-methyl were included in Annex I of the European Commission's Plant Protection Products Directive 91/414.[141] Concerning its environmental behaviour, the hydrolysis rate of chlorpyrifos-methyl was found to be approximately twice that of chlorpyrifos.[145]

Diazinon **51** is a non-systemic insecticide used to control pest insects in soil, on ornamental plants, and on fruit and vegetable field crops. In the USA, all residential uses of diazinon were cancelled in 2004, and agricultural uses on a variety of fruit, vegetable and field crops have been restricted. It is allowed on non-lactating cattle as an ear-tag.[146] WHO classification is II, moderately hazardous, with a LD_{50} of $300\,mg\,kg^{-1}$ (oral, rat).[116] In the EU, authorisations for plant protection products containing diazinon were withdrawn in 2007.[147]

2.3.2.3.5 Heteroarylmethyl Phosphorodithioates. Commercially-available insecticides belonging to this group are included in Figure 2.5. Azinphos-ethyl **53** is a persistent broad spectrum insecticide, effective in the control of chewing and sucking insects of fruits, vegetables, cotton and ornamentals. It is no longer registered for use in many countries due to its extreme acute toxicity in humans. WHO classification is Ib, highly hazardous, with a LD_{50} of $12\,mg\,kg^{-1}$ (oral, rat).[116] The US EPA reregistered the insecticide in 2006, but decided to phase out the remaining uses of the pesticide in 2007, with all uses ending in 2012. It has been banned in the EU since 2002.[104] Azinphos-ethyl is dangerous to bees, fish and harmful to livestock, wild birds and animals. It has a lifetime of hours in the atmosphere, but relatively long persistence (months) in water and in soil where it is biodegraded slowly.[148]

Azinphos-methyl **52**, a broad spectrum persistent insecticide, is used extensively as a foliar application against leaf feeding insects. It works both as a contact insecticide and as a stomach poison. WHO classification is Ib, highly hazardous, with a LD_{50} of 16 mg kg^{-1} (oral, rat).[116] The US EPA reregistered this insecticide in 2006 and it was decided to phase out the remaining uses in 2007, with all uses ending in 2012.[149] It has been banned in the EU since 2002, because it was not approved for inclusion in Annex I to Directive 91/414/EEC.[150]

Phosmet **54** is used to control pests of deciduous fruit. It is a phthalimide-derived, non-systemic insecticide, used on a wide range of fruit trees, ornamentals and vines for the control of aphids, suckers and fruit flies. The compound is also an active ingredient in some dog collars.[151] WHO classification is II, moderately hazardous, with a LD_{50} of 113 mg kg^{-1} (oral, rat). The US EPA issued in 2007 its final decision on the restricted-entry intervals (the time after a pesticide application during which entry into the treated area is restricted) for nine uses of phosmet: apples, apricots, highbush blueberries, grapes, nectarines, peaches, pears, plums and prunes, in order to protect workers and bystanders.[152] In the EU, phosmet was reevaluated (2007), and its uses as an insecticide and acaricide were authorised.[153]

2.3.2.3.6 Miscellaneous Organophosphorus Esters.

Some other different structures of organophosphorus insecticides are included in Figure 2.5. Acephate **55** is a phosphoramidothioate insecticide, with residual systemic activity of about 10–15 days. It kills the insects by direct contact or by ingestion. Therefore, it is used for controlling biting and sucking insects, especially aphids, in vegetables and in horticulture. It also controls leaf miners, lepidopterous larvae (caterpillars), mole crickets, chinch bugs, ants, lacebugs and lawn pests. Used as a turf insecticide, it must be applied 48 hours before irrigation. Wetting the lawn with a 2% soap solution pulls mole crickets and chinch bugs up to the top of the thatch and improves the eradication after spraying. Acephate affects the migratory orientation of adult white-throated sparrows, by affecting the memory of the migratory route.[154] WHO classification is III, slightly hazardous, with a LD_{50} of 945 mg kg^{-1} (oral, rat). In the EU the authorisations for products containing acephate were withdrawn in March, 2003,[155] whereas in 2008 the US EPA imposed new restrictions.[156]

Phoxim **56** is an oxime phosphorothioate insecticide with stomach and contact action, especially useful for controlling stored product pests. WHO classification is II, moderately hazardous, with a LD_{50} of 1975 mg kg^{-1} (oral, rat).[116] It has not been registered in the US EPA, while it is allowed in the EU for use in veterinary medicine for the control of mites, lice and other ectoparasites of pigs and sheep.[157] However, it is banned for use on crops since December 2007.[158]

Trichlorfon **57** is a non-systemic phosphonate insecticide with contact and stomach action, used on golf course turfs, home lawns, food processing plants, ornamental plants and fish ponds, to control arthropod pests such as fleas, flies, ants, cockroaches, water scavenger beetle, water scorpions and giant water bugs, among others. It is a curative grub insecticide, playing an important role in IPM programs for white grubs in turf, allowing turf managers to reduce preventative treatments. WHO classification is II, moderately hazardous, with a LD_{50} of 250 mg kg^{-1} (oral, rat).[116] Trichlorfon is under registration review by the US EPA,[159] while the EU authorisations for plant protection products containing trichlorfon were withdrawn by November 2007.[160]

2.3.2.4 Mode of Action

Organophosphorus insecticides are able to phosphorylate the active site of acetylcholinesterase AChE (EC 3.1.1.7), an enzyme widely distributed in excitable membranes of nerves and muscles, throughout the animal kingdom.[161,162] This enzyme works at the neural synapses to rapidly hydrolyze the neurotransmitter acetylcholine to acetate and choline. When AChE is phosphorylated, acetylcholine accumulates, the nerve remains polarized and there is an over-stimulation at the neuromuscular junctions.[163,164]

AChE is located primarily in the central nervous system of insects, whereas in mammals AChE is found in both the central and peripheral nervous systems. Therefore, differences exist in the symptoms of poisoning by anticholinesterases between the two groups. Those effects (nausea, lacrimation, salivation, sweating and abnormal constriction of the pupil of the eye), associated with poisoning of the parasympathetic nervous system in mammals,[165] are not observed in insects since they do not have a parasympathetic nervous system.[166] Central nervous system poisoning effects (tremors, hyperactivity, ataxia, convulsions paralysis and death) occur in both groups.[167]

The reaction between acetylcholine and AChE takes place in three stages, as shown in Scheme 2.5: (1) formation of the Michaelis complex between the substrate, acetylcholine and AChE; (2) acetylation of the enzyme and formation of choline; and (3) hydrolysis of the acetylated enzyme to give free enzyme and acetic acid. These reactions take place rapidly, so there is no accumulation of acetylcholine across the synapse.[168]

The reaction between the organophosphorus compounds and AChE is analogous, as seen in Scheme 2.5: (1) formation of the complex; (2) phosphorylation of a serine hydroxyl group at the active site of AChE; and

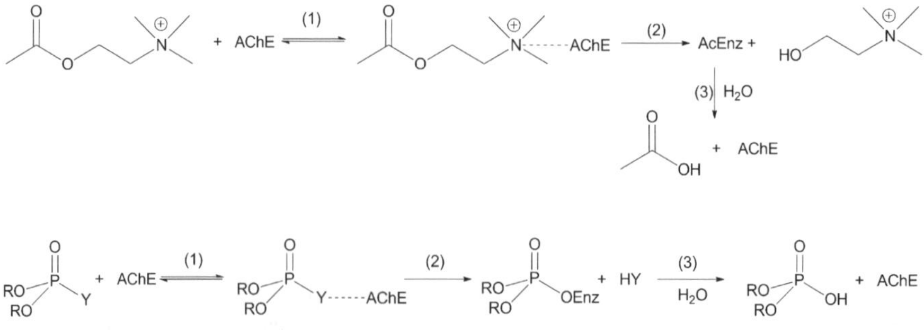

Scheme 2.5 Mode of action of OP insecticides.

(3) dephosphorylation, although this step occurs at an extremely slow rate: from hours (*O,O*-dimethyl phosphates) to days (*O,O*-diethyl phosphates).[90]

The reactivity of organophosphorus insecticides is determined by the electrophilicity of the phosphorus atom; the phosphate esters being much more reactive than the phosphorothionate esters (P=S). The latter are unreactive with acetylcholinesterase, and are active only after *in vivo* oxidation to the phosphate analogues by a microsomal oxidase in the insect gut and in the mammalian liver. For example, parathion **46** is converted into paraoxon by oxidation. The time required for this oxidation (hours), allows treatment with atropine before severe effects are developed. This explains why phosphoro-thionates are less hazardous and have less fatal consequences than their equivalent phosphates.

One of the main drawbacks associated with organophosphorus insecticides is their high toxicity in non-target organisms. Acute poisoning of humans can occur by inhalation, ingestion or dermal exposure. As an example, 120 mg of parathion is known to kill an adult human, and children of five- to six-years old were killed by ingestion of only 2 mg of parathion **46** (about $0.1\,\mathrm{mg\,kg^{-1}}$). Repeated chronic exposure of agricultural workers is also quite dangerous due to the irreversible inhibition of acetylcholinesterase and the need for the enzyme to be resynthesized in the organism. Improper storage, handling and use have caused many cases of poisoning with thousands of deaths occurring each year.[169,170]

In spite of the hazards associated with the use of highly toxic organopho-sphorus compounds, they are still used in great quantities throughout the world. In the USA, OPs account for about half of all insecticides used (by amount sold).[171] In the search for safer OPs, malathion **42** was shown to be about one-hundredth as toxic to humans and higher animals, compared with parathion **46**.[172]

The introduction of a methyl group in the 3-position of the aryl ring of parathion increases the affinity for the insect AChE and decreases it for the mammalian enzyme, due to the stereochemical interaction of the methyl group with the enzyme. This increases the selectivity of insecticides such as feni-trothion or fenthion, which are widely, and safely, used in the control of household and public health pests and animal ectoparasites.

2.3.2.5 Environmental Fate and Toxicity

Organophosphorus insecticides are metabolized readily by oxidation and hydrolysis. Therefore, they do not pose a serious problem of transfer along food chains or risk of bioaccumulation, and subsequently, they are not usually persistent in the environment.

The soil half-life of chlorpyrifos **49** ranges from two weeks to more than one year, depending on soil texture, pH and climate, and it has a very low potential for movement through the soil to groundwater. When applied to moist soils, chlorpyrifos has a volatility half-life of 45 to 163 hours.[173] Diazinon **51** has a low persistence in soil, with a half-life of two to four weeks. It seldom migrates through the soil, but in some instances it may contaminate groundwater.[173] Fenitrothion **44** is degraded in aerobic upland soils with a half-life of less than a week,[174] and it is photochemically degraded under ambient sun light with a half-life range of approximately 5 hours.[175] Malathion **42** displays little persistence in soil, with rapid degradation[176] and reported half-lives in the field range from one to 25 days.[177] In air, malathion is rapidly broken down photochemically, with a half-life of approximately 1.5 days.[176]

Although the OP insecticides are general biocides, toxic to nearly all animals, some of them, such as fenitrothion **44** and malathion **42**, fortunately combine suitable biodegradation with adequate safety to higher animals.[178] The organophosphorus compounds are generally toxic to fish: dichlorvos has a LD_{50} of $51.3\,mg\,l^{-1}$ to zebrafish (24 h exposure, low toxicity), and phoxim has a LD_{50} of $1.28\,mg\,l^{-1}$ to zebrafish (24 exposure, medium-high toxicity).[179] In addition, OP insecticides are highly toxic to beneficial species, such as bees, earthworms, insect predators and insect parasites.[180]

2.3.3 Carbamate Insecticides

2.3.3.1 General Aspects

The insecticides belonging to the carbamate family are esters of carbamic acids, with the general structure depicted in Figure 2.6. There is a strong variability in their spectrum of activity, mammalian toxicity and persistence upon chemical modifications in their general structure.[181]

Carbamate insecticides were developed in the 1950s as a result of investigations on compounds exerting anticholinesterase action on the nervous system similar to that of the organophosphorus insecticides, in an attempt to obtain insecticides with an increased selectively and reduced mammalian toxicity. They are still widely used today.[1,182] The structure of these compounds was inspired by the biologically active alkaloids isolated from plants bearing a carbamato moiety.[183]

As with organophosphorus esters, carbamates exert their toxicological effects against insects by inhibiting nervous tissue AChE found in the synaptic spaces and on the postsynaptic membranes of all neurons. Such activity takes place through a carbamylation reaction in the active site of the enzyme, so a deacylation reaction on acetylcholine does not take place.

Therefore the overall effect of suppressing the normal functioning of such enzyme is an impairment of the conduction of signals in the insect nervous system.[1] Organophosphorus insecticides react irreversibly with acetylcholinesterase (AChE), as indicated in section 2.3.2.4, whereas carbamates act reversibly and their inhibitory effect is so brief that measurements of blood cholinesterase levels in human beings or other animals exposed to them appear to be normal. For this reason, the volume of carbamates needed for getting an insecticidal action exceeds that of organophosphorus insecticides. When used on crops, it is common that several applications of carbamates are required in a growing season.[1]

Some methyl carbamate insecticides proved to exhibit high acute toxicity in mammals, in some cases comparable to that in insects,[65] but they are in general safer than organophosphorus insecticides as a result of the reversibility of their action. The lethal dose of carbamates is much higher than that of organophosphorus insecticides; slight symptoms can appear under long occupational exposure before a dangerous dose is received.[184]

It was also demonstrated that in the presence of ionic substituents, the carbamates exhibited no insecticidal properties despite being, in some cases, potent AChE inhibitors. This is due to the fact that ionic groups prevent the compound from penetrating through the lipophilic sheaths surrounding the nerves.[65]

Carbamates are currently among the most popular insecticides, not only in agriculture, but also for home use, both indoors and on gardens and lawns. These compounds are rapidly detoxified and excreted, so their risk to warm-blooded animals is less than for other agents.[185] They are relatively unstable compounds that break down in the environment within weeks or months, so their persistence is not generally a problem,[182] a major difference from previous organochlorine insecticides.

One significant disadvantage of carbamates is that they are toxic against many beneficial insects, especially honeybees. The widespread use of these compounds has increased the residues in water and foods, especially in fruits, and has led to a higher incidence of accidental poisonings.[186,187] The clinical symptoms in carbamate-mediated poisonings range from the classic cholinergic syndrome, to flaccid paralysis and intractable seizures.

Much effort has also been devoted to the study of the behavior of carbamates in the environment. The identification of degraded products allows the degradation pathways involved to be established, and thus it provides valuable information for potential ways of protecting the environment.[188] Hydrolysis,[189] photodegradation[190] and microbial-mediated biodegradation (*via* carbamate hydrolases)[191] are the main ways of environmental degradation of carbamate insecticides. With some soil-incorporated carbamate pesticides, microbial degradation processes may be so rapid as to limit pesticide effectiveness in controlling targeted pests.[192]

In many occasions, degradation reactions studied under laboratory conditions provide data with significant differences from those obtained in field studies. The reason is the presence in field studies of metal ions,[193] metal oxides[194] or clays[195] that can enhance the transformation rates of the organic pollutants.

In addition, numerous analytical methodologies have been developed for the quantification of carbamate content even at low concentrations in environmental samples, such as soil and water.[196]

The synthesis of carbamates[197] usually involves the reaction of isocyanates (prepared by treatment of amines with hazardous phosgene) with alcohols, or alternatively, the phosgenation of alcohols, followed by reaction with amines (see Scheme 2.6). Due to the environmental problems of the phosgene route, significant attempts have been made to explore a phosgene-free synthesis of carbamates.[198]

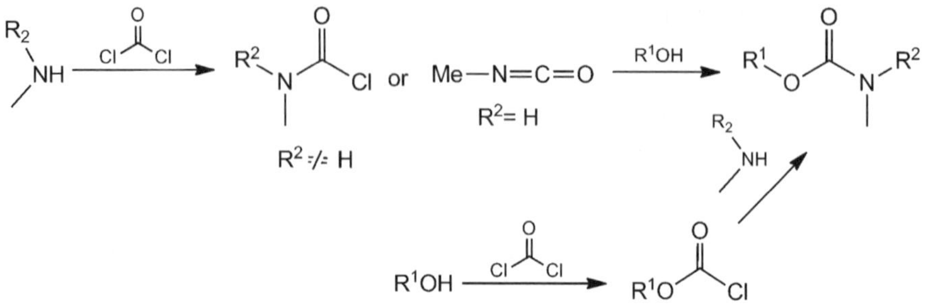

Scheme 2.6   Synthesis of carbamate insecticides.

General formulae for carbamates

Aryl *N*-methylcarbamates

Carbaryl (**66**)

Propoxur (**67**)

N,N-Dimethylcarbamates

Dimetilan (**68**)

Pirimicarb (**69**)

Benzofuranyl *N*-methylcarbamates

Carbofuran (**70**)

Carbosulfan (**71**)

Oxime carbamates

Aldicarb (**72**)

Methomyl (**73**)

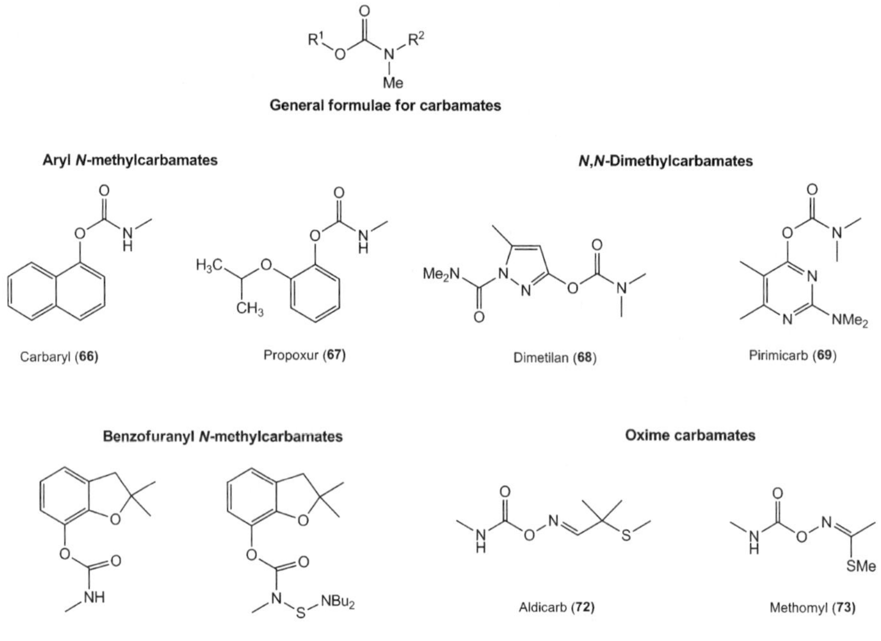

Figure 2.6   General structure and classification of carbamate insecticides.

2.3.3.2   *Classification of Carbamate Insecticides*

Carbamates are usually classified according to the substituent on the NH-moiety, and they fall into two major groups: *N*-methyl and *N*-arylcarbamates, which exhibit insecticidal (both contact and systemic) and herbicidal activities, respectively. A more precise classification of the insecticidal *N*-methyl carbamates is indicated below, and some significant examples from a commercial point of view are depicted in Figure 2.6:

- Aryl *N*-methylcarbamate insecticides (*e.g.* **66**, **67**)
- *N*,*N*-dimethylcarbamate insecticides (*e.g.* **68**, **69**)
- Benzofuranyl *N*-methylcarbamate insecticides (*e.g.* **70**, **71**)
- Oxime carbamate insecticides (*e.g.* **72**, **73**)

Another group of carbamate derivatives is comprised of dithiocarbamates, sulfur-containing carbamates which have little or no esterase-inhibiting action and which are often, but not exclusively, used as fungicides and herbicides. These compounds are not discussed in any further detail in this section.

2.3.3.2.1 Aryl *N*-methylcarbamate Insecticides. Aryl *N*-methyl carbamates became one of the first efficient groups of insecticides in the 1950s, and currently comprise roughly 20 commercial derivatives.[199] They are suitable against small sucking insects such as planthoppers and leafhoppers, and at the same time they lack toxicity against spiders, which act as predators against such insects.[200]

It was demonstrated that the presence of halogens and alkyl residues in either *ortho* or *meta* positions of the phenyl ring led to an increase in insecticidal activity.[201] By contrast, the replacement of the *N*-methyl group with a longer alkyl chain improved the recovery of the insect, and therefore lead to an impairment of insecticidal activity.

One of the most popular carbamates in this groups is carbaryl **66** (1-naphthyl methylcarbamate), commercialized by Bayer under the brand name Sevin[®]. Carbaryl is widely used (granular, liquid, wettable power or dust formulations) to protect food crops, but also for home applications such as gardens,[202] due to its low toxicity in mammals (oral LD_{50} of 500–850 mg kg^{-1} in rats).[203] Propoxur **67** (Baygon[®]) is also quite effective as a home insecticide against cockroaches and mosquitoes,[204] and was introduced as a replacement for DDT. Nevertheless, Propoxur possesses a relatively high toxicity against mammals (LD_{50} of 100 mg kg^{-1} in rats, oral).

As a remarkable example of the lack of an eco-friendly methodology for the production of carbamates, we can mention the accident that occurred on December 3rd, 1984, at the Union Carbide factory in Bhopal (India), where both carbaryl **66** and aldicarb **72** were being synthetized.[205] A gas leak of methyl isocyanate, one of the chemicals used in these syntheses, took place. Methyl isocyanate is slightly heavier than air, so it escaped into the atmosphere and stayed low to the ground. Approximately 40 tons of methyl isocyanate escaped into the air, spreading over a city of nearly 900 000 people. At least 3000 people died in the immediate aftermath, and over half a million people were seriously injured, and since then at least another 20 000 other deaths have occurred from gas-related diseases.

Environmental degradation studies on some of the carbamate insecticides proved that chemical hydrolysis was the major degradation pathway,[189] especially at basic pH values. In the case of phenyl *N*-methylcarbamates, such degradation leads to the formation of methylamine and substituted phenols;

the presence of electron-withdrawing substituents increased the hydrolysis rate. The *in vitro* metabolism of some aryl *N*-methylcarbamates by the fetal tissues of Sprague Dawley rats was also investigated.[206]

Prousalis *et al.* developed[199] a preparation of antibodies from chicken egg yolk to bind aryl *N*-methylcarbamates, which were used as immunosorbent preparations for improving the analytical quantification of environmental residues containing carbamates. Such antibodies exhibited a high binding capacity to carbaryl, trimethacarb, metolcarb, aminocarb and promecarb.

2.3.3.2.2 *N,N*-Dimethylcarbamate Insecticides.

The *N,N*-dimethylcarbamate family of compounds is generally more stable than *N*-methylcarbamates, although their insecticidal properties are considerably lower.[201] Dimetilan **68** and pirimicarb **69** are important examples used in baits for housefly control or as an aphicide on grain crops, respectively.[207]

Most insecticidal carbamates are esters of aromatic and heterocyclic hydroxyl derivatives of *N*-methylcarbamic acid. Reports on aryl *N,N*-dimethylcarbamates are scarce however. In this context, Zhao *et al.*[208] designed and synthesized a series of phthalimido alkyloxyphenyl *N,N*-dimethylcarbamates, in which the phthalimido and phenyl moieties are connected with an alkoxy chain. The variation of the AchE inhibitory activity upon modifying the length of the alkoxy chain and the position of the carbamato moiety on the phenyl ring were analyzed.

Concerning environmental degradation, Huang and Stone[209] studied the hydrolytic cleavage of dimetilan **68** (see Figure 2.7), which can occur *via* nucleophilic attack at the carbamate group or at the substituted ureido moiety. This study showed a synergic effect between hydrolysis rates and the presence of +2 transition metal ions, *e.g.* Ni(II), Cu(II) and Zn(II), but not Pb(II).

2.3.3.2.3 Benzofuranyl *N*-Methylcarbamate Insecticides.

Benzofuranyl-based insecticides are structurally related to aryl *N*-methylcarbamates, where a substituted furan ring is fused to the aromatic residue in the 2,3 positions. Carbofuran **70** is one of the most remarkable examples of benzofuranyl

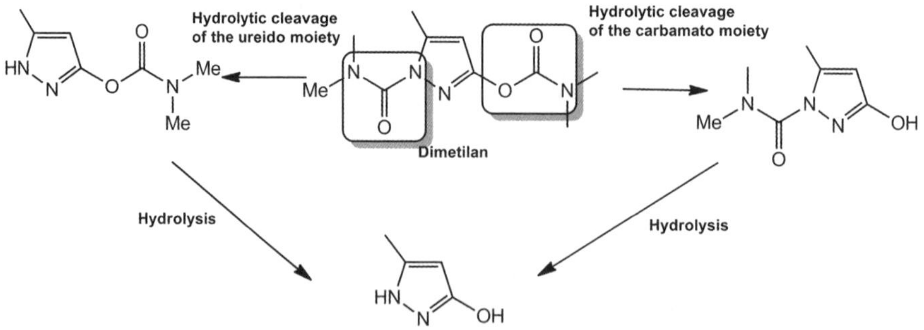

Figure 2.7 Hydrolytic pathways for environmental degradation of dimetilan.

methylcarbamate insecticides, and many aspects of the compound have been studied: characterization of its AChE inhibition,[210] analysis of its influence on mycorrhizal development,[211] stimulation of nitrogenase and the populations of nitrogen-fixing bacteria associated with rice rhizosphere,[212] and its leaching potential.[213]

Carbofuran (with an oral LD_{50} value for rats of $14\,mg\,kg^{-1}$) is one of the most toxic among the carbamate insecticides, and the toxicities of carbofuran and some derivatives have been investigated.[214] Honeybees and earthworms are particularly sensitive to carbofuran, and numerous bird kills have been linked to direct ingestion of carbofuran while sifting sediment.[215] For this reason its granular form has been banned in the USA. Nevertheless, carbofuran does not bioaccumulate to any noteworthy extent.

In 2008, following an announcement by the US EPA banning carbofuran,[216] the only US manufacturer (FMC Corporation) announced that it had voluntarily requested the cancellation of six of the previously-allowed applications of the insecticide (on maize, potatoes, pumpkins, sunflowers, pine seedlings and spinach grown for seed).[217] In May 2009, the EPA cancelled all uses of carbofuran involving crops grown for human consumption.[218]

Much effort has been devoted to the study of the environmental behavior of carbofuran. Base-catalyzed hydrolysis to carbofuran phenol is considered to be the major degradation pathway of this insecticide in both water and sediments.[219] Again, a substantial increase in the hydrolysis rate with increasing pH was observed both in soil and water.[220] By contrast, the slower degradation in acidic and neutral soils was dominated by microbial mechanisms.[221] Thus, a number of hydroxylated carbofuran metabolites produced in soils and in bacterial cultures have been isolated and identified, indicating that hydrolysis is not the only degradation mechanism. Breakdown products in soil include carbofuran phenol,[222] 3-hydroxycarbofuran and 3-ketocarbofuran.[223]

There is a general agreement which establishes that microbial degradation of a soil-applied pesticide may occur when a population of soil microorganisms is repeatedly exposed to a chemical and adapts by developing the ability to catabolize that chemical.[224–226] In particular, for carbofuran, Getzin and Shanks found[227] that enhanced degradation could take place with as little as just one or two applications.

Persistence of carbofuran in several media has also been reported. For instance, Caro *et al.*[228] reported a soil dissipation half-life of 117 days in a cornfield. A low soil pH of 5.3 and low soil moisture content may explain the relatively slow rate of dissipation.

Photolysis is not generally considered a significant degradation pathway in water or soil, like oxidation and volatilization which are generally considered insignificant dissipation pathways for carbofuran in water.[229] On the other hand, because of its water solubility (351 ppm at 25 °C) carbofuran is relatively mobile in soil and in surface runoff. Consequently it has the potential to contaminate lakes, streams and groundwater.

Although carbofuran possesses a low vapor pressure, it has been reported that its volatilization rate is much more rapid under flooded soil conditions than under non-flooded conditions, probably due to co-evaporation with the water of the soil surface.[230]

In addition, several studies have also been reported on the benzofuranyl methylcarbamate-related carbosulfan **71** concerning its degradation,[231] toxicity[232] and metabolism.[233]

2.3.3.2.4 Oxime Carbamate Insecticides. Oxime-based insecticides are the most recent carbamates, whose characteristic feature is the presence of an oximino ester scaffold.[234] The double bond C=N confers rigidity, acting in a similar fashion as the aryl ring in *N*-aryl carbamates. Furthermore, two isomers (*syn* and *anti*) can be obtained which can sometimes be separated.[234] The *Z* configuration was found to confer significantly more activity than the *E* counterpart. Compounds such as aldicarb **72** and methomyl **73**, quite resembling acetylcholine, are commercially-available examples of oxime-based insecticides applied as plant systemic or foliar sprays, respectively.[201]

Aldicarb **72** was the first example of oxime *N*-methylcarbamates, with high water solubility, non-volatility, which was relatively stable under acidic conditions and easily degraded under alkaline conditions. These properties are important determinants of its systemic action in plants and of its problematic environmental behavior. Possible environmental hazards involving this chemical include groundwater contamination and excessive terminal residues in certain foods.[235]

Aldicarb has been registered worldwide to control a wide variety of insect, mite and nematode pests in agriculture. Due to its widespread use, many studies have been devoted to this insecticide. With an oral LD_{50} value for rats of $0.9\,mg\,kg^{-1}$ and classified by the EPA in the highest toxicity category, a strict control for its delivery and use has been established by this environmental agency.[236]

Several reviews concerning the toxicological effects of aldicarb have been published.[237] The first review of aldicarb poisoning circumstances associated with clinical and analytical findings dates from 2000.[238] In this sense, the toxicity and biochemical impact of several oxime carbamates have been tested.[239] Despite its acute toxicity to humans and laboratory animals, aldicarb is readily absorbed through both the gut and the skin, but is rapidly metabolized and excreted in the urine almost completely within twenty-four hours. Moreover, it is not known to be carcinogenic, teratogenic, or to produce other long-term adverse health effects.

Several experiments have been conducted to investigate the degradation of aldicarb, for instance, in sterile, non-sterile and plant-grown soils, and the capability of different plant species to accumulate the compound.[240] This study showed that microorganisms play an important role in the degradation of aldicarb in soil. Likewise, the development of different methods for the separation of aldicarb and its degradation products is also of great interest,[241] allowing for the routine monitoring of aldicarb and its soil derivatives in water at concentrations less than $1\,\mu g\,l^{-1}$.[242]

Given that neither a toxicokinetic model nor an estimate of the target tissue dose of aldicarb and its metabolites in exposed organisms was available, a physiologically based toxicokinetic model was recently developed in rats and humans.[243] The model describes the time-course behavior of the chemical in blood, liver, kidney, lungs, brain and fat. Perkins and Schlenk[244] have also described the *in vivo* acetylcholinesterase inhibition, metabolism and toxicokinetics of aldicarb in Channel catfish.

The use of another important oxime carbamate, methomyl **73**, is also restricted because of its high toxicity to humans.[245] It has been object of several studies related to: the fatal poisoning caused by its inhalation and transdermal absorption;[246] its penetration and fate in insects and two-spotted spider mites;[247] its degradation by reduction in the presence of $Fe(II)$[248] and $Cu(II)$;[249] as well as chromatographic methods for its quantification in blood.[250]

Recently, the risk assessment and chemical decontamination of methomyl from eggplants was investigated, in which it was found that the insecticide was neither appropriate nor effective for application in this vegetable.[251]

2.3.3.3 Current Status

In 2005,[252] the US EPA published a preliminary Cumulative Risk Assessment on *N*-methyl carbamate pesticides, which incorporated exposures from multiple pathways (*i.e.*, food, drinking water and residential/non-occupational exposure to pesticides in air, or on soil, grass and indoor surfaces) for those chemicals. The EPA presented a document[253] in 2007 with its decision regarding the reregistration eligibility of the registered uses of the following carbamates: aldicarb, carbaryl, carbofuran, methomyl, methiocarb, oxamyl and propoxur.

Acknowledgements

We thank the Junta de Andalucía (P08-AGR-03751 and FQM 134), Junta de Extremadura (PRI07A015) and Dirección General de Investigación of Spain (CTQ2008-02813, CTQ2007-66641/BQU) for financial support.

References

1. Ó. López, J. G. Fernández-Bolaños and M. V. Gil, *Green Chem.*, 2005, **7**, 431.
2. J. E. Casida and G. B. Quistad, *Annu. Rev. Entomol.*, 1998, **43**, 1.
3. P. Warrior, *Pest Manag. Sci.*, 2000, **56**, 681.
4. W. Olkowski, S. Daar and H. Olkowski, *The Gardener's Guide to Commonsense Pest Control*, The Taunton Press, 1995, Newtown, CT, USA, p. 22.
5. S. Hörstadius, *Biol. J. Linnean Soc.*, 1974, **6**, 269.
6. E. Panagiotakopulu, P. C. Buckland and P. M. Day, *J. Archaeol. Sci.*, 1995, **22**, 705.
7. R. L. Metcalf and A. Kelma, in *Science in Contemporary China*, ed. L. A. Orleans, Stanford University Press, Standord, CA, USA, 1980, p. 313.

8. Reregistration Eligibility Decision for Inorganic Polysulfides, http://www. epa.gov/oppsrrd1/REDs/inorganic_polysulfides_red.pdf (last accessed October 17th, 2010).

9. http://www.epa.gov/oppsrrd1/REDs/factsheets/0031fact.pdf (last accessed October 17th, 2010).

10. *Encyclopedia of Entomology*, ed. J. L. Capinera, Springer, Dordrecht, The Netherlands, 2nd edn, 2008, vol. 3, p. 2571.

11. S. E. Halcomb, in *Goldfrank's Toxicologic Emergencies*, ed. L. R. Goldfrank, L. S. Nelson, M. A. Howland, N. A. Lemin, N. E. Flomenbaum and R. S. Hoffman, McGraw-Hill, New York, USA, 8th edn, 2006, p. 657.

12. D. B. South, in *Encyclopedia of Pest Management*, ed. D. Pimentel, *online version*, CRC Press, 2002, p. 395.

13. C. Gillott, *Entomology*, Springer, 3rd edn, 2005, p. 746.

14. E. A. Murhy and M. Aucott, *Sci. Total Environ.*, 1998, **218**, 89.

15. *Medical and Biological Effects of Environmental Pollutants: Arsenic*, Committee on Medical and Biological Effects of Environmental Pollutants, National Academy of Sciences, Washington, D. C., USA, 1977, p. 34.

16. E. Hood, *Environ. Health Perspect.*, 2006, **114**, A471.

17. H. L. Haller, *J. Chem. Educ.*, 1942, **19**, 315.

18. V. Bencko and A. Slámová, *J. Public Health*, 2007, **115**, 279.

19. S. T. Omaye, in *Pesticide, Veterinary and Other Residues in Food*, ed. D. H. Watson, Woodhead Publishing Limited, Cambridge, UK, 2004, p. 4.

20. H. Garelick, H. Jones, A. Dybowska and E. Valsami-Jones, in *Reviews of Environmental Contamination and Toxicology-Arsenic Pollution and Remediation: An International Perspective*, ed. H. Garelick and H. Jones, Springer, New York, USA, 2008, vol. 197, p. 46.

21. R. Datta, D. Sarkar, S. Sharma and K. Sand, *Sci. Total Environ.*, 2006, **372**, 39.

22. *Recognition and Management of Pesticide Poisonings*, ed. J. R. Reigart and J. R. Roberts, 5th edn, EPA,. 1999, p. 82, www.epa.gov/oppfead1/safety/healthcare/handbook/handbook.htm (last accessed October 17th, 2010).

23. L. H. Weinstein and A. W. Davison, *Fluorides in the Environment*, CABI Publishing, Wallingford, UK, 2004, p. 7.

24. G. N. Agrios, *Plant Pathol.*, Elsevier Academic Press, 5th edn, 2005, p. 338.

25. W. G. Woods, *Environ. Health Perspect.*, 1994, **102**, 5.

26. P. L. Strong, in *Handbook of Pesticide Toxicology-Principles*, ed. R. I. Krieger and W. C. Krieger, 2nd edn, Academic Press, 2001, p. 1429.

27. G. Nalyanya, D. Liang, R. J. Kopanic, Jr. and C, Schal, *J. Economic Entomol.*, 2001, **94**, 686.

28. L B. Norton, *Ind. Eng. Chem.*, 1948, **40**, 691.

29. R. C. Roark, *Ind. Eng. Chem.*, 1935, **27**, 530.

30. W. Olkowski, S. Daar and H. Olkowski, *The Garderner's Guide to Common-Sense Pest Control*, The Taunton Press, 1995, Newtown, CT, p. 83.

31. R. G. Hollingsworth, *J. Economic Entomol.*, 2005, **98**, 772.
32. M. Elliot, in *Pyrethrum Flowers: Production, Chemistry, Toxicology, and Uses*, ed. J. E. Casida and G. B. Quistad, Oxford University Press, New York, USA, 1995, p. 3.
33. M. Tomizawa and J. E. Casida, *Annu. Rev. Entomol.*, 2003, **48**, 339.
34. M. Cabizza, A. Angioni, M. Melis, M. Cabras, C. V. Tuberoso and P. Cabras, *J. Agric. Food Chem.*, 2004, **52**, 288.
35. P. R. Jefferies, P. Yu and J. E. Casida, *Pest. Sci.* 1997, **51**, 33.
36. I. Ujváry, L. Polgár, B. Darvas and J. E. Casida, *Pest. Sci.* 1995, **44**, 96.
37. G. Brahmachari, *ChemBioChem*, 2004, **5**, 408.
38. R. L. Metcalf, *Agric. Human Values*, 1987, **4**, 15.
39. R. L. Metcalf, *J. Agric. Food Chem.*, 1973, **21**, 511.
40. V. Turusov, V. Rakitsky and L. Tomatis, *Environ. Health Perspect.*, 2002, **110**, 125.
41. A. G. Smith, in *Handbook of Pesticide Toxicology-Principles*, ed. R. I. Krieger and W. C. Krieger, Academic Press, San Diego, California, USA, 2nd edn, 2001, vol. 2, p. 1305.
42. W. H. C. Rueggeberg and D. J. Torrans, *Ind. Eng. Chem.*, 1946, **38**, 211.
43. Gy. Matolcsy, in *Studies in Environmental Science-Pesticide Chemistry*, ed. G. Matolcsy, M. Nádasy and V. Andriska, Elsevier, Amsterdam, Holland, 1988, vol. 32, p. 47.
44. J. Wright, *Environmental Chemistry*, Routledge, New York, USA, 2003, p. 332.
45. T. G. E. Davies, L. M. Field, P. N. R. Usherwood and M. S. Williamson, *IUBMB Life*, 2007, **59**, 151.
46. J. Beard, *Sci. Total Environ.*, 2006, **355**, 78.
47. W. J. Rogan and A. Chen, *Lancet*, 2005, **366**, 763.
48. R. A. Bailey, H. M. Clark, J. P. Ferris, S. Krause and R. L. Strong, *Chemistry of the Environment*, Hartcourt/Academic Press, San Diego, CA, USA, 2nd edn, p. 223.
49. C. P. Chang and F. W. Plapp, Jr., *Pestic. Biochem. Physiol.*, 1983, **20**, 76.
50. J. R. Coats, *Environ. Health Perspect.*, 1990, **87**, 255.
51. A. O. O'Reilly, B. P. S. Khambay, M. S. Williamson, L. M. Field, B. A. Wallace and T. G. E. Davies, *Biochem. J.*, 2006, **396**, 255.
52. E. Zlotkin, *Annu. Rev. Entomol.*, 1999, **44**, 429.
53. J. E. Casida and G. B. Quistad, *J. Pestic. Sci.*, 2004, **29**, 81.
54. P. Kaushik and G. Kaushik, *J. Hazard. Mater.*, 2007, **143**, 102.
55. M. Schlitzer, *ChemMedChem*, 2007, **2**, 944.
56. A. Kumar, N. Valecha, T. Jain and A. P. Dash, *Am. J. Trop. Med. Hyg.*, 2007, **77**, 69.
57. S. Matta, S. L. Kantharia and V. K. Desai, *J. Vector Borne Dis.*, 2005, **42**, 77.
58. S. Ensley, in *Veterinary Toxicology: Basic and Clinical Principles*, ed. R. C. Gupta, Academic Press, New York, USA, 2007, p. 489.
59. S. Abu-El-Haj, M. A. H. Fahmy and T. R. Fukuto, *J. Agric. Food Chem.*, 1979, **27**, 258.

60. W. C. Sagar, R. E. Monroe and M. J. Zabik, *J. Agric. Food Chem.*, 1972, **20**, 1176.
61. J. R. Coats, R. L. Metcalf and I. P. Kapoor, *J. Agric. Food Chem.*, 1977, **25**, 859.
62. D. Zakarya, A. Boulaamail, E. M. Larfaoui and T. Lakhlifi, *SAR QSAR Environ. Res.*, 1997, **6**, 183.
63. A. S. Hirwe, R. L. Metcalf and I. P. Kapoor, *J. Agric. Food Chem.*, 1972, **20**, 818.
64. R. L. Metcalf, I. P. Kappor and A. S. Hirwe, *Bull. World Health Org.*, 1971, **44**, 363.
65. T. R. Fukuto, *J. Pestic. Sci.*, 1988, **13**, 137.
66. G. Holan, W. M. Johnson, C. T. Virgona and R. A. Walser, *J. Agric. Food Chem.*, 1988, **34**, 520.
67. H. van den Berg, *Environ. Health Perspect.*, 2009, **117**, 1656.
68. V. Zitko, in *The Handbook of Environmental Chemistry, Part O: Persistant Organic Pollutants*, ed. H. Fiedler, Springer-Verlag, Berlin, Germany, 2003, vol. 3, p. 47.
69. H.-R. Buser, M. D. Müller, I. J. Buerge and T. Poiger, *J. Agric. Food Chem.*, 2009, **57**, 7445.
70. T. A. Unger, *Pesticide Synthesis Handbook*, Noyes Publications, Saddle River, New Jersey, USA, 1996, p. 809.
71. W. F. Schmidt, S. Bilboulian, C. P. Rice, J. C. Fettinger, L. L. McConnell and C. J. Hapeman, *J. Agric. Food Chem.*, 2001, **49**, 5372.
72. J. E. Casida and M. Tomizawa, *J. Pestic. Sci.*, 2008, **33**, 4.
73. J. R. Bloomquist, *Arch. Insect Biochem. Physiol.*, 2003, **54**, 145.
74. L. M. Hernández, M. A. Fernández and M. J. González, *Bull. Environ. Contam. Toxicol.*, 1991, **46**, 9.
75. J. R. Plimmer, in *Handbook of Pesticide Toxicology-Principles*, ed. R. I Krieger and W. C. Krieger, Academic Press, San Diego, CA, USA, 2nd edn., 2001, p. 95.
76. K. Breivik, J. M. Pacyna and J. Münch, *Sci. Total Environ.*, 1999, **239**, 151.
77. Y. F. Li, *Sci. Total Environ.*, 1999, **232**, 121.
78. P. Bhatt, M. S. Kumar and T. Chakrabarti, *Crit. Rev. Environ. Sci. Technol.*, 2009, **39**, 655.
79. http://www.epa.gov/ttnatw01/hlthef/lindane.html (last accessed October 17th, 2010).
80. V. Andriska, in *Studies in Environmental Science- Pesticide Chemistry*, ed. G. Matolcsy, M. Nádasy and V. Andriska, Elsevier, Amsterdam, Holland, 1988, vol. 32, p. 577.
81. W. Vetter and M. Oehme, in *The Handbook of Environmental Chemistry*, ed. J. Paasivirta, Springer-Verlag, Heidelberg, Germany, 2000, vol. 3, part K, p. 237.
82. http://www.epa.gov/international/toxics/pop.html (last accessed October 17th, 2010).
83. G. W. Ware and D. M. Whitacre, *The Pesticide Book*, 6th edn, Meister Media Worldwide, Willoughby, OH, USA, 2004, p. 47.

84. J. Tucker, *War of Nerves: Chemical Warfare from World War I to Al-Qaeda,* Pantheon Books, New York, USA, 2006.
85. F. Schmaltz, *J. Hist. Neurosci.*, 2006, 15, 186.
86. C. J. Hilmas, J. K. Smart and B. A. Hill, Jr, in *Medical Aspects of Chemical Warfare,* Borden Institute, 2008, p. 43.
87. E. A. Croddy and J. J. Wirtz, *Weapons of Mass Destruction: an Encyclopedia of Worldwide Policy, Technology, and History*, ABC-CLIO, Santa Barbara, USA, 2005, vol. 1, p. 93.
88. P. G. Ghosh, N. A. Sawant, S. N. Patil and B. A. Aglave, *Int. J. Biotechnol. Biochem.*, 2010, **6**, 871.
89. B. C. Gbaruko, E. I. Ogwo, J. C. Igwe and H. Yu, *African J. Biotechnol.*, 2009, **8**, 5137.
90. R. L. Metcalf, in *Ullmann's Encyclopedia of Industrial Chemistry*, Wiley-VCH, Weinheim, Germany, 2002, p. 24.
91. http://www.epa.gov/pesticides/pestsales/01pestsales/usage2001_3.htm (last accessed October 17th, 2010).
92. R. D. Wauchope, T. M. Buttler, A. G. Hornsby, P. W. Augustijn-Beckers and J. P. Burt, *Rev. Environ. Contam. Toxicol.*, 1992, **123**, 1.
93. T. S. S. Dikshith and P. V. Diwan, *Industrial Guide to Chemical and Drug Safety*, John Wiley & Sons, Hoboken, NJ, USA, 2003, p. 150.
94. L. E. Smart, J. H. Stevenson and J. H. H. Walters, *Crop Prot.*, 1989, **8**, 169.
95. M. A. Khan and G. C. Kozub, *Can. J. Comp. Med.*, 1981, **45**, 15.
96. J. S. Ogg, *Vet. Parasitol.*, 1977, **3**, 229.
97. H. W. Chambers, J. S. Boon, R. L. Carr, and J. E. Chambers, in *Handbook of Pesticide Toxicology-Principles*, ed. R. I Krieger and W. C. Krieger, Academic Press, San Diego, CA, USA, 2nd edn., 2001, p. 913.
98. J. E. Casida, *J. Agr. Food Chem.*, 1956, **4**, 772.
99. J. E. Casida, *Science*, 1955, **122**, 587.
100. http://www.pesticideinfo.org/Detail_Chemical.jsp?Rec_Id=PC33331 (last accessed October 17th, 2010).
101. http://www.pesticideinfo.org/Detail_Chemical.jsp?Rec_Id=PC33362 (last accessed October 17th, 2010).
102. http://www.pesticideinfo.org/Detail_Chemical.jsp?Rec_Id=PC35097 (last accessed October 17th, 2010).
103. http://www.epa.gov/pesticides/reregistration/status_op.htm (last accessed October 17th 2010).
104. https://secure.virtuality.net/panukcom/PDFs/Banned%20In%20The%20EU.pdf (last accessed October 17th 2010).
105. http://www.epa.gov/pesticides/reregistration/ddvp/ (last accessed October 17th, 2010).
106. http://ec.europa.eu/food/plant/protection/evaluation/existactive/dichlorvos.pdf (last accessed October 17th, 2010).
107. http://eur-lex.europa.eu/LexUriServ/LexUriServ.do?uri=OJ:L:2010:006:0001:0005:EN:PDF (last accessed October 17th, 2010).
108. http://www.fao.org/docrep/w5715e/w5715e04.htm (last accessed October 17th, 2010).

109. B. Dinham, *The Pesticide Hazard: A Global Health and Environmental Audit*, Zed Books, London, UK, 1993, p. 120.

110. http://www.pesticideinfo.org/Detail_Chemical.jsp?Rec_Id = PC35862 (last accessed October 17th, 2010).

111. http://web.utk.edu/~extepp/napiap/rnn/nw9401pr.htm (last accessed October 17th, 2010).

112. http://www.tca.or.tz/docs/PAN-%20THE%20LIST%20OF%20LISTS.pdf. (last accessed October 17th, 2010).

113. http://www.epa.gov/fedrgstr/EPA-PEST/2008/September/Day-17/p21736. htm (last accessed October 17th, 2010).

114. http://www.fao.org/docrep/w3727e/w3727e0g.htm (last accessed October 17th, 2010).

115. http://www.furs.si/law/EU/ffs/eng/annexI/direktive/RR/dimethoate.pdf (last accessed October 17th, 2010).

116. (a) World Health Organization, The WHO recommended classification of pesticides by hazard and guidelines to classification 2009, Stuttgart, Germany, p. 48; (b) ibid, p. 19.

117. http://www.regulations.gov/search/Regs/home.html#docketDetail?R=EPA-HQ-OPP-2009-0054 (last accessed October 17th, 2010).

118. http://eur-lex.europa.eu/LexUriServ/LexUriServ.do?uri=OJ:L:2002:319: 0003:0011:EN:PDF (last accessed October 17th 2010).

119. M. R. Bonner, J. Coble, A. Blair, L. E. Beane Freeman, J. A. Hoppin, D. P. Sandler and M. C. R. Alavanja, *Am. J. Epidemiol.*, 2007, **166**, 1023.

120. http://www.epa.gov/opp00001/health/mosquitoes/malathion4mosquitoes. htm#malathion (last accessed October 17th, 2010).

121. J. W. Edwards, S. G. Lee, L. M. Heath and D. L. Pisaniello, *Environ. Res.*, 2007, **103**, 38.

122. M. Abramowicz, *Med. Lett. Drugs Ther.*, 1999, **41**, 73.

123. A. M. Downs, K. A. Stafford, I. Harvey and G. C. Coles, *Br. J. Dermatol.*, 1999, **141**, 508.

124. O. Chosidow, B. Giraudeau, J. Cottrell, A. Izri, R. Hofmann, S. G. Mann and I. Burgess, *N. Engl. J. Med.*, 2010, **362**, 896.

125. http://www.epa.gov/pesticides/reregistration/REDs/malathion_red.pdf (last accessed October 17th, 2010).

126. M. F. Bouchard, D. C. Bellinger, R. O. Wright and M. G. Weisskopf, *Pediatrics*, 2010, **125**, e1270.

127. J. C. Franson, E. J. Kolbe and J. W. Carpenter, *J. Wildlife Dis.*, 1985, **21**, 318.

128. http://www.pesticideinfo.org/Detail_Product.jsp?REG_NR=00024100237 &DIST_NR = 000241 (last accessed October 17th 2010).

129. http://www.regulations.gov/search/Regs/home.html#docketDetail?R=EPA-HQ-OPP-2009-0172 (last accessed October 17th, 2010).

130. http://www.hse.gov.uk/pic/reg689-08ann1.pdf (last accessed October 17th, 2010).

131. N. Becker, D. Petric, M. Zgomba, C. Boase, M. Madon, C. Dahl and A. Kaiser, *Mosquitos and their Control*, 2nd edn, Springer, Heidelberg, Germany, 2010, p. 455.

132. http://www.epa.gov/oppsrrd1/REDs/0290ired.pdf (last accessed October 17th, 2010).

133. http://www.pesticideinfo.org/Detail_Chemical.jsp?Rec_Id=PC35122 (last accessed October 17th, 2010).

134. http://yosemite.epa.gov/opa/admpress.nsf/8b75cea4165024c685257359003f 022e/54ccdfd21fb51284852569770053d0dd!OpenDocument (last accessed October 17th, 2010).

135. http://www.raaa.org.pe/documentos/p-list%20of%20lists%202005.pdf (last accessed October 17th, 2010).

136. http://www.pesticideinfo.org/Detail_Chemical.jsp?Rec_Id=PC35110 (last accessed October 17th, 2010).

137. http://www.regulations.gov/search/Regs/home.html#documentDetail?R= 0900006480b55629 (last accessed October 17th 2010).

138. S. B. Halstead, J. A. Suaya and D. S. Shepard, *Lancet*, 2007, **369**, 1410.

139. A Tsuzuki, T. Huynh , T. Tsunoda , L. Luu, H. Kawada and M. Takagi, *Am. J. Trop. Med. Hyg.*, 2009, **80**, 752.

140. http://www.epa.gov/oppsrrd1/REDs/temephos_red.htm#III (last accessed October 17th, 2010).

141. http://eur-lex.europa.eu/LexUriServ/LexUriServ.do?uri=CELEX:32005L0 072:EN:NOT (last accessed October 17th, 2010).

142. http://www.epa.gov/oppsrrd1/REDs/chlorpyrifos_ired.pdf (last accessed October 17th, 2010).

143. V. A. Rauh, R. Garfinkel, F. P. Perera, H. F. Andrews, L. Hoepner, D. B. Barr, R. Whitehead, D. Tang and R. W. Whyatt, *Pediatrics*, 2006, **118**, e1845.

144. http://www.regulations.gov/search/Regs/home.html#documentDetail?R=09 00006480abdc8a (last accessed October 17th, 2010).

145. T. Roberts and D. Hutson, *Metabolic Pathways of Agrochemicals. Part 2: Insecticides and Fungicides,* The Royal Society of Chemistry, Cambridge, UK, 1999, p. 241.

146. http://www.epa.gov/oppsrrd1/reregistration/REDs/factsheets/diazinon_ cancellation_fs.htm (last accessed October 17th, 2010).

147. http://eur-lex.europa.eu/LexUriServ/LexUriServ.do?uri=OJ:L:2007:148:0009: 0010:EN:PDF (last accessed October 17th 2010).

148. http://www.environment-agency.gov.uk/business/topics/pollution/39411. aspx (last accessed October 17th 2010).

149. http://www.epa.gov/oppsrrd1/reregistration/azm/phaseout_fs.htm (last accessed October 17th 2010).

150. http://eur-lex.europa.eu/LexUriServ/LexUriServ.do?uri=OJ:L:2010:006: 0001:0005:EN:PDF (last accessed October 17th, 2010).

151. http://www.epa.gov/oppfead1/endanger/litstatus/effects/phosmet-sumry. pdf (last accessed October 17th, 2010).

152. http://www.epa.gov/oppsrrd1/reregistration/phosmet/phosmet_summary. htm (last accessed October 17th, 2010).

153. http://eur-lex.europa.eu/smartapi/cgi/sga_doc?smartapi!celexplus!prod! DocNumber&lg=en&type_doc=Directive&an_doc=2007&nu_doc=25 (last accessed October 17th, 2010).

154. N. B. Vyas, E. F. Hill, J. R. Sauer and W. J. Kuenzel, *Enviromen. Toxicol Chem.*, 1995, **14**, 1961.
155. http://eur-lex.europa.eu/LexUriServ/LexUriServ.do?uri = OJ:L:2003:082:0040:0041:EN:PDF. (last accessed October 17th 2010).
156. http://www.regulations.gov/search/Regs/home.html#docketDetail?R = EPA-HQ-OPP-2008-0761 (last accessed October 17th, 2010).
157. http://www.ema.europa.eu/docs/en_GB/document_library/Maximum_Residue_Limits_-_Report/2009/11/WC500015646.pdf (last accessed October 17th, 2010).
158. http://eur-lex.europa.eu/LexUriServ/LexUriServ.do?uri = OJ:L:2007:166:0016:0023:EN:PDF (last accessed October 17th 2010).
159. http://www.regulations.gov/search/Regs/home.html#documentDetail?D = EPA-HQ-OPP-2009-0097-0018 (last accessed October 17th, 2010).
160. http://eur-lex.europa.eu/LexUriServ/LexUriServ.do?uri = OJ:L:2007:133:0042:0043:EN:PDF (last accessed October 17th 2010).
161. J. C. Venter, K. di Porzio, A. Robinson, S. M. Shreeve, J. Lai, A. R Kerlavage, S. P. Fracek, Jr., K. U. Lentes and C. M. Fraser, *Prog. Neurobiol.*, 1988, **30**, 105.
162. H. L. Fernandez, R. D. Moreno and N. C. Inestrosa, *J. Neurochem.*, 1996, **66**, 1335.
163. J. P. Walker and S. A. Asher, *Anal. Chem.* 2005, **77**, 1596.
164. D. W. Moss and E. R. Henderson, in *Tietz Textbook of Clinical Chemistry*, ed. C. A Burtis and E. R. Ashwood, 3rd edn, W. B. Saunders Company, Philadelphia, PA, USA, 1999, p. 708.
165. C. N. Pope, *J. Toxicol. Environ. Health, Part B Crit. Rev.*, 1999, **2**, 161.
166. P. A. Dahm, *Bull. World Health Org.*, 1971, **44**, 215.
167. C. F. Golenda, R. A. Wirtz and R. G. Andre, *Comp. Biochem. Physiol., Part C*, 1987, **88**, 61.
168. J. K. Dubey and S. K. Patyal, in *Encyclopedia of Pest Management*, ed. D. Pimentel, CRC Press, New York, USA, 2007, p. 67.
169. L. Karalliedde, M. Eddleston and V. Murray, in *Organophosphates and Health*, ed. L. Karalliedde, F. Feldman, J. Henry and T. Marrs, Imperial College Press, London, UK, 2001, p. 431.
170. M. Eddleston, L. Karalliedde, N. Buckley, R. Fernando, G. Hutchinson, G. Isbister, F. Konradsen, D. Murray, J. C. Piola, N. Senanayake, R. Sheriff, S. Singh, S. B. Siwach and L. Smit, *Lancet*, 2002, **360**, 1163.
171. http://www.deq.state.or.us/lq/pubs/docs/cu/GuidanceEvalResidualPesticides.pdf (last accessed October 17th, 2010).
172. I. M. Jensen and P. Whatling, in *Hayes' Handbook of Pesticide Toxicology*, ed. R. Krieger, Academic Press, Amsterdam, Holland, 3rd edn, 2010, vol. 2, p. 2010.
173. M. A. Kamrin, *Pesticide Profiles: Toxicity, Environmental Impact, and Fate*, CRC Press, Boca Raton, FL, USA, 1997, p. 8.
174. N. Mikami, S. Sakata, H. Yamada and J. Miyamoto, *J. Pesticide Sci.*, 1985, **10**, 491.

175. J. Weber, C. J. Halsall, J. J. Wargent and N. D. Paul, *J. Environ. Monit.*, 2009, **11**, 654.

176. S. Bondarenko and J. Gan, *Environ. Toxicol. Chem.*, 2004, **23**, 1809.

177. R. D. Wauchope, T. M. Buttler, A. G. Hornsby, P. W. M. Augustijn-Beckers and J. P. Burt, *Rev. Environ. Contam. Toxicol.*, 1992, **123**, 1.

178. R. L. Metcalf, in *Introduction to Insect Pest Management*, ed. R. L. Metcalf and W. H. Luckmann, 3rd edn, Wiley, New York, USA, 1994, p. 262.

179. Z.-Y. Zhang, X.-Y. Yu, D.-L. Wang, H.-J. Yan, and X.-J. Liu, *Pest. Manag. Sci.*, 2010, **66**, 84.

180. H. Mehlhorn, *Encyclopedia of Parasitology*, 3rd edn, Springer-Verlag, New York, USA, 2008, vol. 1, p. 419.

181. R. C. Gupta, *Toxicology of Organophosphate and Carbamate Compounds*, ed. R. C. Gupta, Elsevier Academic Press, Amsterdam, Holland, 2006, p. 3.

182. W. B. Wheeler, *J. Agric. Food Chem.*, 2002, **50**, 4151.

183. G. Matolcsy, in *Studies in Environmental Science-Pesticide Chemistry*, ed. G. Matolcsy, M. Nádasy and V. Andriska, Elsevier, Amsterdam, Holland, 1988, vol. 32 p. 90.

184. M. Vandekar, R. Plestina and K. Wilehelm, *Bull. World Health. Org.*, 1971, **44**, 241.

185. M. A. Sogorb and E. Vilanova, *Toxicol. Lett.*, 2002, **128**, 215.

186. K.-S. Koo, Y.-C. Yoo, T. Kubota, S. Ameno and I. Ijiri, *Forens. Sci. Int.*, 1999, **101**, 65.

187. E. Lacassie, P. Marquet, J. M. Gaulier and M. F. Dreyfuss, *Forens. Sci. Int.*, 2001, **121**, 116.

188. H. Parlar, in *Environmental Fate of Pesticides*, ed. D. H. Hutson and T. R. Roberts, Wiley, New York, USA, 1990, p. 246.

189. S. Chiron, J. A. Torres, A. Fernández-Alba, M. F. Alpendurada and D. Barceló, *J. Environ Anal. Chem.*, 1996, **65**, 37.

190. M. Raveton, A. Aajoud, J. C. Willison, H. Aouadi, M. Tissut and P. Ravanel, *Environ. Sci. Technol.*, 2006, **40**, 4151.

191. X. Li, L. Yang, U. Jans, M. E. Melcer and P. Zhang, *Environ. Sci. Technol.*, 2007, **41**, 1635.

192. A. Tal, B. Rubin, J. Katan and N. Aharonson, *Weed Sci.*, 1989, **37**, 434.

193. J. M. Smolen and A. T. Stone, *Environ. Sci. Technol.*, 1997, **31**, 1664.

194. J. M. Smolen and A. T. Stone, *Soil Sci. Soc. Am. J.*, 1997, **62**, 636.

195. M. M. El-Amamy and T. Mill, *Clays Clay Miner.*, 1984, **32**, 67.

196. M.V. Bassett, S. C. Wendelken, B. V. Pepich and D. J. Munch, *J. Chromatogr. Sci.*, 2003, **41**, 100.

197. A. Millar, K.-H. Kim, D. K. Minster, T. Ohgi and S. M. Hecht, *J. Org. Chem.*, 1986, **51**, 189.

198. N. Katada, H. Fujinaga, Y. Nakamura, K. Okumura, K. Nishigaki and M. Niwa, *Catal. Lett.*, 2002, **80**, 47.

199. K. P. Prousalis, G. M. Tsivgoulis and T. Tsegenidis, *Int. J. Environ. Anal. Chem.*, 2007, **87**, 1065.

200. K. Kamoshita, I. Ohno, K. Kasamatsu, T. Fujita and M. Nakajima, *Pestic. Biochem. Physiol.*, 1979, **11**, 104.

201. J. B. Knaak, C. C. Dary, M. S. Okino, F. W. Power, X. Zhang, C. B. Thompson, R. Tornero-Velez and J. N. Blancato, *Rev. Environ. Contam. Toxicol.*, 2008, **193**, 53.

202. http://www.epa.gov/oppsrrd1/reregistration/REDs/carbaryl-red-amended. pdf (last accessed October 17th, 2010).

203. http://www.cdpr.ca.gov/docs/emon/pubs/fatememo/carbaryl.pdf (last accessed October 17th, 2010).

204. C. M. Kamanavalli and H. Z. Ninnekar, *World J. Microbiol. Biotechnol.* 2000, **16**, 329.

205. L. Everest, *Behind the Poison Cloud: Union Carbide's Bhopal Massacre*, Banner Press, Chicago, IL, USA, 1985.

206. L. Wheeler and A. Strother, *Drug Metab. Dispos.*, 1974, **2**, 533.

207. R. L. Metcalf, in *Ullmann's Agrochemicals*, Wiley-VCH Verlag, Weinheim, Germany, 2007, vol. 2, p. 651.

208. Q. Zhao, G. Yang, X. Mei, H. Yuan and J. Ning, *Pestic. Biochem. Physiol.*, 2009, **95**, 131.

209. C.-H. Huang and A. T. Stone, *Environ. Sci. Technol.*, 2000, **34**, 4117.

210. X. Zhang, A. M. Tsang, M. S. Okino, F. W. Power, J. B. Knaak, L. S. Harrison and C. C. Dary, *Toxicol. Sci.*, 2007, **100**, 345.

211. M. P. Sharma and A. Adholeya, *Phytomorphology*, 2005, **55**, 55.

212. P. K. Kanungo, T. K. Adhya and V. Rajaramamohan, *Chemosphere*, 1995, **31**, 3249.

213. R. Diaz-Diaz, K. Loague and J. S. Notario, *J. Contam. Hydrol.*, 1999, **36**, 1.

214. N. Moorthy, A.-H. Lee, I. P. Kapoor and G. J. Hollingshaus, *J. Agric. Food Chem.*, 1994, **42**, 1019.

215. N. Erwin, *J. Pestic. Reform*, 1991, **11**, 15.

216. http://www.epa.gov/pesticides/reregistration/carbofuran/carbofuran_noic. htm. (last accessed October 17th, 2010).

217. B. E. Erickson, *Chem. Eng. News*, 2009, **87**, 18.

218. http://www.ens-newswire.com/ens/may2009/2009-05-11-093.asp (last accessed October 17th, 2010).

219. K. Talebi and C. H. Walker, *Pestic. Sci.*, 1993, **39**, 65.

220. H. C. Bailey, C. DiGiorgio, K. Kroll, J. Miller, D. E. Hinton and G. Starrett, *Environ. Toxicol. Chem.*, 1996, **15**, 837.

221. K. R. Krishna and L. Philip, *J. Environ. Sci. Health,* 2008, **43**, 157.

222. L. W. Getzin, *Environ. Entomol.*, 1973, **2**, 461.

223. W. G. Johnson and T. L. Lavy, *J. Environ. Qual.*, 1995, **24**, 487.

224. K. M. Scow, R. R. Merica and M. Alexander, *J. Agric. Food Chem.*, 1990, **38**, 908.

225. T. B. Parkin and D. R. Shelton, *Pestic. Sci.*, 1994, **40**, 163.

226. N. Singh and N. Sethunathan, *Pestic. Sci.*, 1999, **55**, 740.

227. L. W. Getzin and C. H. Shanks, *J. Environ. Sci. Health*, 1990, **B25**, 433.

228. J. H. Caro, H. P. Freeman, D. E. Glotfelty, B. C. Turner and W. M. Edwards, *J. Agric. Food Chem.*, 1973, **21**, 1010.

229. P. Raha and A. K. Das, *Chemosphere*, 1990, **21**, 99.

230. J. O. Lalah, S. O. Wandiga and W. C. Dauterman, *Bull. Environ. Contam. Toxicol.*, 1996, **56**, 37.
231. A. Sahoo, N. Sethunathan and P. K. Sahoo, *J. Environ. Sci. Health, Part B*, 1998, **33**, 369.
232. B. E. Renzi and R. I. Krieger, *Fundam. Appl. Toxicol.*, 1986, **6**, 7.
233. V. E. Clay and T. R. Fukuto, *Arch. Environ. Contam. Toxicol.*, 1984, **13**, 53.
234. R. L. Metcalf, *Bull. World. Health. Org.*, 1971, **44**, 43.
235. R. L. Baron, *Environ. Health Perspect.*, 1994, **102**, 23.
236. http://www.epa.gov/oppsrrd1/REDs/aldicarb_red.pdf (last accessed October 17th, 2010).
237. J. F. Risher, F. L. Mink and J. F. Stara, *Environ. Health Perspect.*, 1987, **72**, 267.
238. C. Ragoucy-Sengler, *Hum. Exp. Toxicol.*, 2000, **19**, 657.
239. M. A. Radwan, H. B. El-Wakil and K. A. Osman, *J. Environ. Sci. Health*, 1992, **27**, 759.
240. H. Sun, J. Xu, S. Yang, G. Liu and S. Dai, *Chemosphere*, 2004, **54**, 569.
241. A. Dekker and N. W. H. Houx, *J. Environ. Sci. Health*, 1983, **18**, 379.
242. L.-Y. Lin and W. T. Cooper, *J. Chromatogr., A*, 1987, **390**, 285.
243. M. Pelekis and C. Emond, *Environ. Toxicol. Pharmacol.*, 2009, **28**, 179.
244. E. J. Perkins Jr. and D. Schlenk, *Toxicol. Sci.*, 2000, **53**, 308.
245. http://www.regulations.gov/search/Regs/home.html#documentDetail?R= 0900006480b538ad (last accessed October 17th, 2010).
246. A. M. Tsatsakis, G. K. Bertsias, I. N. Mammas, I. Stiakakis and D. B. Georgopoulos, *Bull. Environ. Contam. Toxicol.*, 2001, **66**, 415.
247. A. K. Gayen and C. O. Knowles, *Arch. Environ. Contam. Toxicol.*, 1981, **10**, 55.
248. T. J. Strathmann and A. T. Stone, *Environ. Sci. Technol.*, 2002, **36**, 653.
249. T. J. Strathmann and A. T. Stone, *Environ. Sci. Technol.*, 2001, **35**, 2461.
250. S. Ito, K. Kudo, T. Imamura, T. Suzuki and N. Ikeda, *J. Chromatogr., B*, 1998, **713**, 323.
251. Md.W. Aktar, D. Sengupta, S. Alam and A. Chowdhury, *Environ. Monit. Assess.*, 2010, **168**, 657.
252. http://www.epa.gov/oppsrrd1/REDs/nmc_revised_cra.pdf (last accessed October 17th 2010).
253. http://www.epa.gov/pesticides/reregistration/status_carbamates.htm (last accessed October 17th, 2010).

CHAPTER 3

Pyrethrins and Pyrethroid Insecticides

JEROME J. SCHLEIER III* AND ROBERT K. D. PETERSON

Department of Land Resources and Environmental Sciences, Montana State University, 334 Leon Johnson Hall, Bozeman, MT 59717, USA

3.1   Introduction

Pyrethrum is one of the oldest and most widely used botanical insecticides. Its insecticidal properties have been known for more than 150 years; although the earliest mention of the Chrysanthemum flowers from which it originates comes from early Chinese history, where it is believed that the flower passed into Europe along the silk roads.[1] The term "pyrethrum" refers to the dried and powdered flower heads of a white-flowered, daisy-like plant belonging to the *Chrysanthemum* genus. Pyrethrum's insecticidal properties were recognized in the middle of the 19th century, when an American named Jumticoff discovered that many Caucuses tribes used it for the control of body lice.[1] The earliest cultivation of pyrethrum, also called "Persian pyrethrum" or "Persian powders", was in the region of the Caucuses extending into Northern Persia.[2] The first Persian powders that were processed and commercialized in Europe in the 1820s were most likely prepared from a mixture of *C. roseum* and *C. corneum*. During and after 1876, these preparations were introduced into the USA, Japan, Africa and South America.[3,4] The superior insecticidal properties of *C. cinerariaefolium* were first discovered around 1845 and these species subsequently supplanted previously cultivated species. *Chrysanthemum cinerariaefolium* is currently cultivated in the USA, Japan, Kenya, Brazil, the Democratic Republic of the Congo, Uganda and India.[2,3]

RSC Green Chemistry No. 11
Green Trends in Insect Control
Edited by Óscar López and José G. Fernández-Bolaños
© Royal Society of Chemistry 2011
Published by the Royal Society of Chemistry, www.rsc.org

In 1917, the U.S. military made the first pyrethrum extracts by percolating the ground flower heads with kerosene, which were then incorporated into space sprays for use against house flies and mosquitoes.[1] Since pyrethrins are derived from plants, however, the supply has always been highly variable. A shortage during World War II hastened the search for synthetic insecticides like dichloro-diphenyltrichloroethane (DDT), which could be consistently produced and which was subsequently used by the Allies to manage insect vectors of human pathogens. The introduction of synthetic insecticides like organochlorines, organophosphates and carbamates represented a revolution in insect control because of their high insecticidal toxicity and consistent supply, however, they have been, or are being, phased out of use due to biomagnification, high non-target toxicity, or both.

The commercial limitations of pyrethrum extracts, which are collectively known as pyrethrins and are a mixture of six lipophilic esters, have long been recognized because of their high rate of photodegradation and a short "knockdown" (rapid paralysis) effect. After the discovery of the constituents of pyrethrins, researchers searched for derivatives of pyrethrins that had a higher resistance to photodegradation. This search directly led to the synthesis of pyrethroids. The advantages of pyrethrins and pyrethroids are that they are highly lipophilic, have a short half-life in the environment, have low toxicity to terrestrial vertebrates and do not biomagnify like older chemical classes, such as organochlorines (see Tables 3.1 and 3.2). In her book *Silent Spring*, Rachel Carson recognized that insecticides like pyrethrins offered alternatives to many of the insecticides that were used during the 1940s to 1970s.

Pyrethroids, the synthetic derivatives of pyrethrins, have changed structurally over the past several decades. However, the basic components of pyrethrins, a chrysanthemic acid linked to an aromatic alcohol through an ester linkage, have been conserved (see Figures 3.1 and 3.2). The widespread use of pyrethroids began in the 1970s after the development of photostable pyrethroids like permethrin and fenvalerate. Pyrethroid use has increased substantially throughout the world over the past few decades as organophosphate, carbamate and organochlorine insecticides are being phased out.[5–7] Pyrethrins and pyrethroids are estimated at 23% of the insecticide world market, with more than 3500 registered formulations, and are widely used in agriculture, residential areas, public health and food preparation.[8,9] Permethrin and cypermethrin are the most widely used pyrethroids in the USA, with about

Table 3.1 Bioconcentration factors (BCF) for type I and II pyrethroids and DDT for rainbow trout (*Oncorhynchus mykiss*) from Muir *et al.*[221]

Compound	*BCF*
Cypermethrin	832
Permethrin	1940
Deltamethrin	502
Fenvalerate	403
DDT	72500

Table 3.2 LC_{50} values of pyrethrins, type I (allethrins, permethrin and resmethrin), II (cypermethrin and deltamethrin) and pseudopyrethroid (etofenprox) for mallard duck (*Anas platyrhynchos*), rat (*Rattus norvegicus*) and rainbow trout (*Oncorhynchus mykiss*).

Compound	Mallard Duck[a]	Rat[a]	Rainbow Trout[b]	Source
Pyrethrins	>5620	700	5.1	USEPA[118]
Allethrins	>2000	720	9.7	WHO[222]
Permethrin	>10000	8900	6.43	USEPA[10]; Kumaraguru and Beamish[223]
Resmethrin	>5000	4639	0.28	USEPA[68]
Cypermethrin	>2634	247	0.39	USEPA[10]
Deltamethrin	>4640	128	1.97	WHO[224]
Etofenprox	>2000	>5000	13	USEPA[25]

[a]Acute oral LC_{50} ($mg\,kg^{-1}$).
[b]96-h LC_{50} ($\mu g\,l^{-1}$).

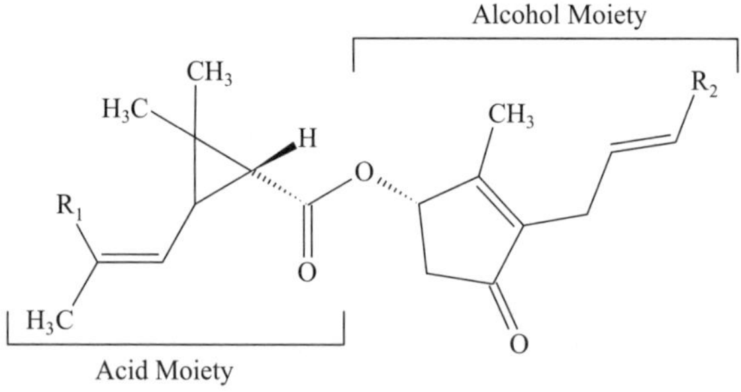

Esters of chrysanthemic acid			Esters of pyrethric acid		
	R$_1$	R$_2$		R$_1$	R$_2$
Pyrethrin I	CH$_3$	CHCH$_2$	Pyrethrin II	CH$_3$OC(O)	CHCH$_2$
Cinerin I	CH$_3$	CH$_3$	Cinerin II	CH$_3$OC(O)	CH$_3$
Jasmolin I	CH$_3$	CH$_2$CH$_3$	Jasmolin II	CH$_3$OC(O)	CH$_2$CH$_3$

Figure 3.1 The chemical structure of the six constituents of pyrethrum extracts which are collectively known as pyrethrins.

910 tonnes of permethrin and 455 tonnes of cypermethrin applied annually.[10,11] Pyrethroids are also used extensively in urban areas, accounting for about 70% of the total usage in California.[6]

3.2 Structure and Chemistry

3.2.1 Pyrethrins

Pyrethrins are prepared from dried *Chrysanthemum cinerariaefolium* and/or *C. cineum* flower heads and are composed of six insecticidally active esters. Pyrethrin extracts are highly viscous liquids with high boiling points, sensitivity to oxidation, and are difficult to store for long periods.[12] Annual world

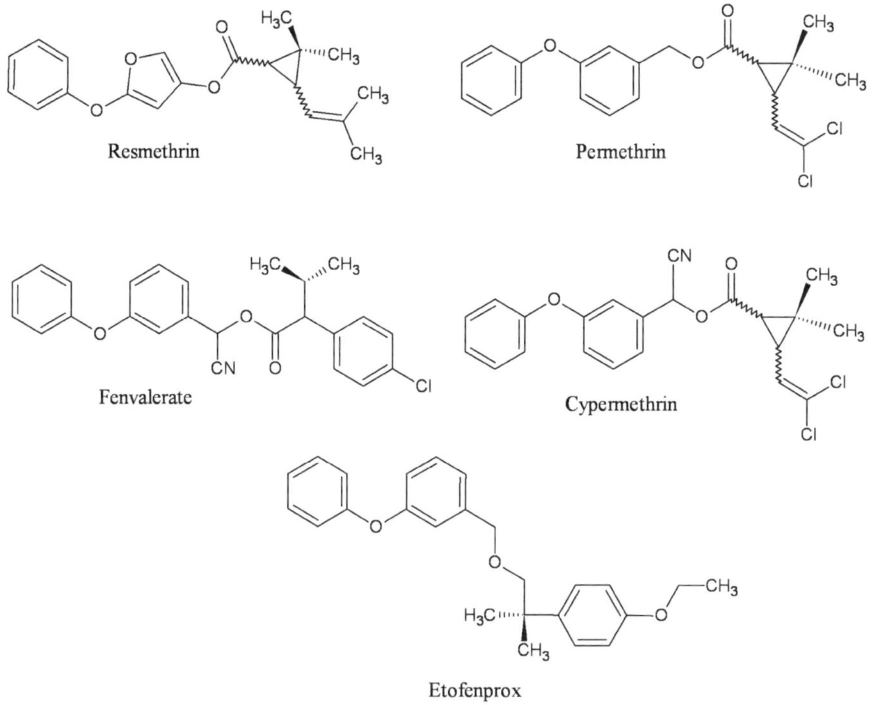

Figure 3.2 The chemical structure of type I (resmethrin and permethrin), type II (fenvalerate and cypermethrin) and pseudopyrethroids.

production of dried flowers has rarely exceeded 20 000 tonnes and, with an average pyrethrins content of 1.5%, the potential yield is 30 kg of 50% extract per tonne. With losses at various processing stages, however, the actual yield is only about 25 kg, giving a potential annual world production of 500 tonnes. Since availability is highly variable, demand often far exceeds supply.[1]

In 1924, Staudinger and Ruzicka[13] elucidated that the active constituents, pyrethrin I and II, are esters of 2,2-dimethyl-3-(2-methyl-l-propenyl)-l-cyclo-propanecarboxylic acid (chrysanthemic) and of 3-(2-methoxycarbonyl-l-prope-nyl)-2,2-dimethyl-l-cyclopropanecarboxylic acid (pyrethric acid), respectively. The six constituents of pyrethrins are pyrethrin I and II, cinerin I and II, and jasmolin I and II. They are collectively known as pyrethrins, which are the esters of two carboxylic acids, chrysanthemic and pyrethric acid (see Figure 3.1). Naming of the six esters of pyrethrins is derived from the alcohol component distinguished by name and number, which are designated by the Roman numeral I and II that represent the esters of the chrysanthemic and pyrethric acid, respectively (see Figure 3.1). There is considerable variation in the proportions of the different constituents of pyrethrins, with the average extract containing 73% pyrethrin I and II, 19% cinerin I and II, and 8% jasmolin I and II.[14] Pyrethrin I and II differ in their insecticidal properties, with pyrethrin I showing greater lethality and pyrethrin II showing greater knockdown.[15]

3.2.2   Pyrethroids

The highly variable availability of pyrethrins encouraged the development and use of synthetic alternatives, which has led to the development of pyrethroids. When the stereochemistry of pyrethrins was elucidated, it formed the model from which pyrethroids were derived; the majority of pyrethroids were derived by modifying the chrysanthemic acid moiety of pyrethrin I and esterifying the alcohols. Synthetic pyrethroids have been developed in order to improve the specificity and activity of pyrethrins, while maintaining the high knockdown and low terrestrial vertebrate toxicity. There is a small group of structural features that pyrethroids require if they are to possess high insecticidal activity, irrespective of the rest of the molecule or the nature of the target species. The pyrethroid active esters are 3-substituted cyclopropanecarboxylic acids which all have a 1R-configuration, a gem-dimethyl substitution at the C-2 of the cyclopropane ring, and only those phenylacetates that contain the corresponding substitute in the 2-position. About 1000 different pyrethroid structures have been synthesized; some are very different from the original structures of the pyrethrin I and II, including structures lacking the dimethylcyclopropane ring and the ester linkages (see Figures 3.1 and 3.2). The level of activity is determined by penetration, metabolism and target site sensitivity, which is in turn determined by the structure of the molecule.

It has been known since the 1840s that pyrethrins are highly photolabile, with a half-life of less than five hours in direct sunlight, greatly limiting their commercial use.[16] The first pyrethroids were synthesized by the replacement of specific structural elements found in pyrethrin I with isosteric moieties to improve metabolic and photochemical stability. Although the synthesis of analogs of pyrethrins began as soon as the active constituents were identified, it was not until 1949 that the first commercially successful pyrethroid, allethrin, was introduced.[17]

The next significant development was through the modification of the alcohol component of pyrethrin I, which was esterified, and this led to the synthesis of resmethrin in 1967.[18] Resmethrin represented the first compound that had an insecticidal activity that was equal to, or greater than, that of pyrethrins, but which exhibited a lower mammalian toxicity. The synthesis of resmethrin and other chrysanthemate esters raised the issue of the stereochemistry of the acid moiety as a determinant of biological activity and metabolism (see Figure 3.2). Pyrethroids have three asymmetric carbon atoms and can have as many as eight possible stereoisomers. The presence of two chiral centers in the cyclopropane ring of chrysanthemic acids produces two pairs of diastereomers, which are designated *cis* and *trans* based on the orientation of the C-1 and C-3 substitutions in relation to the plane of the cyclopropane ring,[19] but only those with the R configuration at the cyclopropane C-1 are insecticidally active.[20]

Despite its positive attributes, resmethrin is not photochemically stable and lacks the degree of persistence needed for agricultural commercialization. In 1973, permethrin was synthesized[21] and was the first compound that exhibited sufficient photostability for agricultural use. Permethrin revolutionized pyrethroids as a class, subsequently leading to their widespread use in pest management applications. Permethrin was synthesized by replacing the methyl groups with chlorine atoms in the acid side-chain, which block photochemical degradation on the adjacent double bond (see Figure 3.2).[21] Permethrin is ten to 100 times more stable in light than resmethrin, yet it is as active against insects as resmethrin while maintaining low mammalian and avian toxicity (see Table 3.2). Like most pyrethroids, the 1*R-trans* isomer of permethrin is rapidly metabolized in organisms, with the 1*R-cis* isomer being more stable and toxic.[22] After the discovery of permethrin, researchers searched for compounds with a higher insecticidal activity than that of pyrethroids and this led to the discovery of the cyano substitute at the benzylic carbon of the 3-phenoxybenzyl group (see Figure 3.2).

Pyrethroids are categorized according to their structure and toxicology, including those lacking the α-cyano group on the phenoxybenzyl moiety (type I) and those with a α-cyano group on the phenoxybenzyl moiety (type II; see Figure 3.2). The next phase of pyrethroid development involved the search for a greater structural variety that could reduce the cost of synthesis and expand the biological activity for new uses.[19] The discovery that less expensive α-substituted phenylacetic acids could be used as substitutes for cyclopropanecarboxylic acids when esterified with the appropriate pyrethroid alcohols, led to the development of pyrethroids like fenvalerate (see Figure 3.2). Pyrethrins and pyrethroids are extremely toxic to many aquatic organisms (see section 3.7), which has led to research looking for pyrethroids that reduce aquatic toxicity while maintaining the favorable properties of the photostable pyrethroids.

The discovery of fenvalerate, which is a α-substituted phenylacetic acid form of the cyclopropanecarboxylic acids, led to the development of the non-ester pyrethroids which are also known as pseudopyrethroids. The common features of permethrin and fenvalerate were used to develop the pseudopyrethroid etofenprox (see Figure 3.2). Pseuodopyrethroids were mainly derived during the 1980s and were found to be substantially different from pyrethrin I and type I and II pyrethroids, so they were not placed into the classical pyrethroid insecticide classification.[23,24] Pseudopyrethoids, such as etofenprox, have approximately 2% of the toxicity to fish of conventional pyrethroids, but they maintain high potency to insects with a characteristic low mammalian toxicity (see Table 3.2).

Pseudopyrethoids have not been widely used in the USA, but etofenprox was recently registered for the control of adult mosquitoes, with crop labeling being currently evaluated.[25] In addition to reducing the toxicity of pyrethroids to aquatic organisms, "green" processes are currently being developed for the preparation of pyrethroids, such as chemoenzymatic synthesis, and the reduction of the 1,2-addition of haloalkanes to polymer-bound olefins has been carried out in solid-phase synthesis to add the dihaloethenylcyclopropane carboxylate moieties.[26]

Due to the similar modes of action of pyrethroids and DDT analogs, researchers have developed "hybrid" pyrethroids that contain the features of both pyrethroids and DDT. The only compound that has been developed for commercial applications is cycloprothrin.[27] Cycloprothrin, a type II pyrethroid, is not as toxic to target organisms as other type II pyrethroids like deltamethrin, but is less toxic to fish.

3.2.3 Physical Properties

Pyrethrins and pyrethroids are highly nonpolar chemicals that have low water solubility and volatility, high octanol–water partition coefficients, and a high affinity to bind to soil and sediment particles (see Table 3.3). Pyrethrins and pyrethroids are rapidly degraded *via* photochemical reactions which result from isomerization of the substituents on the cyclopropane ring, oxidation of the acid and alcohol moieties, dehalogenation of dihalovinyl derivatives, and decarboxylation occurring in type II pyrethroids.[28,29] The non-ester pyrethroids are not subject to hydrolysis, but are broken down *via* oxidation reactions. There is evidence that hydrogen peroxide photochemically produces a hydration reaction with ether cleavage proceeding *via* reaction with a hydroxyl radical.[29] Pyrethroid photodegradation follows first-order kinetics with the main reactions being ester cleavage, photooxidation, photoisomerization and decyanation.[30,31]

In soil under both standard atmospheric and flooded conditions, the photolysis half-life in water ranges from 34.7 to 165 days (see Table 3.4). On soil,

Table 3.3 Physical properties of pyrethroids.

Compound	Molecular Weight	Log P	Water Solubility $(\mu g\,l^{-1})$	Vapor Pressure (mm Hg)	K_{ow} $(\times 10^6)$	K_{oc} $(\times 10^5)$	K_h (atm.m^3 mol^{-1})
Permethrin	391.3[b]	6.1[a]	0.084[a]	2.2×10^{-8b}	1.3[a]	2.8[a]	1.4×10^{-2b}
Bifenthrin	422.9[a]	6.4[a]	0.00014[a]	1.8×10^{-7a}	3[a]	2.4[a]	7.2×10^{-3a}
Cypermethrin	416.3[c]	6.5[a]	0.004[a]	3.1×10^{-9}	3.5[a]	1.4[c]	3.4×10^{-7a}
λ-Cyhalothrin	449.9[a]	7[a]	0.005[a]	1.6×10^{-9a}	10[a]	3.3[a]	1.9×10^{-7a}
Deltamethrin	505.2[a]	4.5[a]	0.0002[a]	9.3×10^{-11a}	3.4[a]	7[a]	3.1×10^{-7a}
Etofenprox	376.4[d]	6.9[e]	0.023[e]	2.5×10^{-8d}	7.9[d]	9.9[e]	3.5×10^{-2e}

Log P = Partition coefficient. K_{ow} = Octanol : Water partition coefficient. K_{oc} = Organic carbon adsorption coefficient. K_h = Henry's law constant.
[a]Laskowski.[33]
[b]USEPA.[10]
[c]USEPA.[11]
[d]USEPA.[25]
[e]Vasquez *et al.*[141]

Table 3.4 Half-lives of pyrethroids in water, light and soil from Laskowski.[33]

Compound	Hydrolysis half-life			Photolysis half-life		Soil Degradation half-life	
	pH 5	pH 7	pH 9	Water	Soil	Aerobic soil	Anaerobic soil
Permethrin	S	S	242	110	104	39.5	197
Bifenthrin	S	S	S	408	96.6	96.3	425
Cypermethrin	619	274	1.9	30.1	165	27.6	55
λ-Cyhalothrin	S	S	8.66	24.5	53.7	42.6	33.6
Deltamethrin	S	S	2.15	55.5	34.7	24.2	28.9

S = Stable.

the photolysis half-life is generally fewer than 55 days, with faster degradation occurring on dry soil. Aerobic degradation occurs rapidly, with the major degradation pathways resulting from ester cleavage, oxidation and hydroxylation.[32,33] The loss of permethrin from water by adsorption on sediment leaves less than 2% in the aqueous phase after seven days[32] with 95% of pyrethroids being adsorbed by sediments within one minute.[34] In natural water–sediment mixtures, ester cleavage is the major degradation process for both type I and II pyrethroids.[32] Pyrethroids are degraded slowly in acidic and neutral pH, but degradation is more rapid in alkaline water.[33] In addition to abiotic reactions, bacteria are capable of degrading pyrethroids and can be specific to both the compound and the stereochemistry.[35,36]

3.3 Mode of Action

Pyrethrins, pyrethroids, DDT and DDT analogs belong to a group of chemicals that are neurotoxic and share a similar mode of action that is distinctive from other classes of insecticides. There are several ways that pyrethrins and pyrethroids can enter the body of an organism to exert their effects. The first mode is non-stereospecific with rapid penetration through the epidermis, followed by uptake by the blood or hemolymph carrier proteins and subsequent distribution throughout the body. Pyrethroid diffusion along the epidermis cells is the main route of distribution to the central nervous system (CNS) after penetration.[37] Pyrethroids also can enter the CNS directly *via* contact with sensory organs of the peripheral nervous system. The sensory structures of both invertebrates and vertebrates are sensitive to pyrethroids.[38] Pyrethroids can also enter the body through the airway in the vapor phase, but such penetration represents only a small contribution due to the low vapor pressure of pyrethroids (see Table 3.3). Pyrethroids can also be ingested, and penetration into the blood–hemolymph through the alimentary canal can play an important role in toxicity.

Pyrethroids have been classified toxicologically into two subclasses based on the induction of either whole body tremors (T syndrome) or a coarse whole body tremor progressing to sinuous writhing (choreoathetosis) with salivation (CS syndrome) following near-lethal dose levels in both rats (*Rattus norvegicus*) and mice (*Mus musculus*), and closely follows the chemical structure of the two types of pyrethroids.[39,40] Type I pyrethroids are characterized by the T-syndrome which consists of aggressive sparring, sensitivity to external stimuli, fine tremors progressing to whole body tremors and prostration. Type I pyrethroids also elevate core body temperature, which is attributed to the excessive muscular activity associated with tremors. Type II pyrethroids are characterized by the CS syndrome which is comprised initially of pawing and burrowing behavior followed by profuse salivation, choreoathetosis, increased startle response, and terminal chronic seizures. Type II pyrethroids decrease core body temperature, which is attributed to excessive salivation and wetting of the ventral body surface. Although salivation typically co-occurs with choreoathetosis, a TS syndrome (tremor with salivation) has also been observed in a few pyrethroids. Multiple lines of evidence show that pyrethroids, as a class, do not act in a similar fashion on the voltage-gated sodium channels, and the classifications of toxicology are not absolute for either invertebrates or vertebrates.[41,42] For example, the type I pyrethroid, bioallethrin, exhibits toxicological symptoms of both type I and II intoxication.

As expected, increasing the dose levels of pyrethrins and pyrethroids results in a proportional increase in motor activity, which is the classic dose–response effect with respect to neurotoxic substances. Pyrethrins and pyrethroids act very quickly to produce symptoms of lost coordination and paralysis which are known as "the knockdown effect", and which are often accompanied by spasms and tremors that induce intense repetitive activation in sense organs and in myelinated nerve fibers. The spasms can be violent and can cause the loss of extremities, such as legs and wings in insects.

The most compelling evidence of a similar mode of action for pyrethrins, pyrethroids, and DDT comes from resistance studies examining knockdown resistance (*kdr*) demonstrating cross resistance. Physiological and biochemical studies of pyrethrins, pyrethroids and DDT show that in both vertebrates and invertebrates the primary mode of action is the binding of the voltage-gated sodium channel.[38,42–44] Mammals, unlike insects, however, have multiple isoforms of the sodium channel that vary by tissue type, as well as biophysical and pharmacological properties.[45]

To understand the primary mode of action, the mechanism by which voltage-gated sodium channels work needs to be reviewed. When the voltage-gated sodium channel is stimulated, it causes a depolarization of the membrane, which changes the nerve cell's permeability to Na^+ and K^+. The excited membrane becomes permeable to Na^+, with a small number of ions acted on when electrical and concentration gradients rush into the membrane causing the depolarization of the membrane. The sodium ions carry a current inward, which is referred to as the "action potential". The inward movement of sodium ions causes the membrane potential to overshoot the membrane potential with the inside becoming positive relative to the outside of the membrane surface. During a spike, the membrane is absolutely refractory, and a stimulus of even greater magnitude cannot cause the gates to open wider or more Na^+ to flow inward. In addition, a neuron is partially refractory for a further few milliseconds and only a strong stimulus will cause a new response.[46] The upper limit of impulses per second is about 100, with each depolarization event lasting only about two to three milliseconds.[46] Pyrethrins and type I pyrethroids modify the sodium channels such that there is a slight prolongation of the open time (*i.e.* sodium tail currents of approximately 20 milliseconds), which results in multiple long action potentials. Type II pyrethroids significantly prolong channel open time (*i.e.* sodium tail currents of 200 milliseconds to minutes), resulting in an increased resting membrane potential and often inducing a depolarization-dependent block of action potentials.

Type I pyrethroids cause multiple spike discharges, while type II pyrethroids cause a stimulus-dependent depolarization of the membrane potential which reduces the amplitude of the action potential, and a loss of electrical excitability in both vertebrates and invertebrates.[38,47] The toxic action is exerted by preventing the deactivation or closing of the gate after activation and membrane depolarization. This results in destabilizing the negative after potential of the nerve due to the leakage of Na^+ ions through the nerve membrane. This causes hyperactivity by delaying the closing sodium channels which allows a persistent inward current to flow after the action potential, causing repetitive discharges that can occur either spontaneously or after a single stimulus. The sodium channel residue that is critical for regulating the action of pyrethroids is the negatively-charged aspartic acid residue at position 802 located in the extracellular end of the transmembrane segment 1 of domain II, which is critical for both the action of pyrethroids and the voltage dependence of channel activation.[48]

The differences between type I and II pyrethroids are expressed in the motor nerve terminals, where type I cause presynaptic repetitive discharges, and type II cause a tonic release of transmitter indicative of membrane depolarization.[38,49] Type II pyrethroids are a more potent toxicant than type I in depolarizing the nerves.[49] Type II pyrethroids are associated with faster activation–deactivation kinetics on the $Na_{v1.8}$ sodium channels than type I pyrethroids in vertebrates.[42] The higher toxicity of type II pyrethroids is mostly attributed to the hyperexcitatory effect on the axons which results from their stronger membrane depolarizing action. Type I pyrethroids modify the sodium channels in the closed state, while type II pyrethroids modify the open but not inactivated sodium channels.[50] However, this relationship does not always hold true; *cis*-permethrin and fenvalerate interact with both closed and open sodium channels, but they bind with greater affinity to the open state.[51–53] Type I repetitive discharges have been shown to be suppressed by cypermethrin, indicating that the two pyrethroid types can interact antagonistically.[53]

Pyrethroids affect the voltage-sensitive calcium channels, γ-aminobutyric acid (GABA) receptors and GABA-activated channels, and voltage-sensitive chloride channel.[43,54] Recent findings suggest that pyrethroids can modulate the activity of voltage-gated calcium (Ca^{2+}) channels.[55] However, these studies report conflicting results on the inhibitory effects of pyrethroids on voltage-gated calcium channels. Neal *et al.*[56] demonstrated that allethrin significantly altered the voltage dependency of activation and inactivation of L-type voltage-gated calcium channels, which suggests that differential modulation of voltage-gated calcium channels subtypes could elucidate some of the conflicting observations of other studies. Type II pyrethroids are more potent enhancers of Ca^{2+} influx and glutamate release under depolarizing conditions than type I pyrethroids.[41,51]

The GABA receptor–chloride ionophore complex is also a target of type II pyrethroids. GABA is an inhibitory transmitter in the synapse of the CNS of both vertebrates and invertebrates. Pretreatment with diazepam (a benzodiazepine anticonvulsant known to act on the GABA receptors) has been shown to selectively delay the onset of toxic symptoms of type II, but not type I, pyrethroids in cockroaches and mice.[38] Radioligand binding studies have shown that deltamethrin, but not its non-toxic α-*R*-cyano epimer, inhibited [³H]dihydropicrotoxin binding to the chloride ionophore in the rat brain GABA receptor complex.[38] Pyrethrins and pyrethroids also inhibit the Cl⁻ channel function at the GABA receptor–ionophore complex.[57]

An additional target proposed for type II pyrethroids is the membrane chloride ion channel.[58] Generally type II pyrethroids decrease the open channel probability of chloride channels, but the type I pyrethroids do not seem to have an effect on the chlorine channel.[42,54,59] Upon further investigation, Burr and Ray[59] found that the type I pyrethroid bioallethrin, and type II pyrethroids β-cyfluthrin, cypermethrin, deltamethrin and fenpropathrin, significantly decreased the probability that the ligand-gated chloride channel would be an open channel. However, they found that the type I pyrethroids, bifenthrin, bioresmethrin, *cis*-permethrin and *cis*-resmethrin, and the type II pyrethroids, cyfluthrin, lambda-cyhalothrin, esfenvalerate and tefluthrin, did not. Interestingly, the type I pyrethroid, bioallethrin, significantly alters the probability of opening the ligand-gated chloride channel, but has generally a weaker response than type II pyrethroids.[42] One hypothesis was that bioallethrin may be a mixed-type pyrethroid.[43,59] The blockade of the voltage-sensitive chloride channels is associated with salivation, which is a hallmark of type II pyrethroid intoxication and could contribute to the enhanced excitability of the CNS.[43]

Pyrethroids inhibit the Ca-ATPase, Ca-Mg ATPase neurotransmitters and the peripheral benzodiazepine receptors,[60] but their action on these sites is minor compared with the voltage-gated sodium channels. The effects on these sites could, however, enhance the uncontrolled convulsions and tremors.[43]

3.3.1 Enantioselective Toxicity

Formulations of pyrethroids are mixtures of the 1*R*-*cis*- and 1*R*-*trans*-isomers. Only the cyclopropanecarboxylic acid esters that have the *R* absolute configuration at the cyclopropane C-1 and α-cyano-3-phenoxybenzyl esters with the *S* absolute configuration at the C-α are toxic.[38] Of the four stereoisomers of permethrin and resmethrin, the highest acute toxicity is observed in the 1*R*-*cis*- and 1*R*-*trans*-isomers, which contribute 94 to 97% of the toxic dose, while the 1*S*-*trans*- and 1*S*-*cis*-isomers contribute insignificantly to the toxicity.[19,61,62] Studies of the non-toxic isomers of pyrethroids found that they were less than 1% as toxic as the corresponding toxic isomer.[63] Chronic toxicity tests in *Daphnia magna* with respect to survival and fecundity for 1*R*-*cis*-bifenthrin have been shown to have 80-fold greater toxicity than 1*S*-*cis*- bifenthrin after 14 days.[64] The difference in toxicity can be attributed to the absorbed dose of 1*R*-*cis*-isomer which was approximately 40-fold higher than that of 1*S*-*cis*- isomer. Pyrethroids exhibit significant enantioselectivity in oxidative stress, with the *trans*-permethrin exhibiting 1.6 times greater cytotoxicity than *cis*-permethrin at concentrations of $20\,\mathrm{mg\,l^{-1}}$ in rat adrenal pheochromocytoma cells.[65] It should be noted that effects on neurotoxicity at both the cellular and visible level occur at doses 2000 times greater than exposures seen in the environment.

3.3.2 Effects of Sex, Age and Size on Toxicity

Studies of the toxicity of insecticides have shown a significant difference in sensitivity between sexes, with invertebrate males generally more sensitive than females, but the opposite is true for mammals. This may be due to size differences, lipid content, and enzyme activity, but differences between sexes may not always be observed.[10,11,66–68] Age and size are the most important factors influencing the susceptibility of organisms to insecticides, because these factors are related to increases in body fat content and enzymatic activity.[67,69] Adult males and gravid females of German cockroaches (*Blattella germanica*) were generally found to be more sensitive than non-gravid females to pyrethroids.[70] However, whilst the body mass of gravid and non-gravid females did not differ, that of the males was smaller. In contrast to invertebrates, female rats are more sensitive to pyrethroids than males; this difference in sensitivity is most likely due to hormone differences. A larger body mass does not necessarily mean that a higher dose is required to kill insects. This was demonstrated by Antwi and Peterson,[71] who showed that house crickets (*Acheta domesticus*) were more sensitive to pyrethroids than adult convergent lady beetles (*Hippodamia convergens*) and larval fall armyworms (*Spodoptera frugiperda*).

Younger invertebrates and vertebrates are generally more sensitive than older immature organisms, with susceptibility decreasing with each successive stage.[72,73] The sensitivity of the younger developmental stages is most likely due to age-related differences in pharmacokinetics and pharmacodynamics. These differences may be a result of the lower enzymatic activity, particularly of the esterases and cytochrome P450 monooxygenases (CYP), of younger organisms and in insects where the cuticle has not hardened, allowing more of the insecticide to be absorbed. In vertebrates, however, the evidence is unclear as to whether the differences in sensitivity of the voltage-gated sodium channel isoforms are due to the isoform of the sodium channel since they differ between fetal and post-natal rats. Therefore, regulatory agencies do not assume an increased toxicity of pyrethroids to juveniles based on pharmacokinetic dynamics.[74]

3.3.3 Temperature

DDT and pyrethroids share of a number of similar properties in their mode of action. The temperature and mode of action of pyrethroids have been connected since Vinson and Kearns[75] found that DDT had a higher toxicity at lower temperatures because of an intrinsic susceptibility of some physiological systems, rather than penetration or metabolism which was subsequently confirmed for pyrethrins.[76] A strict negative temperature correlation is not always observed because type II pyrethroids are in some cases positively correlated with temperature (see Table 3.5).[77,78] The same temperature dependent toxicities have been observed in warm-blooded animals.[37] The decreased toxicity at higher temperatures is mostly the result of desorption of the pyrethroid from the target site. In contrast to insects, mites show a positive temperature effect to both type I and II pyrethroids.[79]

Table 3.5 The lethal concentration that kills 50% of a population (LC$_{50}$) for 3-phenoxybenzyl pyrethroids and DDT against third instar larvae of tobacco budworm (*Heliothis virescens*) and Asian citrus psyllid (*Diaphorina citri*) at 37.8, 26.7 and 15.6 °C from Sparks *et al.*[77] and Boina *et al.*[78]

Compound	Temp °C	LD$_{50}$ ($\mu g\,g^{-1}$)	LD$_{50}$ Ratio 37.8–15.6 °C
Permethrin	37.8	1.94	−9
	26.7	1.44	
	15.6	0.22	
Sumithrin	37.8	4.64	−24.2
	26.7	2.51	
	15.6	0.19	
Cypermethrin	37.8	0.51	−1.81
	26.7	0.24	
	15.6	0.28	
Deltamethrin	37.8	0.016	1.55
	26.7	0.044	
	15.6	0.088	
Fenvalerate	37.8	0.22	2.29
	26.7	0.39	
	15.6	0.51	
Bifenthrin	17	1.94	−9
	27	1.44	
	37	0.22	
DDT	37.8	62.36	−15.35
	26.7	31.49	
	15.6	4.06	

3.4 Metabolism

In mammals and birds, pyrethroids augment the electrical activity in the brain, spinal column and peripheral neurons which underlie the induced paresthesia, convulsions, and tremors.[38] The low toxicity of pyrethroids is attributed to their rapid metabolism in the blood and liver, with more than 90% of pyrethroids being excreted as metabolites in urine within 24 hours after exposure.[80–82] Indeed, although extensively used, there are relatively few reports of human, domestic animal or wild animal pyrethroid poisonings.[83,84]

Cytochrome P450s are extremely important in the metabolism of xenobiotics and endogenous compounds. Cytochrome P450s can metabolize a large number of substrates because they exist in numerous different isoforms and they have several functional roles, including growth, development and metabolism of xenobiotics. The two types of primary metabolic enzymes involved in the detoxification of pyrethroids are microsomal monooxygenases and esterases. The detoxification of pyrethrins and pyrethroid insecticides is primarily through oxidative metabolism by CYP, which yields metabolites with hydroxyl groups substituted in both the acidic and basic moieties.[85] The presence of a *cis*-substituted acid moiety and a secondary alcohol moiety indicates that hydrolytic metabolism would be limited, and subsequent studies in mammals have found hydrolysis to be minimal.[86] The metabolic pathway of *cis*- and *trans*-permethrin is displayed in Figure 3.3 and shows the different CYP involved in the metabolism of pyrethroids. The initial biotransformation of pyrethroids is through attack by either esterases at the central ester bond, or by CYP-dependent monooxygenases at one or more of the acid or alcohol moieties, and this generally achieves detoxification of the compound (see Figure 3.3).

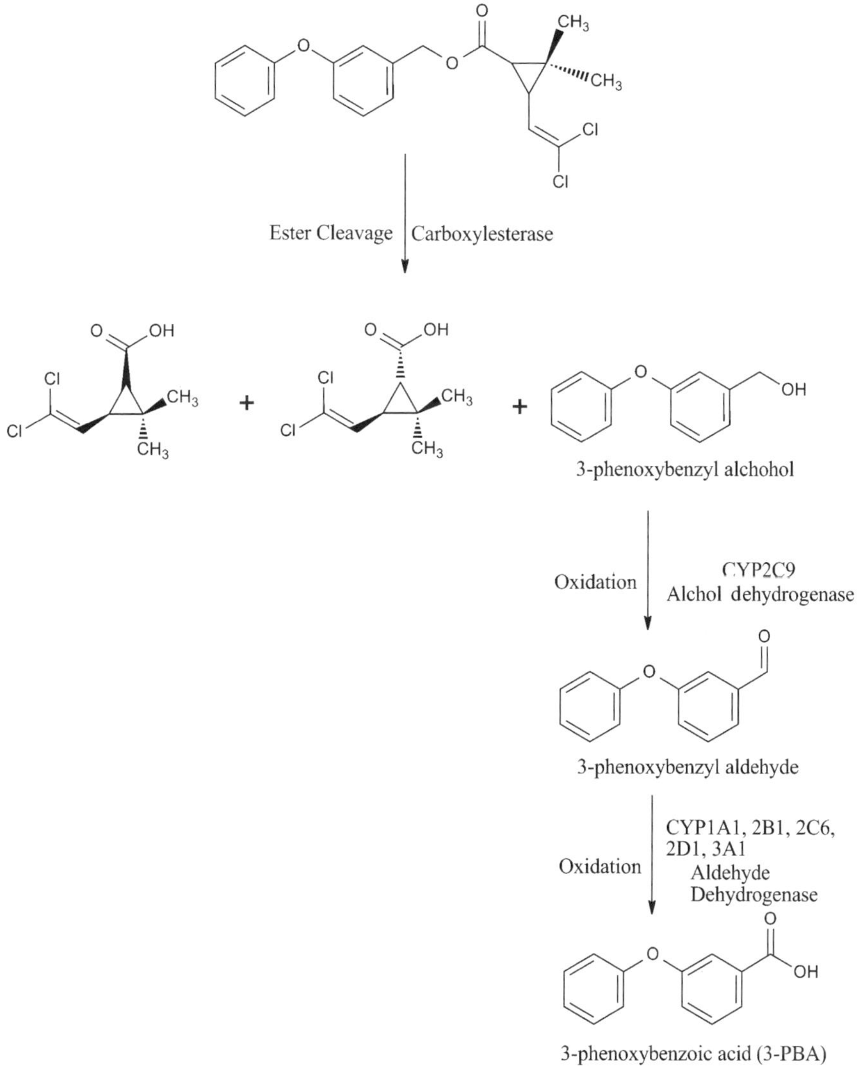

Figure 3.3 Metabolism of permethrin in mammals. Abbreviation CYP: cytochrome P450.

CYPs are not, however, involved in the hydrolysis or in the oxidation of the *trans*-isomers of pyrethroids to phenoxybenzyl alcohol and phenoxybenzoic acid, the main forms of pyrethroids that are excreted (see Figure 3.3).[87] The human alcohol and aldehyde dehydrogenases are the enzymes involved in the oxidation of phenoxybenzyl alcohol to phenoxybenzoic acid (see Figure 3.3).[87]

For type I pyrethroids, following ester cleavage, the primary alcohol moieties undergo further oxidation *via* the aldehyde to carboxylic acids. However, type II alcohols lose the cyanide non-enzymatically to form the aldehyde.[43] The principal sites of oxidation for pyrethrins I in rats are the terminal double bond and the *trans* methyl group of the isobutenyl substituent of the acid moiety, which undergoes sequential oxidation to a carboxylic acid.[88] In mammals, as in insects, the *cis*-isomers are generally more toxic than the corresponding *trans*-isomers. This phenomenon may be because the liver fractions are poor at metabolizing *cis*-isomers, while the *trans*-isomers are readily metabolized by esterases.[87] The *cis*-isomers are also less readily absorbed by the stomach hence limiting their toxicity.[37] For reference, technical-grade mixtures of permethrin contain 30% of *cis*-isomer, while formulations contain about 35%.

Pyrethroids are metabolized predominantly by esterases. The first stage involves cleavage of the ester bond, generating 3-phenoxybenzaldehyde, 3-phenoxybenzoic acid, and (2,2-dichlorovinyl)-3,3-dimethylcyclopropanecarboxylic acid as major metabolites (see Figure 3.3).[89] The major metabolites detected in the urine of mammals (see Figure 3.3) are 3-phenoxybenzoic acid (3PBA; the product of the oxidation of the hydrolytic product of many of pyrethroids), 4-fluoro-3-phenoxybenzoic acid (4F3PBA; a metabolite of the fluorine-substituted pyrethroid insecticides), and *cis*- and *trans*-(2,2-dichlorovinyl)-3,3-dimethylcyclopropane-1-carboxylic acid (*cis*- and *trans*-DCCA; metabolites of chlorinated pyrethroids, such as permethrin, cypermethrin and cyfluthrin).[80,82,90]

There are also specific metabolites for certain pyrethroids. For example, *cis*-(2,2-dibromovinyl)-3,3-dimethylcyclopropane-1-carboxylic acid (DBCA) is the main metabolite of deltamethrin.[90] The ratio of *trans* : *cis* DCCA can be used to determine the exposure pathway *via* dermal and oral routes.[91] Other, more minor, metabolites include those resulting from hydroxylation at the acidic gem dimethyl group and at the phenoxy group of the alcohol and from oxidation, which results in carboxylic acids and phenols.[92] Once these oxidations occur, the resulting carboxylic acids and phenols may be conjugated by a variety of enzymes, and are subsequently excreted as either free metabolites or conjugated with sugars or amino acids which are rapidly excreted.

3.5 Synergists

Yamamoto[93] defined synergism as where the interaction of two or more toxins is such that their combined effect is greater than simply the sum of their individual toxicities. For pesticide formulations, synergists are typically nontoxic compounds at the dosage applied, but which enhance the toxicity of the active pesticide ingredient. The main route of detoxification of insecticides is through

CYP-mediated detoxification. The CYP enzymes bind molecular oxygen and receive electrons from NADPH to introduce an oxygen molecule into the toxicant, thus catalyzing the oxidation of toxicants. *N*-Octyl bicycloheptene dicarboximide (MGK-264) and piperonyl butoxide (PBO) are the most commonly used synergists and are incorporated into insecticide formulations to inhibit the CYP.[94,95] The enhancement of toxicity for pyrethroids is not as great as it is for pyrethrins.[96] Synergists are mixed at a concentration of two to 50 times that of the insecticide, and enhance the toxicity one to 100 times.

Piperonyl butoxide and MGK-264 have been shown to increase the toxicity of pyrethroids to aquatic organisms, but there is no indication that PBO acts as a synergist in mammals.[52,94,97–99] In addition to inhibiting mixed function oxidases, PBO has also been shown to enhance the penetration rates of pyrethroids through the cuticle of insects.[100]

3.6 Resistance

The fact that pyrethrins, pyrethroids and DDT share a common mode of action, and therefore a common binding domain on the sodium channel, has important implications for the continued use of pyrethrins and pyrethroids in pest management. Resistance to insecticides may cost more than $1.4 billion per year in the USA alone.[101] Selection for resistance to either class of insecticides will lead to resistance to both, which has been extensively documented in mosquitoes.[102] Pyrethroids are currently the most widely used insecticides for the indoor and outdoor control of mosquitoes and are the only chemical recommended for the treatment of mosquito nets (the main tool for preventing malaria in Africa). However, mosquito-borne diseases are emerging and re-emerging in parts of the world and it is thought to mainly be due to widespread mosquito resistance to pyrethroids and the drug resistant strains of vector-borne pathogens.

Insect resistance is dependent on the volume and frequency of applications of insecticides and the inherent characteristics of the insect species. Resistance to pyrethroids comes in two forms: (1) non-metabolic resistance through the decreased sensitivity or reduction in the number of voltage-gated sodium channels, and (2) metabolic resistance *via* detoxifying enzymes, oxidases and decreased cuticle penetration. There are four mechanisms by which resistance is expressed: (1) decreased sensitivity of the sodium channels due to altered structure, (2) decreased sensitivity to pyrethroids through a change in the kinetics of the channel, (3) reduced number of channels available for pyrethroids to bind, and (4) altered lipid membrane around the nerve.[103]

The main form of non-metabolic resistance is the *kdr* and *super-kdr* mutations.[104] Farnham[105] first demonstrated that *kdr* resistance is caused by a recessive gene, and characterized it as resistance to the knockdown effect (*i.e.*, it lowers the sensitivity of the sodium channel). German cockroaches that demonstrate knockdown resistance take about twice as long as susceptible ones to express toxic symptomology.[106] Resistant strains subsequently recover two to four hours after the knockdown and appear normal within 24 hours. The *kdr* gene has been mapped to the autosome 3 that confers an enhanced level of resistance that is designated *super-kdr*.[105,107] *Kdr* resistance is genetically linked to the *para*-homologous sodium channel gene, but correlation between presence of the *para* mutation and knockdown resistance has been infrequently observed and depends on the strain of the insect being studied.[108,109] An important attribute of *kdr* resistance is that synergists do not appreciably alter the toxicity.[110] Decreased cuticle penetration of pyrethroids has been demonstrated in a number of insect species and is generally found in addition to other resistance mechanisms like increased enzyme activity.[111,112]

The detoxification of insecticides through the action of CYP is one of the more important resistance mechanisms. Metabolic resistance can be reduced through the use of a synergist, but non-metabolic resistance cannot.[113] The CYP binds molecular oxygen and receive electrons from NADPH to introduce an oxygen molecule into the substrate. Resistance *via* CYP is associated with overtranscription of a single CYP gene, *Cyp6g1*, in *Drosophila melanogaster*.[114] Cytochrome P450 resistance slows female emergence time and produces smaller body size and lowers energy reserves (glycogen and lipids), which when combined affect the fitness of resistance compared to non-resistant female mosquitoes.[115] Resistance can also be associated with decreased cuticle penetration.[116] Resistance has also been associated with increases in carboxylester hydrolases and glutathione transferases, but these pathways most likely do not confer a large resistance because they play a small role in the detoxification of pyrethrins and pyrethroids.[117]

3.7 Risk Assessment

3.7.1 Human Health Risk Assessment

Human health risk assessments have been performed for pyrethrins and pyrethroids by the United States Environmental Protection Agency (USEPA), other government regulatory agencies around the world and university researchers. The USEPA has found that dietary exposure to pyrethrins and pyrethroids is below reference doses;[10,11,68,118] these results are supported by the current weight of evidence based on urinary metabolites (see section 3.8). Permethrin has been observed in animal models to be a carcinogen and the USEPA estimated the worst-case lifetime average daily exposure based on a tier 1 conservative model to be $0.117\,\mu g\,kg^{-1}\,day^{-1}$. This exposure does not result, however, in an increase in incidents of cancer to the general US population.[10]

One commonly overlooked use of pyrethrins and pyrethroids is with ultra-low volume (ULV) application techniques, which are applied from trucks, helicopters or airplanes and are used for the control of public health pests such as adult mosquitoes and midges. Ultra-low volume applications are commonly used in and around residential areas, so exposure to bystanders within the spray area may occur. This type of application utilizes small droplets of 5 to 25 μm which produce aerosol clouds that are designed to stay aloft to impinge on flying pests and can thus travel considerable distances. Carr *et al.*[119] conducted a dietary risk assessment for aerial ULV applications of resmethrin above agricultural fields as a result of a public health emergency, and found that exposures would result in negligible human dietary risk.

Peterson *et al.*[120] performed a tier-1 deterministic human health risk assessment for acute and subchronic exposures to pyrethroids used in mosquito management after truck-mounted ULV applications. They found that the acute and subchronic risks to humans would result in negligible risk. Schleier III *et al.*[121] followed up with a probabilistic risk assessment and found that Peterson *et al.*[120] overestimated risk, as expected, by about ten-fold. Subsequent risk assessments of acute and subchronic exposures to pyrethroids used during ULV applications in and around military bases revealed that the risks would not exceed their respective reference dose.[122,123] However, Macedo *et al.*[123] found that aggregate exposure to permethrin from ULV applications and impregnated battle dress uniforms exceeded the standard threshold of an excess of one-in-a-million cancer risk above background levels. Schleier III *et al.*[122] completed a probabilistic risk assessment for exposure to indoor residual sprays using cypermethrin and lambda-cyhalothrin which exceeded their respective reference doses. Although these studies found risks that exceeded their respective toxic endpoint, more realistic exposures would reduce the risk estimates, which is common when using higher tiered risk assessments.[121,124]

3.7.2 Ecological Risk Assessment

Pyrethroids are extremely toxic to many aquatic organisms, and thus could pose a substantial ecological risk (see section 3.10). In this section, we have employed a species sensitivity distribution from the USEPA's Ecotox Database[125] for permethrin using 41 aquatic species based on 96 hour LC_{50}s (see Figure 3.4). Species sensitivity distributions are used to calculate the concentrations at which a specified proportion of species will be affected, referred to as the hazardous concentration (HC) for p% of the species (HCp). The resulting HC5 is $0.047\,\mu g\,l^{-1}$, which amounts to approximately 33% of the maximum concentrations seen in the environment (see section 3.9). The minimum concentration observed in the environment is $0.0054\,\mu g\,l^{-1}$, which would result in 0.35% of the species reaching their respective LC_{50} value. At the maximum concentrations seen in the environment ($3\,\mu g\,l^{-1}$), which are rarely observed, 65% of the species would be affected (see Figure 3.4). These results are supported by aquatic risk assessments performed for pyrethroids. The toxicity of pseudopyrethroids, like etofenprox, is lower with respect to aquatic organisms than other pyrethroids currently used; thus they represent a lower risk to the aquatic environment.[126]

A probabilistic aquatic risk assessment conducted by Maund *et al.*[124] for cotton-growing areas focused on pyrethroid exposure in static water bodies as a worst-case scenario. They found that exposures were several orders of

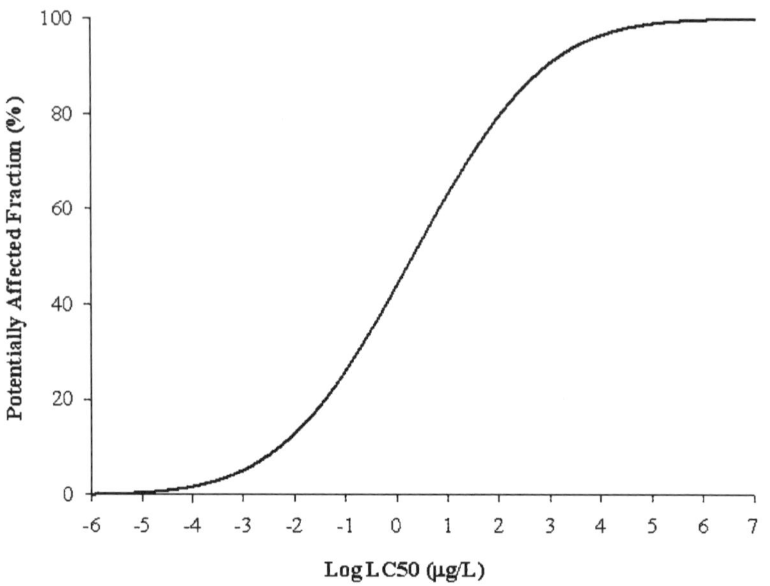

Figure 3.4   Acute species sensitivity distribution constructed from the lethal concentrations that kill 50% of a population (LC$_{50}$) for permethrin, demonstrating the proportion of species affected for aquatic organisms.

magnitude lower than those that would cause effects based on laboratory and field studies. Davis *et al.*[127] conducted a deterministic ecological risk assessment for truck-mounted ULV applications of pyrethroids, and found that the risks to mammals, birds, aquatic vertebrates, and aquatic and terrestrial invertebrates were negligible. These results were subsequently supported using actual environmental concentrations after aerial and truck-mounted ULV applications.[128–131] Studies by Schleier III and Peterson[131] using caged house crickets as a surrogate for medium- to large-sized terrestrial invertebrates showed that ULV applications of permethrin did not result in increased mortality. These results can most likely be applied to smaller insects as well, because house crickets have been found to be more sensitive to pyrethroids than adult convergent lady beetles and larval fall armyworms.[71]

3.8   Biomonitoring and Epidemiology

Current human biomonitoring and epidemiological studies show that pyrethroid exposures to the general population are low and adverse effects are highly unlikely. The main route of exposure for the general public to pyrethrins and pyrethroids is through dietary intake.[7] Urinary metabolite data from both the USA and Germany show that exposure to pyrethroids in the general population is similar, with the highest exposure coming from the most commonly used pyrethroids, permethrin and cypermethrin,[7,90,132–134] with infrequent exposure to pyrethroids like cyfluthrin and deltamethrin.[90] The average daily intake of permethrin in the USA due to diet has been estimated at about $3.2\,\mu g\,day^{-1}$, which is approximately 0.1% of the acceptable daily intake.[135] Children have been found to have higher levels, which may be attributed to their higher ingestion rates of household dust.[132] Even when residential areas have been treated directly with truck-mounted ULV applications of pyrethroids, urinary metabolites have shown no statistical difference when compared with results from untreated areas.[136] The use of biomarkers to monitor pyrethroid exposure may be problematic because the estimation of daily absorbed doses of pyrethroids from volume-weighted or creatinine-adjusted concentrations can lead to substantial under- or over-estimation when compared with doses reconstructed directly from amounts excreted in urine during a period of time.[137]

Occupational application of pyrethroids resulting in the highest concentrations of metabolites in urine samples are from indoor pest-control operators. However, occupational exposures to pyrethroids do not seem to lead to adverse effects.[138] Weichenthal *et al.*[74] reviewed the epidemiological evidence relating to occupational exposures and cancer incidence in agricultural workers applying permethrin, and found an increased odds ratio, but the associations were small and imprecise because of small sample sizes and clear exposure–response relationships were not observed. This is most likely because pyrethroids are slowly absorbed across the skin which prevents high levels of exposure. Karpati *et al.*[139] found no increase in asthma cases after truck-mounted ULV applications in residential neighborhoods in New York, New York, USA. Epidemiologically, the USEPA found that the weight of evidence shows no clear or consistent pattern of effects to indicate an association between pyrethrins or pyrethroid exposure and asthma and allergies.[140]

3.9 Environmental Fate

Commercially available pyrethroids are effective in the field at rates of $0.2\,kg\,ha^{-1}$ or less, with the most active compounds, such as cypermethrin, being effective at rates of $0.015\,kg\,ha^{-1}$. The main routes that pyrethroids enter aquatic systems are *via* bound soil surface movement (run-off) or through drift. When considering the potential for run-off, it is important to remember that pyrethrins, pyrethroids and pseudopyrethroids rapidly degrade in most soil types, under both aerobic and anaerobic conditions, and are strongly absorbed to soil (see Table 3.3).[33,141] The strong adsorption of pyrethroids to soil suggests that when such aquatic contamination does occur, it will most likely be in the form of erosion of soil particles through high wind or large rain events. If pyrethroids drift into water bodies, they are rapidly absorbed by the sediments and organic content in the water column, so they will only be present in the water phase for a short time. The most frequently detected pyrethroids in irrigation, storm water run-off and sediments are bifenthrin, lambda-cyhalo-thrin, cypermethrin and permethrin, with bifenthrin measured at the highest concentrations because it is commonly used for residential pest control.[142–145] Concentrations of pyrethroids in the environment range from 0.0054 to $0.015\,\mu g\,l^{-1}$ in the dissolved phase and 0.0018 to $0.870\,\mu g\,l^{-1}$ in suspended sediment in urban or agricultural areas.[146]

Run-off losses of pesticides from treated fields have been extensively studied, with losses ranging from less than 1% to 10% of the applied product entering waterways.[147] Run-off studies after single and multiple applications of pyrethroids found $\leq 1\%$ of the applied chemical is present throughout the year.[148,149] Residue analysis of water and sediment samples following the final application of a cumulative seasonal exposure, simulated with 12 drift applications and six run-off events of lambda-cyhalothrin and cypermethrin, showed that pyrethroid residues were rapidly lost from the water column with residues of lambda-cyhalothrin and cypermethrin of less than $0.002\,\mu g\,l^{-1}$.[150] Residues in sediment reached a maximum level of approximately $25\,\mu g\,kg^{-1}$, subsequently declining to $<9\,\mu g\,kg^{-1}$ within four months.[150]

The greatest amount of pesticide run-off occurs when severe rain events occur soon after application. The size of the draining catchment has been shown to be negatively correlated with the concentration of insecticide present after agricultural pesticide application.[147] Concentrations of pyrethroids after run-off events ranged from 0.01 to $6.2\,\mu g\,l^{-1}$ in the aqueous phase, and non-point sediment loads were 1 to $300\,mg\,kg^{-1}$.[147] Insecticide concentrations of less than $10\,\mu g\,l^{-1}$ were only observed in catchment sizes of less than $100\,km^2$, with the majority of detections occurring in catchments less than $10\,km^2$.[147]

Pesticide spray drift is defined as the physical movement of a pesticide through air at the time of application or soon thereafter to any site other than that intended for the application.[151,152] Models and field studies show that as the distance from the spray sources increases the concentration deposited decreases, resulting in a concentration gradient in the water.[128,153–157] Generally, aerial applications result in higher levels of spray drift compared with ground applications, which can be partly attributed to the equipment used on ground sprayers that reduces drift.

Measurements of pyrethroid concentrations in farm ponds following spray drift from aerial applications found 0.2 to 7% of the product deposited in the water, which was dependent on the distance from the spray source.[158] Vineyards treated with cypermethrin *via* mistblowers resulted in surface deposits from spray drift ranging from 0.04 to $0.45\,mg\,m^{-2}$ and concentrations in subsurface water ranging from 0.4 to $1.7\,\mu g\,l^{-1}$ soon after spraying, decreasing to $<0.1\,\mu g\,l^{-1}$ within a few hours.[159] Shires and Bennett[160] observed peak concentration after aerial applications to winter wheat of $0.03\,\mu g\,l^{-1}$ of cypermethrin in subsurface water samples. The concentrations declined rapidly after spraying and generally resulted in little to no adverse effects on invertebrates and caged fish, but there was a slight increase of invertebrate drift.[159,160]

Jensen *et al.*[161] found no detectable concentrations of pyrethrins and permethrin in water samples from wetlands before and after truck-mounted ULV. Weston *et al.*[162] found no detectable concentrations of pyrethrins ten and 34 hours after application in suburban streams after airplane ULV applications over Sacramento, California, USA. Schleier III *et al.*[129] found no detectable concentrations of pyrethrins one hour after airplane ULV applications in irrigation ditches and static ponds. Concentrations of resmethrin is Suffolk County, New York, USA were detected in 11% of water samples taken with concentrations ranging from non-detectable to $0.293\,\mu g\,l^{-1}$, and no concentrations were detected after two days.[163] Zulkosky *et al.*[164] measured concentrations of resmethrin ranging from non-detectable to $0.98\,\mu g\,l^{-1}$ and non-detectable concentrations of sumithrin one hour after truck-mounted ULV application. Schleier III and Peterson[128] measured concentrations of permethrin after truck-mounted ULV applications ranging from 0.0009 to $0.005\,\mu g\,cm^{-2}$, depending on the distance from the spray source. The lower concentrations of pyrethrins and pyrethroids measured after ULV applications are most likely due to the lower use rate, which is < 5% of agricultural applications.

3.10   Ecotoxicology

Pyrethrins and pyrethroids are broad spectrum insecticides, and as such they may also impact on beneficial insects, such as parasitoids, predators and bees.[37] They are also highly toxic to aquatic organisms which are generally more susceptible to pyrethroids than terrestrial organisms (see Table 3.2).[165,166] Birds rapidly eliminate pyrethroids *via* ester hydrolysis and oxidation, and generally eliminate the insecticides two to three times faster than mammals.[166] The lower toxicity and higher elimination rate is most likely a function of the higher metabolic rates of birds.

Pyrethroids are highly toxic to fish and aquatic invertebrates, excluding mollusks, and are slightly less toxic to amphibians. Symptoms of intoxication in fish include hyperactivity, loss of balance and the development of darkened areas on the body. Generally the toxicity of pyrethroids to fish increases with an increasing octanol–water partition coefficient.[167] The sensitivity of fish is mainly due to their poor ability to metabolize pyrethroids, with the only major metabolite recovered from a variety of pyrethroids in rainbow trout (*Oncorhynchus mykiss*) being the 4′-hydroxy metabolite which is produced through oxidation.[168–170] Fenvalerate and permethrin exposure to trout shows little to no esterase activity or ester hydrolysis, which is the main detoxification route for mammals and birds.[171–173] In trout exposed to cypermethrin, low levels of ester hydrolysis were observed, but at levels which were nevertheless lower than those in other vertebrates.[169]

The higher acute toxicity of pyrethroids to fish can be accounted for by the uptake and reduced metabolism with higher brain sensitivities compared with that of other vertebrates.[174] Pyrethrins and pyrethroids are most toxic to trout species, but the differences between fish species are less than a half an order of magnitude.[175] Trout are two to three times more sensitive to pyrethroids than bluegill sunfish (*Lepomis macrochirus*) and fathead minnows (*Pimephales promelas*), and three to six times more sensitive than southern leopard frogs (*Rana sphenocephala*) and boreal toads (*Bufo boreas boreas*).[176] Pyrethroid toxicity to amphibians has not been extensively studied and there is limited knowledge of the mechanism by which they are less sensitive than other aquatic organisms. This is important given that amphibians are generally the most sensitive organisms to environmental pollutants, with many of the declines in numbers attributed to their high sensitivity to environmental toxins.[177]

Gills are the most likely route of exposure for fish to anthropogenic agents because of their large surface area, countercurrent flow and thin epithelial layer.[178] However, fish exposure to pyrethroids through gills only results in 20% to 30% of the total absorbed dose.[172] It is unclear as to what the main route of uptake is for fish. The *trans*-permethrin is 110 times more toxic to rainbow trout than to mice by both intravenous and intraperitoneal administration.[179] The half-life of pyrethroids in mammals and birds is six to 12 hours, but in trout the half-life is greater than 24 hours.[171] Lethal brain residues in rainbow trout of permethrin, cypermethrin and fenvalerate were 6% to 33% of the lethal brain residues in mice and quail, indicating that the mode of action and metabolism of pyrethroids are important factors in the increased toxicity to fish.[169,179–181] The difference is due to microsomal oxidation, with the metabolism of *trans*-permethrin being 35 times greater in mice compared with that in rainbow trout, which is most likely the cause of the increased sensitivity of fish.[179,180] Pyrethroids may also affect the respiratory surfaces and renal ion regulation which can contribute to the increased toxicity in fish.[182,183]

A large number of toxicity tests have been carried out on a wide range of terrestrial and aquatic organisms under laboratory conditions, in the presence and absence of sediment and dissolved organic matter. Type II pyrethroids have a greater toxicity than pyrethrins and type I pyrethroids to both aquatic and terrestrial invertebrates.[165,184] The difference in toxicity has been attributed to the decreased degradation of the cyano-substituted pyrethroids by both hydrolases and oxidases.

Pyrethroids undergo minimal biomagnification in vertebrates and invertebrates because of their rapid metabolism and excretion.[185] For example the octanol–water partition coefficient (K_{ow}) for permethrin is about six, while for the major metabolites it is approximately three. These characteristics are often associated with a propensity to biomagnify, much like DDT. However when pyrethroids are compared with the bioconcentration factor of DDT which is known to biomagnify in organisms they are only less than 3% of DDT (see Table 3.1). There have been limited studies on the bioconcentration factor of pyrethrins because they are rapidly metabolized, but it has been estimated to be 11 000 based on the K_{ow}.[16]

Although pyrethroids display very high acute toxicities to aquatic organisms when in the aqueous phase, the presence of suspended sediment substantially reduces the freely dissolved concentration of pyrethroids, and therefore their bioavailability. Pyrethrins and pyrethroids have little mobility in soils and are associated with sediments in natural water; consequently, they will only be in the water phase for a relatively short time, limiting their exposure to many organisms.[186,187] In addition, the half-lives of many pyrethroids in aquatic systems that are not bound to sediment are one to five days, which suggests that small streams are more likely to show effects on non-target organisms because of the lower dilution of the insecticides. Therefore, chronic exposures to organisms that do not have a benthic lifestyle will most likely not result in observed effects because pyrethroids dissipate rapidly (dissipation half-life in the water column is generally less than one day). The rapid dissipation of pyrethroids makes it difficult to reconcile field exposures with those used in laboratory studies that maintain constant concentrations from ten to 100 days. The pH of the water used does not influence the toxicity of pyrethrins or pyrethroids, but hard or saline water can increase the toxicity to aquatic organisms.[171,175] In addition, the values obtained for aquatic organisms in the laboratory can be difficult to reproduce because pyrethroids strongly bind to solvents or surfaces such as glass, which can cause an overestimation of the toxicity values.[188]

Concerns have been expressed that pesticide mixtures, especially pyrethroids with their widespread use, may have greater than additive toxicity in the environment. Brander *et al.*[189] found that type I and II pyrethroids can be antagonistic to one another, lowering the toxicity of the mixture to *D. magna*. This is most likely due to the competitive binding at the voltage-gated sodium channels, but there have been few studies to examine the physiological and biochemical mechanisms involved.[53] In addition to other pyrethroids present in the environment, other pesticides such as fungicides have been shown to interact synergistically with pyrethroids.[190,191] However the concentrations that have been observed to increase the toxicity are greater than those seen in the environment, and given the physicochemical properties of pyrethroids, these exposures are unlikely to result in a substantial increase in toxicity.

In addition to the concern of increased toxicity due to the addition of pyrethroids to aquatic systems, research has shown that the pyrethroid synergist PBO can increase the toxicity of pyrethroids already present in the environment.[97–99,162] However the concentration needed to significantly affect populations of organisms is unknown and is strongly dependent on the amounts and types of pyrethroids already present. Concentrations of PBO in both irrigation ditches and static ponds rapidly decreased to $0.012\,\mu g\,l^{-1}$ within 36 h after applications of synergized formulations, greatly reducing the exposure of organisms to both the pyrethroid and PBO.[129]

3.10.1 Formulation Toxicity

There is contradictory evidence about the differences in toxicity between pyrethroids with emulsifiers (*i.e.*, formulated products) and technical-grade pyrethroids. Emulsifiers are designed to keep the pyrethroid in solution, but they can also inhibit the uptake of the active ingredient into organisms. Coats and O'Donnell-Jeffery[192] found that emulsifiable concentrate formulations of permethrin, fenvalerate and cypermethrin were two to nine times more toxic to rainbow trout than technical-grade materials. However, there was no significant difference in uptake in rainbow trout found between emulsified formulations and technical-grade fenvalerate. Mosquito larvae are 67 times more susceptible to technical-grade fenvalerate than the formulated product.[171,193] Beggel *et al.*[194] observed larger sublethal effects of formulations than technical-grade bifenthrin on fathead minnows. Technical-grade fenvalerate was more toxic to fathead minnows than the emulsifiable concentrate formulation at 96 h, but by 168 h the two formulations were of similar toxicity.[170] Schleier III and Peterson[131] found that the toxicity of technical-grade permethrin was about 10-fold greater than an emulsifiable concentrate to house crickets. The influence of the formulation on the toxicity is more important after spray drift because it lands directly in the water body. Oil-based formulations could retain more of the insecticide because of their high hydrophobicity, which could result in the insecticide being bioavailable longer, while emulsified concentrates could disperse faster in the water.

3.11   Ecological Field Studies

Experiments with cypermethrin showed that its concentration in *D. magna* and *Chironomus tentans* decreased as the dissolved organic carbon content of the water increased.[195] Acute pyrethroid toxicity decreased 60% to 92% depending on the concentration of suspended sediments.[187] Yang *et al.*[186] found that pyrethroids adsorbed on particles or dissolved organic matter were completely unavailable for uptake by *D. galeata mendotae* after a 24-hour exposure period.

Pyrethroids have even been observed to have beneficial effects on aquatic organisms. A concentration of $0.005\,\mu g\,l^{-1}$ of fenvalerate resulted in an increase in longevity of *D. galeata mendotae* adults.[196] The intrinsic rate of increase was not affected by fenvalerate until the concentration reached $0.05\,\mu g\,l^{-1}$, however concentrations of $0.01\,\mu g\,l^{-1}$ caused the net reproductive rate and the generation time to decrease.[196] After 21 days of continuous and pulsed exposures to fenvalerate over a concentration range of 0.1 to $1\,\mu g\,l^{-1}$, recovery of *D. magna* to reproduction was similar to controls.[197] Reynaldi *et al.*[198] found that acute exposures of $0.3\,\mu g\,l^{-1}$ of fenvalerate resulted in reduced feeding activity and smaller body size in *D. magna*, and exposure to concentrations of $0.6\,\mu g\,l^{-1}$ or greater resulted in delayed maturation.

Changes in aquatic communities have mostly been found at concentrations of 5 to $10\,\mu g\,l^{-1}$ of pyrethroid in water with recovery occurring within weeks, which is 0.2 to 10 times the concentrations found in the environment.[199–201] Hill[158] reviewed approximately 70 freshwater field studies in natural or farm ponds, streams and rivers, rice paddies, and microcosms and mesocosms and found that there were little to no acute effects on fish and aquatic invertebrates. However, environmental concentrations could affect some of the most sensitive species. Kedwards *et al.*[202] applied cypermethrin aerially adjacent to a farm pond and observed that Diptera were most affected by increasing concentrations, but the populations quickly recovered after the application. The sediment-dwelling invertebrates Gammaridae and Asellidae were adversely affected by direct applications of cypermethrin and lambda-cyhalothrin in experimental ponds, but increases in populations of Planorbidae, Chironomidae and Lymnaeidae were also observed.[203] Van Wijngaarden *et al.*[204] reviewed 18 microcosm and mesocosm studies using eight different pyrethroids with single and multiple exposures. They found that Amphipoda and Hydacarina were the most sensitive to pyrethroid exposure, and recovery to populations occurred within two months after the final application. Roessink *et al.*[14] compared lambda-cyhalothrin applied three times at one-week intervals at concentrations of 10, 25, 50, 100 and 250 ng l^{-1} in mesotrophic (macrophyte dominated) and eutrophic (phytoplankton dominated) ditch microcosms (approximately 0.5 m^3). At concentrations of 25 ng l^{-1} and greater, population and community responses were measured with indirect effects on rotifers and microcrustaceans more pronounced in the plankton-dominated systems. At concentrations of 100 and 250 ng l^{-1}, which is 100-fold higher than concentrations observed in the environment, the rate of recovery of the macroinvertebrate community was lower in the macrophyte-dominated systems, most likely due to the prolonged decline of the amphipods.

Dabrowski *et al.*[205] found that mayfly nymphs are more likely to be affected by spray-drift exposure than by run-off exposure because of the reduced bioavailability of sediment-bound pyrethroids. Schulz and Liess[206] observed chronic toxicity to *Limnephilus lunatus* after pulsed exposures to fenvalerate. Soil with low organic matter content has a greater toxicity than soil with high organic matter content.[207] However, soil aging was not found to exert any effect on lambda-cypermethrin toxicity in the springtail *Folsomia candida.*[207]

Researchers have provided evidence that the 1999 lobster (*Homarus americanus*) die-off in Long Island Sound was not caused by the use of ULV resmethrin and sumithrin (also known as delta-phenothrin) insecticides in response to the introduction of West Nile virus.[208–210] Jensen *et al.*[161] showed that the use of truck-mounted ULV above wetlands had no significant impact on aquatic macroinvertebrates and *Gambusia affinis*, but did have a significant impact on flying insects. However, flying insect abundance recovered 48 hours after application. Milam *et al.*[211] found less than 10% mortality for *Pimephales promelas* and *Daphnia pulex* after truck-mounted and airplane ULV applications of permethrin. Davis and Peterson[212] found little impact on aquatic and terrestrial invertebrates after single and multiple applications of either permethrin or sumithrin by truck-mounted ULV. After airplane ULV applications of pyrethrins, Boyce *et al.*[213] found no impact on large-bodied insects within the spray zone, but they did observe an impact on smaller-bodied insects.

3.12 Conclusions

Pyrethroids have become widely used because they are highly effective against many pests, have low mammalian and avian toxicity, and lack environmental persistence. The discovery of the first photostable pyrethroid, permethrin, revolutionized pyrethroids as a class and subsequently led to their increased use in pest management. Pyrethroids represent an incredibly diverse set of compounds that are currently used for all major pest control applications. However, they are broad-spectrum insecticides that are highly toxic to non-target terrestrial insects and many aquatic organisms.

Currently, the mode of action of pyrethroids is well understood and is characterized by either fine tremors (T-syndrome/Type I pyrethroids) or choreoathetosis and salivation (CS-syndrome/Type II pyrethroids), although not all steps between cellular changes in excitability and behavior are well understood. The secondary modes of action for pyrethroids are not well understood and more research is needed in this area.

The environmental fate and physical properties of pyrethrins and pyrethroids are well understood. Pyrethroids are persistent in soils and sediment with half-lives greater than 30 days, but their half-lives are substantially lower than legacy pesticides such as DDT. Pyrethroids are rapidly biodegraded and, contrary to their high K_{ow} values, they do not biomagnify through higher trophic levels of the food chain. Due to their long half-lives in sediment, certain sediment-dwelling invertebrates may be affected by pyrethroids, especially in urban areas where the insecticides are heavily used. A surprising finding is that pyrethroid exposure of fish through their gills results in only 30% or less of the total absorbed dose, which is contrary to many other anthropogenic agents. Although the environmental effects of fenvalerate and esfenvalerate on aquatic organisms have been studied extensively, more research is required on the effects of pseudopyrethroids on aquatic organisms.

Commercial research and development efforts in the discovery of novel pyrethroids have largely ceased since the late 1990s; however, work is still being done to introduce single- or enriched-isomer mixtures of compounds like cypermethrin and cyhalothrin. With the voluntary cancellation of fenvalerate and esfenvalerate[214] and the end of major development of pyrethroids by many manufacturers (with the exception of companies like Sumitomo which recently developed metofluthrin for commercial use in Japan),[215] pyrethroid development seems to be well past its peak. The key to the continued commercialization of pyrethroids in Europe and the USA may lie with pseudopyrethroids like etofenprox that have yet to be widely used, but display lower acute toxicity to aquatic organisms.

About 80 species of arthropods are resistant to pyrethroids around the world.[216] Across the USA, there has been an increase in *kdr* resistance in bed bugs (*Cimex lectularius*), which is thought to have led to their reappearance in many cities.[217] Pyrethroids suffer from an inherent disadvantage because at the outset *kdr* resistance also confers resistance to DDT analogs, and prior resistance to DDT has already selected for this mechanism of resistance. This is of great concern with respect to pyrethroid resistance among *Anopheles gambiae* in West Africa, which could render the use of pyrethroid-impregnated bed nets ineffective in the prevention of malaria.[218] This trend could be exacerbated because of the renewed use of DDT as an indoor residual spray. The increasing cost for the discovery of insecticides with novel modes of action, in conjunction with other insecticides like organophosphates losing registration, could render pyrethrins and pyrethroids less effective.

Pyrethrins, pyrethroids and their synergists that were registered after 1984 are currently undergoing registration reviews in the USA to evaluate the effectiveness of recent regulatory decisions and to consider new data.[219] The registration review is focused on developmental neurotoxicity, because recent studies have shown decreases in rat pup weight, pup weight gain, and/or brain weight.[74] In addition, the USEPA has recently updated spray drift regulations for pyrethroids, increasing the buffer between sprayed areas and aquatic environments.[220]

Even though pyrethroids are not pest-specific insecticides and have been used for the past 40 years, they continue to be commonly used. This is because they target a wide variety of pests, have low application rates, have low mammalian toxicity and have a favorable environmental fate profile. Provided that they are used appropriately, pest resistance to them is managed effectively and regulations for them are based on scientific evidence, the pyrethrins and pyrethroids will continue to be used well into the foreseeable future.

References

1. A. Glynne-Jones, *Biopesticides*, 2001, **12**, 195.
2. B. K. Bhat, in *Pyrethrum Flowers: Production, Chemistry, Toxicology, and Uses*, ed. J. E. Casida and G. B. Quistad, Oxford University Press, New York, NY, USA, 1995, pp. 69.
3. A. Prakash and J. Rao, *Botanical Pesticides in Agriculture*, CRC Press Inc., Boca Raton, FL, USA, 1997.
4. J. M. G. Wainaina, in *Pyrethrum Flowers: Production, Chemistry, Toxicology, and Uses*, ed. J. E. Casida and G. B. Quistad, Oxford University Press, New York, NY, USA, 1995, pp. 49.
5. USEPA, *Qualitative Assessment of Impacts of Risk Management Strategies for Endosulfan on Multiple Crops: Extending REIs and Cancellation*, U.S. Environmental Protection Agency, Washington, D.C., USA, 2010; http://www.regulations.gov/search/Regs/home.html#documentDetail? R=0900006480afefa1.
6. F. Spurlock and M. Lee, in *Synthetic Pyrethroids*, ed. J. Gan, F. Spurlock, P. Dendley and D. P. Weston, American Chemical Society, Washington, DC, 2008, pp. 3.
7. USDHHS, *Toxicological Profile for Pyrethrins and Pyrethroids*, U.S. Department of Health and Human Services, Agency for Toxic Substances and Disease Registry, Atlanta, GA, USA, 2007; http://www.atsdr.cdc.gov/ToxProfiles/tp155-p.pdf.
8. USEPA (U.S. Environmental Protection Agency), *Pyrethroids and Pyrethrins*; http://www.epa.gov/oppsrrd1/reevaluation/pyrethroids-pyrethrins.html.
9. J. E. Casida and G. B. Quistad, *Annu. Rev. Entomol.*, 1998, **43**, 1.
10. USEPA, *Reregistration Eligibility Decision (RED) for Permethrin* EPA 738-R-09-306, U.S. Environmental Protection Agency, Washington D.C., USA, 2009; http://www.epa.gov/oppsrrd1/REDs/permethrin-red-revised-may2009.pdf.

11. USEPA, *Reregistration Eligibility Decision for Cypermethrin (revised 01/ 14/08)* EPA OPP-2005-0293, U.S. Environmental Protection Agency, Washington D.C., USA, 2008; http://www.epa.gov/oppsrrd1/ REDs/cypermethrin_revised_red.pdf.

12. L. Crombie, in *Pyrethrum Flowers: Production, Chemistry, and Uses*, ed. J. E. Casida and G. B. Quistad, Oxford University Press, New York, NY, USA, 1995, pp. 121.

13. H. Staudinger and L. Ruzicka, *Helv. Chim. Acta*, 1924, **7**, 177.

14. I. Roessink, G. H. P. Arts, J. D. M. Belgers, F. Bransen, S. J. Maund and T. C. M. Brock, *Environ. Toxicol. Chem.*, 2005, **24**, 1684.

15. A. W. Farnham, *Pestic. Sci.*, 1973, **4**, 513.

16. D. G. Crosby, in *Pyrethrum Flowers: Production, Chemistry, Toxicology, and Uses*, ed. J. E. Casida and G. B. Quistad, Oxford University Press, New York, NY, USA, 1995, pp. 194.

17. M. S. Schechter, N. Green and F. B. LaForge, *J. Am. Chem. Soc.*, 1949, **71**, 3165.

18. M. Elliott, A. W. Farnham, N. F. Janes, P. H. Needham and B. C. Pearson, *Nature*, 1967, **213**, 493.

19. D. M. Soderlund, *Xenobiotica*, 1992, **22**, 1185.

20. M. Elliott, A. W. Farnham, N. F. Janes, P. H. Needham and D. A. Pulman, in *Mechanisms of Pesticide Action*, ed. G. K. Kohn, American Chemical Society, Washington D.C., USA, 1974, pp. 80.

21. M. Elliott, A. W. Farnham, N. F. Janes, P. H. Needham, D. A. Pulman and J. H. Stevenson, *Nature*, 1973, **246**, 169.

22. R. L. Holmstead, J. E. Casida, L. O. Ruzo and D. G. Fullmer, *J. Agric. Food Chem.*, 1978, **26**, 590.

23. Y. Katsuda, *Pestic. Sci.*, 1999, **55**, 775.

24. M. Elliott, in *Pyrethrum Flowers: Production, Chemistry, Toxicology, and Uses*, ed. J. E. Casida and G. B. Quistad, Oxford University Press, New York, NY, USA, 1995, pp. 3.

25. USEPA, *Etofenprox (also Ethofenprox) Summary Document Registration Review: Initial Docket August 2007*, U.S. Evironmental Protection Agency, Washington D.C., USA, 2007; http://www.regulations.gov/search/ Regs/contentStreamer?objectId=090000648027aac3&disposition=attachment &contentType=pdf.

26. Ó. López, J. G. Fernández-Bolaños and M. V. Gil, *Green Chem.*, 2005, **7**, 431.

27. G. Holan, D. F. O'Keefe, C. Virgona and R. Walser, *Nature*, 1978, **272**, 734.

28. J.-P. Demoute, *Pestic. Sci.*, 1989, **27**, 375.

29. T. Katagi, *J. Agric. Food Chem.*, 1991, **39**, 1351.

30. P. Y. Liu, Y. J. Liu, Q. X. Liu and J. W. Liu, *J. Environ. Sci. (China)*, 2010, **22**, 1123.

31. I. Mukherjee, R. Singh and J. N. Govil, *Bull. Environ. Contam. Toxicol.*, 2010, **84**, 294.

32. J. P. Leahey, in *The Pyrethroid Insecticides*, ed. J. P. Leahey, Taylor & Francis Inc., Philadelphia, PA, USA, 1985, pp. 263.

33. D. A. Laskowski, *Rev. Environ. Contam. Toxicol.*, 2002, **174**, 49.

34. M. S. Sharom and K. R. Solomon, *J. Agric. Food Chem.*, 1981, **29**, 1122.

35. P. Guo, B. Z. Wang, B. J. Hang, L. Li, S. P. Li and J. He, *Int. J. Syst. Evol. Microbiol.*, 2010, **60**, 408.

36. C. Zhang, L. Jia, S. H. Wang, J. Qu, K. Li, L. L. Xu, Y. H. Shi and Y. C. Yan, *Bioresour. Technol.*, 2010, **101**, 3423.

37. K. Naumann, *Synthetic Pyrethroid Insecticides: Structures and Properties*, Springer-Verlag, New York, NY, USA, 1990.

38. D. M. Soderlund and J. R. Bloomquist, *Annu. Rev. Entomol.*, 1989, **34**, 77.

39. R. D. Verschoyle and W. N. Aldridge, *Arch. Toxicol.*, 1980, **45**, 325.

40. L. J. Lawrence and J. E. Casida, *Pestic. Biochem. Physiol.*, 1982, **18**, 9.

41. X. Wang, F. Xue, A. Hua and F. Ge, *Physiol. Entomol.*, 2006, **31**, 190.

42. C. B. Breckenridge, L. Holden, N. Sturgess, M. Weiner, L. Sheets, D. Sargent, D. M. Soderlund, J. S. Choi, S. Symington, J. M. Clark, S. Burr and D. Ray, *Neurotoxicology*, 2009, **30**, S17.

43. D. M. Soderlund, J. M. Clark, L. P. Sheets, L. S. Mullin, V. J. Piccirillo, D. Sargent, J. T. Stevens and M. L. Weiner, *Toxicology*, 2002, **171**, 3.

44. J. R. Bloomquist, *Annu. Rev. Entomol.*, 1996, **41**, 163.

45. A. L. Goldin, *Ann. N. Y. Acad. Sci.*, 1999, **868**, 38.

46. J. L. Nation, *Insect Physiology and Biochemistry*, CRC Press LLC, Boca Rotan, FL, USA, 2002.

47. S. J. Lozano, S. L. O'Halloran, K. W. Sargent and J. C. Brazner, *Environ. Toxicol. Chem.*, 1992, **11**, 35.

48. Y. Z. Du, W. Z. Song, J. R. Groome, Y. Nomura, N. G. Luo and K. Dong, *Toxicol. Appl. Pharmacol.*, 2010, **247**, 53.

49. V. L. Salgado, S. N. Irving and T. A. Miller, *Pestic. Biochem. Physiol.*, 1983, **20**, 169.

50. D. M. Soderlund, *Pestic. Biochem. Physiol.*, 2010, **97**, 78.

51. P. J. Forshaw, T. Lister and D. E. Ray, *Toxicol. Appl. Pharmacol.*, 2000, **163**, 1.

52. F. Cantalamessa, *Arch. Toxicol.*, 1993, **67**, 510.

53. J. G. Scott and Z. Wen, *Pest Manag. Sci.*, 2001, **57**, 958.

54. S. B. Symington, A. G. Zhang, W. Karstens, J. Van Houten and J. M. Clark, *Pestic. Biochem. Physiol.*, 1999, **65**, 181.

55. D. A. Meyer, J. M. Carter, A. F. M. Johnstone and T. J. Shafer, *Neurotoxicology*, 2008, **29**, 213.

56. A. P. Neal, Y. K. Yuan and W. D. Atchison, *Toxicol. Sci.*, 2010, **116**, 604.

57. J. E. Casida and L. J. Lawrence, *Environ. Health Perspect.*, 1985, **61**, 123.

58. J. J. Van Hemmen and D. H. Brouwer, *Sci. Total Environ.*, 1995, **168**, 131.

59. S. A. Burr and D. E. Ray, *Toxicol. Sci.*, 2004, **77**, 341.

60. F. Matsumura and S. M. Ghiasuddin, *J. Environ. Sci. Health B.*, 1983, **18**, 1.
61. J. Miyamoto, *Environ. Health Perspect.*, 1976, **14**, 15.
62. W. Liu, J. Gan, D. Schlenk and W. A. Jury, *Proc. Natl. Acad. Sci. U. S. A.*, 2005, **102**, 701.
63. A. E. Lund and T. Narahashi, *Pestic. Biochem. Physiol.*, 1983, **20**, 203.
64. M. Zhao, C. Wang, K. K. Liu, L. Sun, L. Li and W. Liu, *Environ. Toxicol. Chem.*, 2009, **28**, 1475.
65. F. Hu, L. Li, C. Wang, Q. Zhang, X. Zhang and M. Zhao, *Environ. Toxicol. Chem.*, 2010, **29**, 683.
66. K. A. Usmani and C. O. Knowles, *J. Econ. Entomol.*, 2001, **94**, 874.
67. S. J. Yu, *The Toxicology and Biochemistry of Insecticides*, CRC Press, Boca Raton, FL, USA, 2008.
68. USEPA, *Reregistration Eligibility Decision for Resmethrin* Case No. 0421, U.S. Environmental Protection Agency, Washington D.C., USA, 2006; http://www.epa.gov/oppsrrd1/REDs/resmethrin_red.pdf.
69. F. Matsumura, *Toxicology of Insecticides*, Plenum Press, New York, NY, 1985.
70. S. F. Abd-Elghafar, A. G. Appel and T. P. Mack, *J. Econ. Entomol.*, 1990, **83**, 2290.
71. F. B. Antwi and R. K. D. Peterson, *Pest Manag. Sci.*, 2009, **65**, 300.
72. N. Prabhaker, S. J. Castle and N. C. Toscano, *J. Econ. Entomol.*, 2006, **99**, 1805.
73. K. Haya, *Environ. Toxicol. Chem.*, 1989, **8**, 381.
74. USEPA, *Pyrethroids: Evaluation of Data from Developmental Neurotoxicity Studies and Consideration of Comparative Sensitivity* Decision No.: 407265; DP Barcode: D371723, U.S. Environmental Protection Agency, Washington D.C., USA, 2010; http://www.regulations.gov/search/Regs/home.html#documentDetail?R=0900006480aa79fc.
75. E. B. Vinson and C. W. Kearns, *J. Econ. Entomol.*, 1952, **45**, 484.
76. M. S. Blum and C. W. Kearns, *J. Econ. Entomol.*, 1956, **49**, 862.
77. T. C. Sparks, A. M. Pavloff, R. L. Rose and D. F. Clower, *J. Econ. Entomol.*, 1983, **76**, 243.
78. D. R. Boina, E. O. Onagbola, M. Salyani and L. L. Stelinski, *J. Econ. Entomol.*, 2009, **102**, 685.
79. N. M. Eesa and L. E. Moursy, *Exp. Appl. Acarol.*, 1990, **10**, 77.
80. C. V. Eadsforth and M. K. Baldwin, *Xenobiotica*, 1983, **13**, 67.
81. G. Leng, A. Leng, K.-H. Kuhn and J. Lewalter, *Xenobiotica*, 1997, **27**, 1273.
82. C. V. Eadsforth, P. C. Bragt and N. J. Van Sittert, *Xenobiotica*, 1988, **18**, 603.
83. T. Mitsche, H. Borck, B. Horr, N. Bayas, H. W. Hoppe and F. Diel, *Allergy*, 2000, **55**, 93.
84. L. G. Costa, G. Giordano, M. Guizzetti and A. Vitalone, *Front. Biosci.*, 2008, **13**, 1240.
85. D. M. Soderlund and D. C. Knipple, *Insect Biochem. Mol. Biol.*, 2003, **33**, 563.

86. D. Thomas, M. Weedermann, L. Billings, J. Hoffacker and R. A. Washington-Allen, *Ecol. Soc.*, 2009, **14**, 15.
87. J. Choi, R. L. Rose and E. Hodgson, *Pestic. Biochem. Physiol.*, 2002, **74**, 117.
88. K. R. Mohan and C. P. Weisel, *J. Expo. Sci. Environ. Epidemiol.*, 2010, **20**, 320.
89. M. A. Sogorb and E. Vilanova, *Toxicol. Lett.*, 2002, **128**, 215.
90. D. B. Barr, A. O. Olsson, L.-Y. Wong, S. Udunka, S. E. Baker, R. D. Whitehead, Jr., M. S. Magsumbol, B. L. Williams and L. L. Needham, *Environ. Health Perspect.*, 2010, **118**.
91. B. H. Woollen, J. R. Marsh, W. J. D. Laird and J. E. Lesser, *Xenobiotica*, 1992, **22**, 983.
92. E. Hodgson and R. L. Rose, *Drug Metab. Rev.*, 2005, **37**, 1.
93. I. Yamamoto, in *Pyrethrum – Natural Insecticide International Symposium Proceedings. Mode of Action of Synergists*, Academic Press, New York, NY, USA, 1973, pp. 195.
94. USEPA, *Reregistration Eligibility Decision for Piperonyl Butoxide (PBO)* Case No. 2525, U.S. Environmental Protection Agency, Washington D.C., 2006; http://www.epa.gov/oppsrrd1/reregistration/REDs/piperonyl_red.pdf.
95. USEPA, *Reregistration Eligibility Decision for n-Octyl Bicycloheptene Dicarboximide (MGK-264)* EPA 738-R-06-006 U.S. Environmental Protection Agency, Washington D.C., 2006; http://www.epa.gov/oppsrrd1/reregistration/REDs/mgk_red.pdf.
96. A. W. Farnham, in *Piperonyl Butoxide: the Insecticide Synergist*, ed. D. G. Jones, Academic Press, London, UK, 1998, pp. 199.
97. E. L. Amweg, D. P. Weston, C. S. Johnson, J. You and M. J. Lydy, *Environ. Toxicol. Chem.*, 2006, **25**, 1817.
98. E. A. Paul and H. A. Simonin, *Bull. Environ. Contam. Toxicol.*, 1995, **55**, 453.
99. E. A. Paul, H. A. Simonin and T. M. Tomajer, *Arch. Environ. Contam. Toxicol.*, 2005, **48**, 251.
100. L. Kennaugh, D. Pearce, J. C. Daly and A. A. Hobbs, *Pestic. Biochem. Physiol.*, 1993, **45**, 234.
101. D. Pimentel, H. Acquay, M. Biltonen, P. Rice, M. Silva, J. Nelson, V. Lipner, S. Giordano, A. Horowitz and M. D'Amore, *Bioscience*, 1992, **42**, 750.
102. N. Liu, Q. Xu, F. Zhu and L. Zhang, *Insect Science*, 2006, **13**, 159.
103. J. Baas, T. Jager and B. Kooijman, *Sci. Total Environ.*, 2010, **408**, 3735.
104. T. A. Miller and V. L. Salgado, in *The Pyrethroid Insecticides*, ed. J. P. Leahey, Taylor & Francis Inc., Philadelphia, PA, USA, 1985, pp. 43.
105. A. W. Farnham, *Pestic. Sci.*, 1977, **8**, 631.
106. Y. H. Zhang, S. S. Liu, H. L. Liu and Z. Z. Liu, *Pest Manag. Sci.*, 2010, **66**, 879.
107. D. G. Cochran, in *Pyrethrum Flowers: Production, Chemistry, and Uses*, ed. J. E. Casida and G. B. Quistad, Oxford University Press, New York, NY, USA, 1995, pp. 234.

108. J. W. Pridgeon, A. G. Appel, W. J. Moar and N. Liu, *Pestic. Biochem. Physiol.*, 2002, **73**, 149.
109. K. Dong, S. M. Valles, M. E. Scharf, B. Zeichner and G. W. Bennett, *Pestic. Biochem. Physiol.*, 1998, **60**, 195.
110. M. Y. Liu, C. N. Sun and S. W. Huang, *J. Econ. Entomol.*, 1982, **75**, 965.
111. R. V. Gunning, C. S. Easton, M. E. Balfe and I. G. Ferris, *Pestic. Sci.*, 1991, **33**, 473.
112. S. F. Abd-Elghafar and C. O. Knowles, *J. Econ. Entomol.*, 1996, **89**, 590.
113. I. Ishaaya, *Arch. Insect Biochem. Physiol.*, 1993, **22**, 263.
114. P. J. Daborn, J. L. Yen, M. R. Bogwitz, G. Le Goff, E. Feil, S. Jeffers, N. Tijet, T. Perry, D. Heckel, P. Batterham, R. Feyereisen, T. G. Wilson and R. H. ffrench-Constant, *Science*, 2002, **297**, 2253.
115. M. C. Hardstone, X. Huang, L. C. Harrington and J. G. Scott, *J. Med. Entomol.*, 2010, **47**, 188.
116. J. A. Ottea, S. A. Ibrahim, A. M. Younis and R. J. Young, *Pestic. Biochem. Physiol.*, 2000, **66**, 20.
117. R. L. Rose, L. Barbhaiya, R. M. Roe, G. C. Rock and E. Hodgson, *Pestic. Biochem. Physiol.*, 1995, **51**, 178.
118. USEPA, *Reregistration Eligibility Decision for Pyrethrins* Case No. 2580, U.S. Environmental Protection Agency, Washington D.C., 2006; http://www.epa.gov/oppsrrd1/REDs/pyrethrins_red.pdf.
119. W. C. Carr, P. Iyer and D. W. Gammon, *Scientific World Journal*, 2006, **6**, 279.
120. R. K. D. Peterson, P. A. Macedo and R. S. Davis, *Environ. Health Perspect.*, 2006, **114**, 366.
121. J. J. Schleier III, P. A. Macedo, R. S. Davis, L. M. Shama and R. K. D. Peterson, *Stoch. Environ. Res. Risk Assess.*, 2009, **23**, 555.
122. J. J. Schleier III, R. S. Davis, L. M. Barber, P. A. Macedo and R. K. D. Peterson, *J. Med. Entomol.*, 2009, **46**, 693.
123. P. A. Macedo, R. K. D. Peterson and R. S. Davis, *J. Toxicol. Environ. Health A.*, 2007, **70**, 1758.
124. S. J. Maund, K. Z. Travis, P. Hendley, J. M. Giddings and K. R. Solomon, *Environ. Toxicol. Chem.*, 2001, **20**, 687.
125. USEPA (United States Environmental Protection Agency), ECOTOX database; http://cfpub.epa.gov/ecotox/ecotox_home.cfm.
126. L. Yameogo, K. Traore, C. Back, J.-M. Hougard and D. Calamari, *Chemosphere*, 2001, **42**, 965.
127. R. S. Davis, R. K. D. Peterson and P. A. Macedo, *Integr. Environ. Assess. Manag.*, 2007, **3**, 373.
128. J. J. Schleier III and R. K. D. Peterson, *Arch. Environ. Contam. Toxicol.*, 2010, **58**, 105.
129. J. J. Schleier III, R. K. D. Peterson, P. A. Macedo and D. A. Brown, *Environ. Toxicol. Chem.*, 2008, **27**, 1063.
130. J. J. Schleier III, Thesis, Montana State University, Bozeman, MT, 2008.
131. J. J. Schleier III and R. K. D. Peterson, *Ecotoxicology*, 2010, **16**, 1140.

132. K. Becker, M. Seiwert, J. Angerer, M. Kolossa-Gehring, H. W. Hoppe, M. Ball, C. Schulz, J. Thumulla and B. Seifert, *Int. J. Hyg. Environ. Health*, 2006, **209**, 221.
133. U. Heudorf and J. Angerer, *Environ. Health Perspect.*, 2001, **109**, 213.
134. T. Schettgen, U. Heudorf, H. Drexler and J. Angerer, *Toxicol. Lett.*, 2002, **134**, 141.
135. CDC, *Fourth National Report on Human Exposure to Environmental Chemicals*, Department of Health and Human Services; Centers for Disease Control and Prevention, Atlanta, GA, USA, 2009; http://www.cdc.gov/exposurereport/.
136. M. Currier, M. McNeill, D. Campbell, N. Newton, J. S. Marr, E. Perry, S. W. Berg, D. B. Barr, G. E. Luber, S. M. Kieszak, H. S. Rogers, S. C. Backer, M. G. Belson, C. Rubin, E. Azziz-Baumgartner and Z. H. Duprey, *MMWR Morb. Mortal. Wkly Rep.*, 2005, **54**, 529.
137. M. C. Fortin, G. Carrier and M. Bouchard, *Environ. Health*, 2008, **7**, 13.
138. J. Hardt and J. Angerer, *Int. Arch. Occup. Environ. Health*, 2003, **76**, 492.
139. A. M. Karpati, M. C. Perrin, T. Matte, J. Leighton, J. Schwartz and R. G. Barr, *Environ. Health Perspect.*, 2004, **112**, 1183.
140. USEPA, *A review of the relationship between pyrethrins, pyrethroid exposure and asthma and allergies*, U.S. Environmental Protection Agency, Washington, D.C., USA, 2009; http://www.epa.gov/oppsrrd1/reevaluation/pyrethrins-pyrethroids-asthma-allergy-9-18-09.pdf.
141. M. E. Vasquez, A. S. Gunasekara, T. M. Cahill and R. S. Tjeerdema, *Pest Manag. Sci.*, 2010, **66**, 28.
142. D. P. Weston, R. W. Holmes, J. You and M. J. Lydy, *Environ. Sci. Technol.*, 2005, **39**, 9778.
143. E. L. Amweg, D. P. Weston, J. You and M. J. Lydy, *Environ. Sci. Technol.*, 2006, **40**, 1700.
144. J. L. Domagalski, D. P. Weston, M. Zhang and M. Hladik, *Environ. Toxicol. Chem.*, 2010, **29**, 813.
145. W. Lao, D. Tsukada, D. J. Greenstein, S. M. Bay and K. A. Maruya, *Environ. Toxicol. Chem.*, 2010, **29**, 843.
146. M. L. Hladik and K. M. Kuivila, *J. Agric. Food Chem.*, 2009, **57**, 9079.
147. R. Schulz, *J. Environ. Qual.*, 2004, **33**, 419.
148. S. Smith, T. E. Reagan, J. L. Flynn and G. H. Willis, *J. Environ. Qual.*, 1983, **12**, 534.
149. B. R. Carroll, G. H. Willis and J. B. Graves, *J. Environ. Qual.*, 1981, **10**, 497.
150. S. T. Hadfield, J. K. Sadler, E. Bolygo, S. Hill and I. R. Hill, *Pestic. Sci.*, 1993, **38**, 283.
151. I. Craig, N. Woods and G. Dorr, *Crop Prot.*, 1998, **17**, 475.
152. Y. Gil, C. Sinfort, S. Guillaume, Y. Brunet and B. Palagos, *Biosyst. Eng.*, 2008, **100**, 184.
153. A. J. Bilanin, M. E. Teske, J. W. Barry and R. B. Ekblad, *Trans. Am. Soc. Agric. Eng.*, 1989, **32**, 327.

154. B. Z. Duan, W. G. Yendol, K. Mierzejewski and R. Reardon, *Pestic. Sci.*, 1992, **36**, 19.
155. M. E. Teske and J. W. Barry, *Trans. Am. Soc. Agric. Eng.*, 1993, **36**, 27.
156. K. Baetens, Q. T. Ho, D. Nuyttens, M. De Schampheleire, A. M. Endalew, M. Hertog, B. Nicolai, H. Ramon and P. Verboven, *Atmos. Environ.*, 2009, **43**, 1674.
157. M. E. Teske, S. L. Bird, D. M. Esterly, T. B. Curbishley, S. L. Ray and S. G. Perry, *Environ. Toxicol. Chem.*, 2002, **21**, 659.
158. I. R. Hill, *Pestic. Sci.*, 1989, **27**, 429.
159. U. Norum, N. Friberg, M. R. Jensen, J. M. Pedersen and P. Bjerregaard, *Aquat. Toxicol.*, 2010, **98**, 328.
160. N. O. Crossland, S. W. Shires and D. Bennett, *Aquat. Toxicol.*, 1982, **2**, 253.
161. T. Jensen, S. P. Lawler and D. A. Dritz, *J. Am. Mosquito Control Assoc.*, 1999, **15**, 330.
162. D. P. Weston, E. L. Amweg, A. Mekebri, R. S. Ogle and M. J. Lydy, *Environ. Sci. Technol.*, 2006, **40**, 5817.
163. I. J. Abbene, S. C. Fisher and S. A. Terracciano, *Concentrations of Insecticides in Selected Surface Water Bodies in Suffolk County, New York, before and after Mosquito Spraying, 2002–04* Open-File Report 2005-1384, U.S. Geological Survey, New York, NY, USA, 2005; http://ny.water.usgs.gov/pubs/of/of051384/.
164. A. M. Zulkosky, J. P. Ruggieri, S. A. Terracciano, B. J. Brownawell and A. E. McElroy, *J. Shellfish Res.*, 2005, **24**, 795.
165. B. D. Siegfried, *Environ. Toxicol. Chem.*, 1993, **12**, 1683.
166. S. P. Bradbury and J. R. Coats, *Rev. Environ. Contam. Toxicol.*, 1989, **108**, 133.
167. V. Zitko, W. G. Carson and C. D. Metcalfe, *Bull. Environ. Contam. Toxicol.*, 1977, **18**, 35.
168. A. H. Glickman, A. A. R. Hamid, D. E. Rickert and J. J. Lech, *Toxicol. Appl. Pharmacol.*, 1981, **57**, 88.
169. R. Edwards and P. Millburn, *Pestic. Sci.*, 1985, **16**, 201.
170. S. P. Bradbury, J. R. Coats and J. M. McKim, *Environ. Toxicol. Chem.*, 1985, **4**, 533.
171. J. R. Coats, D. M. Symonik, S. P. Bradbury, S. D. Dyer, L. K. Timson and G. J. Atchison, *Environ. Toxicol. Chem.*, 1989, **8**, 671.
172. S. P. Bradbury and J. R. Coats, *Environ. Toxicol. Chem.*, 1989, **8**, 373.
173. M. M. Mumtaz and R. E. Menzer, *J. Agric. Food Chem.*, 1986, **34**, 929.
174. R. Edwards, P. Millburn and D. H. Hutson, *Toxicol. Appl. Pharmacol.*, 1986, **84**, 512.
175. V. Dev, K. Raghavendra, K. Barman, S. Phookan and A. P. Dash, *Vector Borne Zoonot. Dis.*, 2010, **10**, 403.
176. C. M. Bridges, F. J. Dwyer, D. K. Hardesty and D. W. Whites, *Bull. Environ. Contam. Toxicol.*, 2002, **69**, 562.

177. D. J. Marcogliese, K. C. King, H. M. Salo, M. Fournier, P. Brousseau, P. Spear, L. Champoux, J. D. McLaughlin and M. Boily, *Aquat. Toxicol.*, 2009, **91**, 126.

178. D. J. Lauren, in *Aquatic Toxicology and Risk Assessment: Fourteenth Volume, ASTM STP 1124*, ed. M. A. Mayes and M. G. Barron, American Society for Testing and Materials, Philadelphia, PA, USA, 1991, pp. 223.

179. A. H. Glickman, S. D. Weitman and J. J. Lech, *Toxicol. Appl. Pharmacol.*, 1982, **66**, 153.

180. A. H. Glickman and J. J. Lech, *Toxicol. Appl. Pharmacol.*, 1982, **66**, 162.

181. S. P. Bradbury and J. R. Coats, *J. Toxicol. Environ. Health*, 1982, **10**, 307.

182. S. P. Bradbury, J. M. McKim and J. R. Coats, *Pestic. Biochem. Physiol.*, 1987, **27**, 275.

183. D. M. Symonik, J. R. Coats, S. P. Bradbury, G. J. Atchison and J. M. Clark, *Bull. Environ. Contam. Toxicol.*, 1989, **42**, 821.

184. L. E. Mokry and K. D. Hoagland, *Environ. Toxicol. Chem.*, 1990, **9**, 1045.

185. I. R. Hill, in *The Pyrethroid Insecticides*, ed. J. P. Leahey, Taylor & Francis Inc., Philadelphia, PA, USA, 1985, pp. 151.

186. W. C. Yang, F. Spurlock, W. P. Liu and J. Y. Gan, *Environ. Toxicol. Chem.*, 2006, **25**, 1913.

187. W. C. Yang, J. Y. Gan, W. Hunter and F. Spurlock, *Environ. Toxicol. Chem.*, 2006, **25**, 1585.

188. C. E. Wheelock, J. L. Miller, M. J. Miller, B. M. Phillips, S. J. Gee, R. S. Tjeerdema and B. D. Hammock, *Aquat. Toxicol.*, 2005, **74**, 47.

189. S. M. Brander, I. Werner, J. W. White and L. A. Deanovic, *Environ. Toxicol. Chem.*, 2009, **28**, 1493.

190. K. B. Norgaard and N. Cedergreen, *Environ. Sci. Pollut. Res.*, 2010, **17**, 957.

191. N. Cedergreen, A. Kamper and J. C. Streibig, *Aquat. Toxicol.*, 2006, **78**, 243.

192. J. R. Coats and N. L. O'Donnell-Jeffery, *Bull. Environ. Contam. Toxicol.*, 1979, **23**, 250.

193. S. P. Bradbury, J. R. Coats and J. M. McKim, *Environ. Toxicol. Chem.*, 1986, **5**, 567.

194. S. Beggel, I. Werner, R. E. Connon and J. P. Geist, *Sci. Total Environ.*, 2010, **408**, 3169.

195. S. J. Maund, M. J. Hamer, M. C. G. Lane, E. Farrelly, J. H. Rapley, U. M. Goggin and W. E. Gentle, *Environ. Toxicol. Chem.*, 2002, **21**, 9.

196. K. Day and N. K. Kaushik, *Environ. Pollut.*, 1987, **44**, 13.

197. S. Reynaldi and M. Liess, *Environ. Toxicol. Chem.*, 2005, **24**, 1160.

198. S. Reynaldi, S. Duquesne, K. Jung and M. Liess, *Environ. Toxicol. Chem.*, 2006, **25**, 1826.

199. L. M. Cole and J. E. Casida, *Pestic. Biochem. Physiol.*, 1983, **20**, 217.

200. J. M. Giddings, K. R. Solomon and S. J. Maund, *Environ. Toxicol. Chem.*, 2001, **20**, 660.
201. J. R. Bloomquist, *Rev. Pestic. Toxicol.*, 1993, **2**, 185.
202. T. J. Kedwards, S. J. Maund and P. F. Chapman, *Environ. Toxicol. Chem.*, 1999, **18**, 158.
203. C. H. Walker, in *Organic Pollutants: an Ecotoxicological Perspective*, ed. C. H. Walker, CRC Press, Boca Raton, FL, USA, 2009, pp. 231.
204. R. P. A. Van Wijngaarden, T. C. M. Brock and P. J. Van den Brink, *Ecotoxicology*, 2005, **14**, 355.
205. J. M. Dabrowski, A. Bollen, E. R. Bennett and R. Schulz, *Agric. Ecosyst. Environ.*, 2005, **111**, 340.
206. R. Schulz and M. Liess, *Environ. Toxicol. Chem.*, 2001, **20**, 185.
207. B. Styrishave, T. Hartnik, P. Christensen, O. Andersen and J. Jensen, *Environ. Toxicol. Chem.*, 2010, **29**, 1084.
208. R. E. Landeck-Miller, J. R. Wands, K. N. Chytalo and R. A. D'Amico, *J. Shellfish Res.*, 2005, **24**, 859.
209. J. Pearce and N. Balcom, *J. Shellfish Res.*, 2005, **24**, 691.
210. M. Levin, B. Brownawell and S. De Guise, *J. Shellfish Res.*, 2007, **26**, 1161.
211. C. D. Milam, J. L. Farris and J. D. Wilhide, *Arch. Environ. Contam. Toxicol.*, 2000, **39**, 324.
212. R. S. Davis and R. K. D. Peterson, *J. Am. Mosquito Control Assoc.*, 2008, **24**, 270.
213. W. M. Boyce, S. P. Lawler, J. M. Schultz, S. J. McCauley, L. S. Kimsey, M. K. Niemela, C. F. Nielsen and W. K. Reisen, *J. Am. Mosquito Control Assoc.*, 2007, **23**, 335.
214. USEPA, *Fenvalerate; Product Cancellation Order* EPA–HQ–OPP–2008–0263; FRL–8371–8, U.S. Environmental Protectition Agency, Washington D.C., USA, 2008; http://www.regulations.gov/search/Regs/home.html#documentDetail?R=090000648066029d.
215. M. Noritada, U. Kazuya, S. Yoshinori, I. Tomonori, S. Masayo, Y. Tomonori and U. Satoshi, *Sumitomo Kagaku*, 2005, 4.
216. M. Whalon, D. Mota-Sanchez. R. M. Hollingworth, P. Bills and L. Duynslager, *The Database of arthropods resistant to pesticides*, Michigan state University; http://www.pesticideresistance.org/DB/index.php.
217. F. Zhu, J. Wigginton, A. Romero, A. Moore, K. Ferguson, R. Palli, M. F. Potter, K. F. Haynes and S. R. Palli, *Arch. Insect Biochem. Physiol.*, 2010, **73**, 245.
218. J. Hemingway, L. Field and J. Vontas, *Science*, 2002, **298**, 96.
219. USEPA, *Registration Review: Summary of Planned Schedule for Opening Registration Review Dockets by Fiscal Year 2010 to 2013*, U.S. Environmental Protection Agency Washington D.C., USA 2010; http://www.epa.gov/oppsrrd1/registration_review/2010-13-schedule-summary.pdf.

220. USEPA, *Letter from: George LaRocca. Re: Updated spray drift language for pyrethroid agricultural use products*, U.S. Environmental Protection Agency, Washington D.C., USA, 2008.
221. D. C. G. Muir, B. R. Hobden and M. R. Servos, *Aquat. Toxicol.*, 1994, **29**, 223.
222. WHO, *Environmental Health Criteria 87: Allethrins*, World Health Organization, Geneva, Switerland, 1989.
223. A. K. Kumaraguru and F. W. H. Beamish, *Water Res.*, 1981, **15**, 503.
224. WHO, *Environmental Health Criteria 97: Deltamethrin*, World Health Organization, Geneva, Switzerland, 1990.

Basic and Applied Aspects of Neonicotinoid Insecticides

R. NAUEN*[1] AND P. JESCHKE[2]

[1] Bayer CropScience AG, Research Insecticides Biology, Building 6220, Alfred-Nobel Str., 50, D-40789 Monheim, Germany; [2] Bayer CropScience AG, Research Insecticides Chemistry, Building 6240, Alfred-Nobel Str., 50, D-40789 Monheim, Germany

4.1 Introduction

Sustainable agriculture aims to supply sufficient food for the world population while minimizing environmental impact. Part of sustainable agriculture is the application of insecticides in order to protect crops from deleterious invertebrate pests, feeding on plants cultivated for human consumption. The discovery of neonicotinoid insecticides can be considered as a milestone in insecticide research.[1] Neonicotinoids represent the fastest-growing class of insecticides introduced to the market since the commercialization of pyrethroids.

The market share for neonicotinoids of the total global market for insecticides (€6.330 billion) was 24% in 2008. Further worldwide success of neonicotinoids over the next few years will result from the growth of established neonicotinoid compounds replacing older, and environmentally less benign substance classes, such as organophosphates and carbamates. Neonicotinoids are potent broad spectrum insecticides possessing contact, stomach and systemic activity. They are especially active on hemipteran pest species such as aphids, whiteflies and planthoppers, but neonicotinoids have also been commercialized in order to control different coleopteran and some lepidopteran

RSC Green Chemistry No. 11

Green Trends in Insect Control

Edited by Óscar López and José G. Fernández-Bolaños

© Royal Society of Chemistry 2011

Published by the Royal Society of Chemistry, www.rsc.org

pest species.[1] The physicochemical properties of neonicotinoids mean that they can be used with a wide range of different application techniques, including foliar, seed treatment, soil drench and stem application in several crops. Due to their favorable mammalian safety characteristics (see section 4.2), some neonicotinoids, like imidacloprid, are also important in the control of subterranean pests and in veterinary use.[1,2]

All commercial neonicotinoid insecticides bind selectively to insect nicotinic acetylcholine receptors (nAChRs) and evoke the same effect as the natural neurotransmitter acetylcholine, *i.e.* agonistic activation of the receptors by causing a transient inward-current leading to the generation of action potentials. As with acetylcholine, neonicotinoid binding to nAChRs is reversible, as shown by their rapid desensitization/recovery during short-term exposure in electrophysiological whole-cell voltage clamp assays on isolated neurons from insects. Radioligand binding studies conducted with tritiated imidacloprid also revealed saturable, specific and reversible binding with fast kinetics.[1–3]

In this chapter, some of the aspects contributing to the commercial success of neonicotinoid insecticides will be briefly reviewed. These include the unique chemical and biological properties of neonicotinoids; their broad-spectrum insecticidal activity at low application rates; the development of new and refined application technologies related to the systemic characteristics of neonicotinoids; as well as their favorable safety profile and selectivity. The chapter also summarizes the latest results on pollinator safety, and discusses the biochemical mode of action of neonicotinoids with special reference to target-site selectivity.

4.2 Target-Site Selectivity of Neonicotinoids

In contrast to the naturally occurring alkaloid (*S*)-(–)-nicotine [isolated as botanical insecticide from *Nicotiana* species; IRAC (Insecticide Resistance Action Committee) Mode of Action (MoA) classification: *n*AChR agonists, group 4B], neonicotinoid insecticides act selectively on the insect central nervous system (CNS) as an agonist of the post-synaptic *n*AChRs, their molecular target site.[1–3] There is a distinct difference in the selective toxicity between (*S*)-(–)-nicotine and neonicotinoid insecticides. Nicotine sulfate is very toxic to bees and warm-blooded animals (*e.g.* LD_{50} of $3\,mg\,kg^{-1}$ for mice, and $<5\,mg\,kg^{-1}$ can be the lethal oral dosage for adult humans of $70\,kg$), and can be readily absorbed through the skin.[4,5] It is sold as a 40% nicotine sulfate concentrate under trade names that include Black Leaf 40 or Tender Leaf Plant Insect spray. (*S*)-(–)-Nicotine is protonated at neutral or lower pH, forming a water-soluble ammonium ion that decreases its insect toxicity, although this ammonium form is recognized by *n*AChRs.

In contrast to (*S*)-(–)-nicotine, the neonicotinoids have a low mammalian toxicity (*see* the data of acute oral and dermal rat toxicity in Table 4.1) and low use rates, that minimize toxicological risks for the consumer by dietary intake of residues with food.

Table 4.1 Toxicological profiles of (*S*)-(−)-nicotine and neonicotinoid insecticides.

Compounds	Acute oral (rat)[a]	Acute dermal (rat)[a]
(*S*)-(−)-Nicotine	50–60	n.d.[b]
Neonicotinoids:		
[A] Ring systems:		
Imidacloprid	>450	>5,000
Thiacloprid	836 (m), 444 (f)	>2.000
Thiamethoxam	1.563	>2.000
[B] Non-cyclic compounds:		
Nitenpyram	1680 (m), 1575 (f)	>2.000
Acetamiprid	217 (m), 146 (f)	>2,000
Clothianidin	>5.000 (m, f)	>2.000 (m, f)
Dinotefuran	2.450 (m, f)	>2.000

[a]LD_{50} mg a.i. kg^{-1} body weight; m = males, f = females.
[b]n.d. = not detected; nicotine readily passes into the bloodstream from dermal contact.

Neonicotinoid insecticides show either little or almost no binding affinity to mammalian nicotinic acetylcholine receptors (nAChRs) due to fundamental differences between the nAChRs of insects and mammals.[3] The binding affinity of imidacloprid to the common mammalian neuronal α4β2 nAChRs is at least 1000-fold lower than that to insect nAChRs. This is due to the interaction of neonicotinoids (such as imidacloprid) with a unique subsite consisting of cationic amino acid residue(s) in insect receptors, and which are absent in mammalian nAChRs.[3,6]

Electrophysiology, computational chemistry and site-directed mutagenesis, in conjugation with homology modeling of the *n*AChR ligand binding domain (*n*AChR LBD)–imidacloprid complexes have been used to elucidate the nature and the diversity of neonicotinoid actions.[6] In order to understand the structural factors involved in target-selectivity, the crystal structures of the acetylcholine-binding protein (AChBP) from the mollusk *Lymnaea stagnalis* (Ls), in complex with the neonicotinoid insecticides, imidacloprid and clothianidin,[7] as well as *Aplysia californica* (AC)-AChBP in complex with imidacloprid and thiacloprid[8] were elucidated. From both crystal structures, a common concept for the nAChR LBD–neonicotinoid interactions was identified, which clearly differs from the binding modes of nicotinoids.

4.3   Chemical Structure of Neonicotinoids

As a result of the efficient MoA given by the *n*AChR as target there is no cross-resistance to conventional long-established insecticide classes, such as chlorinated hydrocarbons, organophosphates, carbamates, pyrethroids and several other chemical classes of insecticides used to control insect pests on major crops. Since the introduction of imidacloprid in 1991, neonicotinoids represent a fairly new class of chemistry classified in the same MoA class (*n*AChR agonists, group 4A) by IRAC.[1]

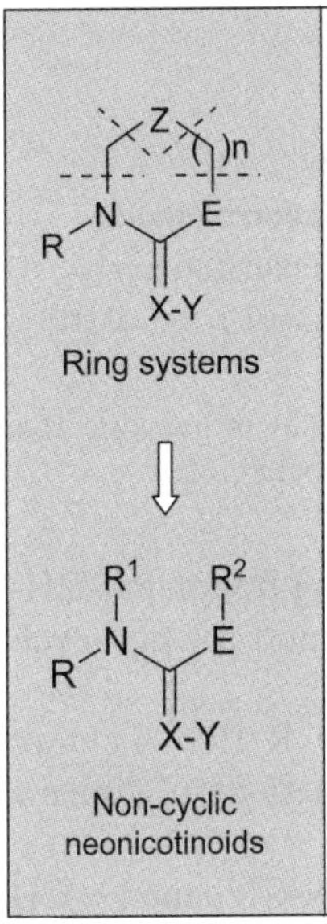

Neonicotinoids:

Ring systems

a Imidacloprid — R = CPM, n = 0, Z = CH_2, E = NH, X-Y = NNO_2

b Thiacloprid — R = CPM, n = 0, Z = CH_2, E = S, X-Y = NCN

c Thiamethoxam — R = CTM, n = 1, Z = O, E = NMe, X-Y = NNO_2

Non-cyclic neonicotinoids

d Nitenpyram — R = CPM; R^1 = Et, E-R^2 = NHMe, X-Y = $CHNO_2$

e Acetamiprid — R = CPM; R^1 = Me, E-R^2 = Me, X-Y = NCN

f Clothianidin — R = CTM; R^1 = H, E-R^2 = NHMe, X-Y = NNO_2

g Dinotefuran — R = TFM; R^1 = H, E-R^2 = NHMe, X-Y = NNO_2

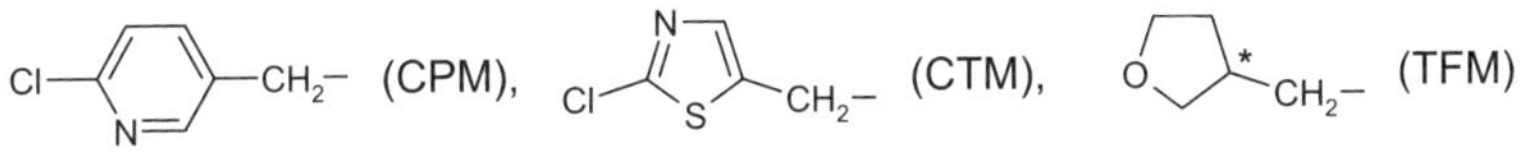

$Cl-\!\!\!<\!\!>\!\!-CH_2-$ (CPM), $Cl-\!\!\!<\!\!>\!\!-CH_2-$ (CTM), $O\!\!<\!\!>\!\!*CH_2-$ (TFM)

Figure 4.1 Commercial neonicotinoid insecticides: ring systems (a–c) *versus* non-cyclic neonicotinoids (d–g). Abbreviations: CPM (chloropyridyl), CTM (chlorothiazolyl) and TFM (tetrahydrofuryl), * mixture or (*R*)- and (*S*)-enantiomers.

Today, seven neonicotinoid insecticides are on the market: three cyclic compounds [neonicotinoids with five-membered ring systems, such as imidacloprid[8,9] and thiacloprid (Bayer CropScience),[10] and the six-membered ring neonicotinoid thiamethoxam (Syngenta)][11] and four non-cyclic compounds [nitenpyram (Sumitomo Chemical Takeda Agro Company),[12] acetamiprid (Nippon Soda),[13] clothianidin (Sumitomo Chemical Takeda Agro Company/ Bayer CropScience)[14] and dinotefuran (Mitsui Chemicals)].[15]

4.3.1 Structural Diversity of Neonicotinoids

Considering their pharmacophore moieties [-N-C(E) = X-Y], neonicotinoid insecticides can be classified as *N*-nitro-guanidines (imidacloprid, thiamethoxam, clothianidin and dinotefuran), nitromethylenes (nitenpyram) and *N*-cyano-amidines (acetamiprid and thiacloprid).[16]

The overall chemical structure of both ring systems and non-cyclic commercial neonicotinoids consists of different segments (1)–(3) (see Figure 4.1).[17–19] This chemical structure affects some of the principles of green chemistry in this substance class, such as their high target selectivity, remarkable physico-chemistry, as well as additional effects like phytotonic behavior (see Figure 4.2 and section 4.7).

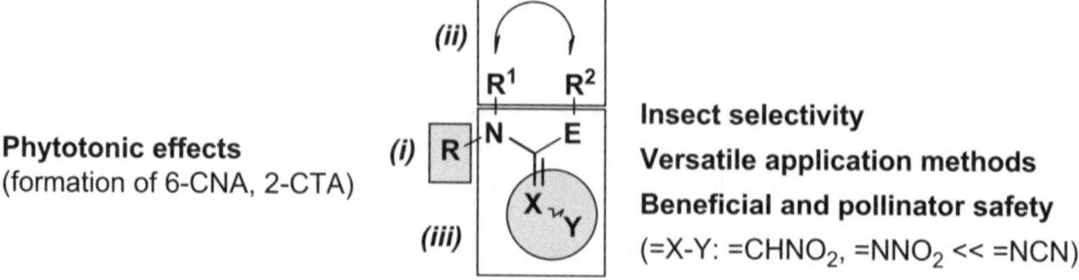

Figure 4.2 Structural segments (*i–iii*) in neonicotinoid insecticides influencing their target selectivity, physico-chemistry and phytotonic behavior.

(1) For five- and six-membered ring systems, the bridging fragment [-CH$_2$-Z-(CH$_2$)$_n$-: $n = 0$; Z $=$ CH$_2$ and $n = 1$; Z $=$ O, NMe] and for non-cyclic neonicotinoids the separate substituents (R^1, R^2);

(2) The heteroarylmethyl or heterocyclylmethyl group R [R $=$ 6-chloro-pyrid-3-ylmethyl (CPM), 2-chloro-1,3-thiazol-5-ylmethyl (CTM) and ($\pm$)-6-tetrahydro-fur-3-ylmethyl (TFM)]; and

(3) The functional group [$=$ X-Y] (*e.g.* [$=$ N-NO$_2$], [$=$ N-CN] and [$=$ CH-NO$_2$]) as part of the different pharmacophore types [-N-C(E) $=$ X-Y].

In comparison to the corresponding ring systems (*i.e.* imidacloprid, thiacloprid or thiamethoxam), the non-cyclic neonicotinoids show similar broad insecticidal activity by forming a so-called *quasi*-cyclic conformation when binding to the insect *n*AChRs.[20,21] In this context, the partial ring cleavage of the six-membered ring system thiamethoxam into the non-cyclic clothianidin (see Figure 4.1; transformation of c into f) in insect and plant tissues has been demonstrated and discussed in detail.[22,23]

In total, *N*-nitro-guanidines (functional group [$=$ X-Y]: [$=$ N-NO$_2$]) are the most prominent subclass, they account for around 85% of the neonicotinoid insecticide market ($2236 million in 2009).

4.3.2   Green Technologies for Manufacturing of Neonicotinoids

Green chemistry utilizes a set of 12 principles that reduces or eliminates the use or generation of hazardous substances in the design, manufacture and application of chemical products.[24]

The design of *greener* processes is also under consideration in the manufacture of neonicotinoid insecticides and their intermediates. This can be exemplified in case of the two *N*-nitro-guanidines dinotefuran and clothianidin.

4.3.2.1   *Dinotefuran – (R,S)-3-(Hydroxymethyl)-tetrahydrofuran Intermediate*

From a green chemistry standpoint, the introduction of a formyl group in 7,12-dioxaspiro[5,6]dodec-9-ene (**4**) by rhodium-catalyzed hydroformylation using syngas (RhHCO(TPP)$_3$; CO/H$_2$; 130 bar; 100 °C; toluene) and its reduction has

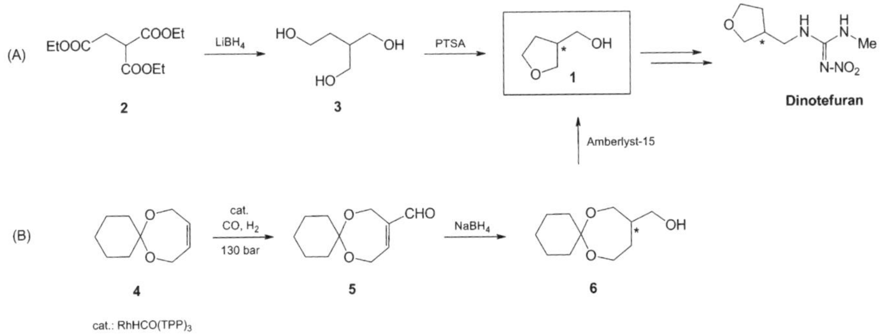

Scheme 4.1 Synthesis of 3-hydroxymethyl)-tetrahydrofuran: pathway (A) by reduction of diethyl 2-carboethoxysuccinate (**2**), and (B) by rhodium-catalyzed hydroformylation of 7,12-dioxaspiro[5,6]dodec-9-ene (**4**) using syngas. *mixture of (*R*)- and (*S*)-enantiomers (adapted from Potluri *et al.*, 2005).[25]

been proposed as a useful alternative and environmentally friendly preparation process (*via* intermediates **5** and **6**; yields are $\geq$ 90%) for the dinotefuran intermediate (*R,S*)-3-(hydroxymethyl)-tetrahydrofuran (**1**; scheme 4.1B).[25]

Previous synthesis of the intermediate (**1**) involved LiBH$_4$ reduction of diethyl 2-ethoxycarbonylsuccinate (**2**), followed by cyclocondensation of the resulting 2-hydroxymethylbutane-1,4-diol (**3**) using *para*-toluenesulfonic acid (PTSA; see scheme 4.1a). This method did have some disadvantages, however, such as:

(1) The difficult separation of the triol (**3**) from aqueous solution during workup because of its very high water solubility;
(2) The large quantities of LiBH$_4$ required to reduce the three ester groups of (**2**), forming large quantities of salts; and
(3) The separation of PTSA from the reaction mixture, before isolation of (**1**) by solvent extraction or distillation.

4.3.2.2 *Clothianidin – O-Methyl-N-nitroisourea Intermediate*

Recently, *O*-methyl-*N*-nitroisourea (**8**) was described as a key building block for the synthesis of *N*-[(2-chloro-5-thiazolyl)methyl]-*N'*-nitro-carbaminic acid methyl ester (**9**), which can be easily transformed into clothianidin by treatment with *N*-methylamine (see scheme 4.2).[26]

It was found that an organic solvent was not essential for either reaction step. Both, the replacement of the amino group in (**8**) and that of the methoxy group in (**9**) could be carried out in water as a very safe solvent under cleavage of ammonia or methanol.[27]

Twenty years ago, at the beginning of the neonicotinoid chemistry, the methylthio group in *N*-[(2-chloro-5-thiazolyl)methyl]-*N'*-nitro-carbamimi-dothioic acid methyl ester (**7**) was the preferred leaving group for coupling with the 5-aminomethyl-2-chloro-1,3-thiazole (CTM-NH$_2$) intermediate, resulting in clothianidin and methyl mercaptan waste, respectively.[28]

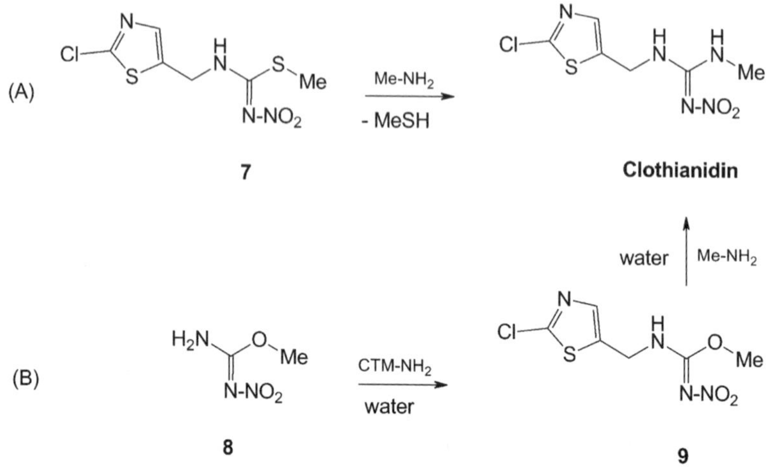

7   **Clothianidin**

8   **9**

Scheme 4.2   Syntheses of clothianidin: pathway (A) by replacement of the methylthio group in N'-[(2-chloro-5-thiazolyl)methyl]-N-nitro-carbamimidothioic acid methyl ester (**7**) by methylamine, and pathway (B) by replacement of the amino group in O-methyl-N-nitro-isourea (**8**) by 5-aminomethyl-2-chloro-1,3-thiazole (CTM-NH$_2$) and following replacement of the methoxy group in (**9**) by methylamine in water (adapted from patent applications).

4.3.3   Physico-Chemical Properties of Neonicotinoids

The physico-chemical properties of the five- and six-membered ring systems and non-cyclic neonicotinoids played an important role in their successful development as modern insecticides for sustainable agriculture.

In this context, photostability is a significant factor in their agricultural field performance. As described earlier,[29–31] the energy gap for the different functional groups [$=$X-Y] from the ground state to the single state increased in the order [$=$CH-NO$_2$]$<$[$=$N-NO$_2$]$<$[$=$N-CN].

For technical application methods for neonicotinoids in the field, such as for soil drench, seed treatment or foliar application (see section 4.5), their uptake, good translaminar (see section 4.5.1) and acropetal distribution in plants is crucial for their insecticidal activity against numerous sucking pests. Therefore, not only the bioisosteric fragments (CPM *vs* CTM substituents or pharmacophore moieties [-N-C(E)$=$X-Y]) but also the whole molecular shape, including the resulting water solubility, has to be considered.[32] As so-called "push–pull" olefins, neonicotinoids can form conjugated coplanar electron-donating and electron-accepting groups. In comparison with other non-polar insecticide classes, the polar, non-volatile neonicotinoids have greater water solubilities (*e.g.* in case of nitenpyram: 840 g l^{-1} at pH 7) and lower logP$_{OW}$ values (*e.g.* dinotefuran: -0.644 at 25 °C) (see Table 4.2).

From these observations, the following conclusions can be drawn:

(1) Generally, non-cyclic neonicotinoids are less lipophilic than the corresponding five- and six-membered ring systems;

Table 4.2 Physico-chemical properties of (*S*)-(−)-nicotine and neonicotinoid insecticides.

Compounds	Color and physical state	Melting point (°C)	Density (g ml^{-1} at 20 °C)	Solubility in water (g l^{-1} at 20 °C)	LogP_{OW} (at 25 °C)
(*S*)-(−)-Nicotine		247[a]	1.01	miscible	1.2
Neonicotinoids:					
[A] Ring systems					
Imidacloprid	colorless crystals	144	1.54	0.61	0.57[b]
Thiacloprid	yellow crystalline powder	136	1.46	0.185	1.26[c]
Thiamethoxam	slightly creamy crystalline powder	139.1	1.57	4.1	−0.13
[B] Non-cyclic compounds:					
Nitenpyram	pale-yellow crystals	83–84	1.40[d]	840[e]	−0.64
Acetamiprid	colorless	98.9	1.330	4.20[f]	0.8
Clothianidin	clear colorless solid powder	176.8	1.61	0.327	0.7
Dinotefuran	white crystalline solid	94.6–101.5	1.33	54.3 ± 1.3	0.644

[a]boiling point.
[b]at 22 °C.
[c]at 20 °C.
[d]at 26 °C
[e]at pH 7.
[f]at 25 °C.

(2) Water solubility is influenced by the functional group [=X-Y] within the pharmacophore moiety [-N-(C(E))=X-Y] and increases in the order [=N-NO$_2$] < [=N-CN] < [=CH-NO$_2$]; and

(3) Regarding E, the lipophilicity increases in the order NH < O < C < S.

According to Briggs *et al.*,[33] somewhat more lipophilic neonicotinoid insecticides should be favorable for seed treatment application (see section 4.5.1), because their root uptake and translocation is more effective than in the case of more hydrophilic compounds (see Figure 4.3).

Due to their higher lipophilicity, thiacloprid and clothianidin show the highest root uptake, whereas nitenpyram and dinotefuran are considered to be more xylem mobile than the other neonicotinoids.

4.3.3.1   Penetration and Translocation of Neonicotinoids

For good leaf coverage in foliar applications, absorption of the ai (active ingredient) through the leaf cuticle and its movement within the leaves from the upper to the lower surface (translaminar) and towards the tip of the leaf (acropetal) is essential for optimal control.

The good penetration, translaminar distribution and xylem mobility of neonicotinoid insecticides in plants can be demonstrated for example by the

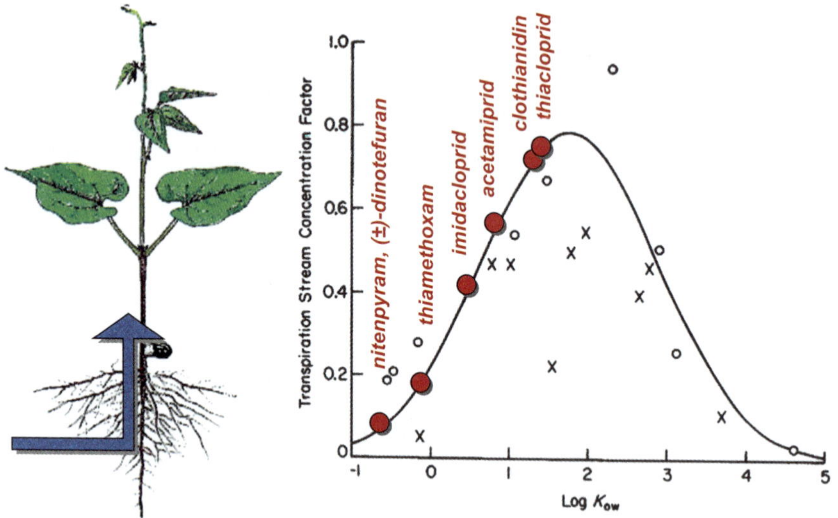

Figure 4.3   Systemicity of commercial neonicotinoid insecticides. Relationship between the translocation of neonicotinoids to barley shoots following uptake by the roots (expressed as the Transpiration Stream Concentration Factor), and the octan-1-ol/water partition coefficients (log K_{ow}). Due to their higher lipophilicity thiacloprid and clothianidin show the best root uptake, whereas nitenpyram and dinotefuran are more xylem mobile than the other neonicotinoids. (Diagram and experimental data taken and adapted from Briggs *et al.*, 1982).[33]

physico-chemical behavior of thiacloprid in cabbage, which is comparable to imidacloprid.[34]

The amounts of thiacloprid stripped off from cabbage leaf application sites were 23% and 17% at one day and seven days after application, respectively. Levels in the true leaf increased from 63% to 77% as measured one day and seven days after application, respectively. These results demonstrate that thiacloprid is readily taken up by cabbage leaves, thus providing good systemic control of leaf-sucking pests. The visualization of the translocation pattern of ^{14}C-labelled thiacloprid equivalents by phosphor imaging technology revealed xylem mobility, *i.e.* translocation of the neonicotinoid in the upwards direction, even one day after application to cabbage leaves (see Figure 4.4a). This xylem mobility is further highlighted in Figure 4.4b, which shows an excellent distribution of thiacloprid in cucumber leaves after spray application.

4.4 Biological Profile of Neonicotinoids

Neonicotinoid insecticides exhibit extensive biological activity with widespread agricultural uses, and numerous reviews, articles and book chapters have been published over the past decade on the subject.[35,36] It is beyond the scope of this chapter to provide a comprehensive overview of the agronomic and horticultural uses of neonicotinoid insecticides, since more than 140 and 115 crop uses have been described and commercialized for imidacloprid and thiamethoxam alone.[36]

The unique properties of neonicotinoids include high intrinsic acute and residual activity against sucking and some chewing insect species; high efficacy against susceptible and insecticide-resistant populations (such as aphids,

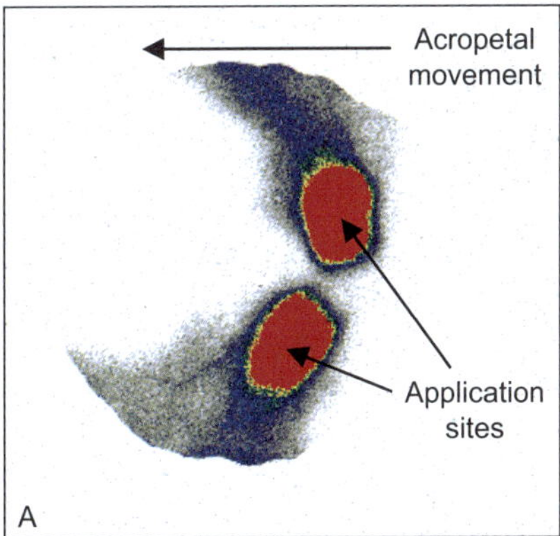

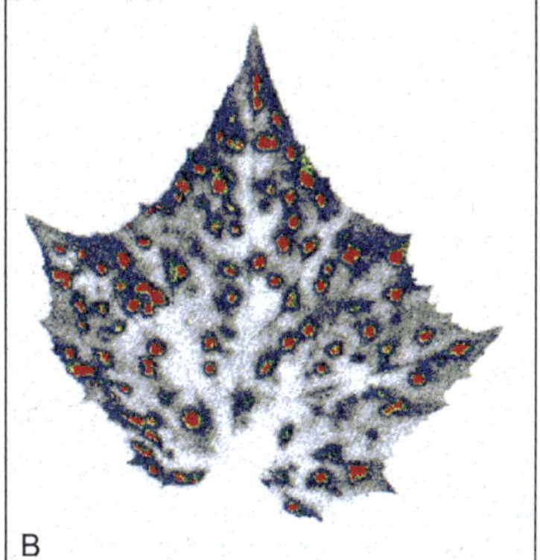

Figure 4.4   (A) Translocation of [14C]-thiacloprid in cabbage plants 1 day after application of 2 × 5 µl droplets onto the first true leaf. Radioactivity on the surface of the cuticle (residue) was already removed by cellulose-acetate stripping, (B) translocation pattern of thiacloprid sprayed to a cucumber leaf. (Pictures taken by Nauen).

whiteflies, leafhoppers and planthoppers, Colorado potato beetles and others); and excellent acropetal translocation. These properties make neonicotinoids suitable for use on a wide variety of crops, including aphids in vegetables, sugar beet, cotton, pome fruit, cereals and tobacco; leafhoppers, planthoppers and water weevil in rice; whiteflies in vegetables, cotton and citrus; lepidopteran leafminer in pome fruit and citrus; and wireworms in sugar beet and corn. Termites and turf pests such as white grubs are also important indications covered by imidacloprid.[9,36]

Due to their high systemicity, diverse application methods for neonicotinoids are feasible which have been introduced into practice. Soil treatments can be done by incorporation of granules, injection, application with irrigation water, spraying, use of tablets, *etc.* Plants, or parts of plants, are treated by seed dressing, pelleting, implantation, dipping, injection and painting. Such methods have led to a more economic and environmentally friendly use of these products which fits well into various IPM (Integrated Pest Management) programmes.[36]

4.4.1   Plant Virus Vector Control by Neonicotinoids

Phytopathogenic viral and bacterial diseases are of increasing importance worldwide, impacting on the quality and the yield of infested crops. For example, it has been calculated that $1\,cm^2$ of potato leaf can contain around 20×10^9 potato X viruses.

Plant viruses can be transmitted by plant-feeding sucking insects. Common virus vectors are aphids, whiteflies, thrips and leafhoppers, whereas bacterial diseases are particularly transmitted by leafhoppers and psyllids. Aphids and other sucking pests cause direct damage to plants when they feed on plants, but far more important is the indirect damage caused through the transmission of viral diseases. For example the green peach aphid, *Myzus persicae,* is known to transmit more than 150 viral diseases. Transmission can occur after just a few seconds (non-persistent) or after a longer period (persistent) of feeding, depending on the type of virus. Aphids and leafhoppers get contaminated with viral particles by feeding on infested plants ("acquisition"), they remain infested for variable times and by different modes ("retention"), and finally they pass virus particles onto other non-infested plants ("inoculation"). The relation between plant virus and virus vector is often specific.

Due to their excellent plant systemic properties, neonicotinoid insecticides such as imidacloprid, thiamethoxam and clothianidin also control important vectors of plant viral diseases, thereby suppressing the secondary spread of viruses in various crops.

For imidacloprid, this control was discovered during its early development supported by the antifeedant effects of sublethal doses of imidacloprid on the whitefly, *Bemisia tabaci,*[37] and later observations, *e.g.* for the persistant barley yellow dwarf virus (BYDV) transmitted by the oat bird-cherry aphid, *Rhopalosiphum padi* L., and the grain aphid, *Sitobion avenae.*[38] Applied as a pellet of 90 g ai (active ingredient) unit^{-1}, beet mild yellows virus (BMYV) was effectively controlled in sugar beet seed, as demonstrated in field trials conducted in the UK between 1989 and 1991.[39,40] An outstanding crop protection was achieved with seed treatment or foliar applications in cereals against aphids and the spread of BYDV,[41–43] in tobacco against thrips and tomato spotted wilt virus (TSWV),[44] in tomato against whiteflies and tomato yellow leaf curl virus (TYLCV), and in citrus against glassy-winged sharpshooters as vectors for the bacterium *Xylella fastidiosa*. A prominent example of virus vector control is described for TSWV which became a serious disease in Georgia flue-cured tobacco in the 1990s.[35] Insecticides registered for the control of the thrips vector had been shown to be ineffective in most cases. After the introduction of imidacloprid for the control of aphids and other insect pests of tobacco, a reduction in TSWV was observed. Data from Mexico, Brazil, Greece, Guatemala and Italy further confirm that imidacloprid can be used to reduce damage caused to tobacco by viruses. In addition to TSWV, Potato virus Y, tobacco etch virus and beet curly top virus are known to be reduced by imidacloprid treatment. With all except TSWV, high levels of insect vector control were observed.[35] In all cases imidacloprid treatment reduced, but did not prevent, viral infections; the neonicotinoid reduced both the incidence and severity of viral infections of tobacco.

4.4.2 Integrated Pest Management – Beneficial and Pollinator Safety

An important aspect of neonicotinoid insecticides is their suitability for integration into Integrated Pest Management (IPM) systems. Depending on the application method and timing, non-target organisms are not affected by neonicotinoids, unlike some older chemical classes of insecticides.

Safety for beneficials and pollinators has especially been optimized in neonicotinoids by selectivity in space.[45] Application into the soil by different methods allows the transport of the compound to the pest within the plant and without harming beneficial organisms.[46] On the other hand, selectivity in time allows for example, foliar application against starting pest populations when beneficial arthropods are still absent.

Safety of neonicotinoids to pollinators, especially honeybees, has been the subject of many scientific studies and much public debate in recent years. Whereas *N*-cyano-amidines (*e.g.* acetamiprid and thiacloprid) are characterized by a low toxicity to honeybees, and can therefore even be sprayed into flowering crops during times of bee activity,[47–49] the *N*-nitro-guanidines (such as imidacloprid, thiamethoxam and clothianidin) are intrinsically highly toxic to honeybees.[48–51] Therefore, spray applications have to be restricted to periods when the treated crops are not in bloom in order to avoid exposure of pollinators.

An application type that effectively minimizes exposure of bees and other pollinators is seed treatment, which is very common with neonicotinoid products in crops such as corn, canola, cereals, sugar beets and sunflowers.

As shown by many studies, translocated systemic residues of the active ingredients of the applied products occur, if at all, at only trace level in bee-relevant plant matrices (such as nectar or pollen), and the detected residue levels are far below concentrations that could harm bees.[50–55] Nevertheless, neonicotinoid seed-dressing products have frequently been suspected to be a potential causative factor involved in increased bee colony mortalities, as have been reported in recent decades from most parts of the world, and hypotheses have been put forward that chronic exposure to nectar and pollen containing neonicotinoid residues at sublethal concentrations could weaken or otherwise affect bee colonies.[56–58] Concerns like this in certain individual countries have even led to the suspension of certain seed-dressing uses (*e.g.* imidacloprid in sunflower in France in 1999). However, after more than fifteen years of intense research and numerous studies, no scientifically sound evidence has been found that insecticidal seed-treatment products, when correctly applied, can damage bee colonies under realistic field conditions.[49,50,52,53,59–65] On the contrary, there is more and more evidence that there is no correlation between exposure to neonicotinoid-treated crops and bee colonies,[66–72] and that the observed bee health problems are primarily caused by a combination of biotic and abiotic factors, of which the most important ones are parasites (especially the *Varroa* mite) and diseases.[69,70,72–80]

In 2008, an incident took place in Southwestern Germany, where a large number of bee colonies were damaged by abraded dust from a neonicotinoid product that was used as seed-dressing in corn, which was emitted to the environment during the drilling process.[81,82] The investigation of the issue soon revealed that the incident was caused by faulty seed treatment in combination with a couple of other coinciding unfavorable factors.[55] Since then, considerable and successful efforts have been undertaken in order to minimize the potential exposure of non-target organisms to dust from seed-dressing products by optimization of the seed-dressing process, as well as the sowing machinery, so that damage to honeybees by insecticidal dusts can be reliably excluded when the appropriate measures are implemented.[55,83–85]

Another potential pathway of exposure of honeybees to neonicotinoid seed-dressing products that was recently under discussion was "guttation" (a mechanism of active excretion of water by vascular plants). It was shown in laboratory studies that the guttation liquid of seedlings grown from neonicotinoid-treated seeds may contain intrinsically bee-toxic concentrations of the applied substances.[86] A comprehensive series of large-scale field studies conducted under realistic conditions to evaluate the potential risks of guttation in treated crops to honeybee colonies showed, however, that guttation is, under practical conditions, not normally a relevant water source for honeybees, and that adverse effects to bee colonies exposed to guttating fields of treated crops are not to be expected.[87]

In summary, neonicotinoid applications are safe to pollinators when the relevant safety measures are complied with. Neonicotinoid seed treatment is also a very environmentally friendly type of pesticide application, especially in comparison with insecticide spray treatments, as it greatly reduces total substance output and effectively reduces the exposure of beneficial organisms, in particular pollinators.

4.5   Versatile Application Methods for Neonicotinoids

Due to their remarkable physico-chemical properties (see section 4.3.3), neonicotinoid insecticides have a high degree of versatility, not seen to the same extent in other insecticidal classes.

Most neonicotinoids can be used as foliar sprays, seed-coating treatments (see section 4.5.1) and *via* soil application. At present, approximately 60% of all neonicotinoid applications are soil/seed treatments and most spray applications are especially targeted against pests attacking crops such as cereals, corn, rice, vegetables, sugar beet, potatoes, cotton and others. However, the increasing success of neonicotinoid insecticides also relies on versatile application methods (see Table 4.3).

In addition, other important application methods with irrigation water in drip or drench systems for vegetables or in floating box systems for tobacco seedlings, enable a long-lasting control of aphids and whiteflies to be achieved.[88]

For example, imidacloprid (Confidor) can be applied with a "drop nozzle" (a bell shaped spraying gun) so that the product drops on the leaves with a regulated droplet size (avoidance of spray drifting), pours down the stem and the remaining portion gets down to the soil. By this method initial efficacy is achieved on the leaves and long lasting efficacy provided by uptake *via* the stem and root system. Soil applications, *e.g.* by drip irrigation, are safe for a wider range of beneficial insects such as *Pseudaphycus flavidulus*, *Leucopis sp.* and *Sympherobius maculipennis* Kimmins which are predators of mealybugs and mites.

Table 4.3   Overview of versatile application methods of neonicotinoids (examples).

Application type	*Pest species (example)*	*Crops (selection)*
Trunk irrigation	Woolly apple aphid *Eriosoma lanigerum* (Hausmann)	Apple trees
Drip irrigation	Coffee leafminer *Perileucoptera coffeella* (Guérin-Méneville)	Coffee
Soil drench	Citrus leafminer *Phyllocnistis citrella* (Stainton)	Citrus trees
Stem injection	Emerald ash borer *Agrilus planipennis* (Fairmaire)	Riparian trees
Float system	Tobacco aphid *Myzus nicotianae* (Blackman)	Tobacco
In-furrow	Colorado potato beetle *Leptinotarsa decemlineata* (Say)	Potato
Bud injection	Banana weevil borer *Cosmopolites sordidus* Germar	Banana

The in-furrow application of imidacloprid (Admire) and thiamethoxam (Actara) is commonly used in potato and vegetable crops. To ensure root uptake, the neonicotinoid is applied at specified doses as in-furrow spray during planting or as a narrow band directly below the seed row some days before planting in a bedding operation. The float system technology is used primarily in tobacco, but also in vegetables. This application method provides considerable savings as neither soil disinfection nor weed control is required and economic benefits result from the higher yields and uniform growth. Neonicotinoids like imidacloprid (Confidor) are ideally suited to this technology, as they can either be sprayed or rinsed on the plants in the float boxes or added directly to the floating water. Direct stem injection-based application of neonicotinoids (*e.g.* Confidor) within a synchronized procedure (no remains of spray mixtures) is commonly used in citrus, hops or various trees, such as riparian trees.[89] A good uptake has been observed if the product is applied on young, green bark. Some neonicotinoids can be either applied hydroponically (*e.g.* seedling boxes in floating systems) or foliarly (*e.g.* pots) to protect the seedlings in nursery boxes. This application has been proven to be very economic and effective with rates of 10 g ai 50 m^{-2} (Confidor).

Soil injection to protect emerging vegetable seedlings against soil-inhabitating and sucking pests is common practice in the U.S.A. On the other hand, banana weevil and thrips are controlled *via* trunk and bud injection.

4.5.1 Seed Treatment Application with Neonicotinoids

Protection of the environment is an important motivator for continuing to develop new products and methods for seed treatment. This innovative method has additional benefits which make it not only interesting for economic reasons but also from an ecological point of view. The advantages are self evident: the ai is only found where activity is desired.

Compared with broadcast spraying or in-furrow treatment with granules, seed treatment effectively reduces the treated area considerably, *e.g.* from a 10 000 m^2 plot of arable land to around 60 m^2 or 1% of the corresponding field area (see Figure 4.5).

Moreover, seed treatment has less impact on non-target organisms and is not prone to drift, since the product is applied under controlled conditions (often in closed systems), which are not only independent of the weather, but also user safe, because the user does not come into direct contact with the ai during application.

In addition to seed dressing, film coating (for field crops and vegetables), pelleting (for sugar beets and fodder beets), pelleting and coating (sequential application of different film coatings) or multilayer coating (incorporation of insecticides and fungicides) also allow an effective, environmentally safe and ideal protection of young plants against insect attack and a variety of fungal diseases.

Applying the ai directly to the seed disinfects its halo surface. Soon after sowing, the plant absorbs the ai from the disinfectant halo through the roots

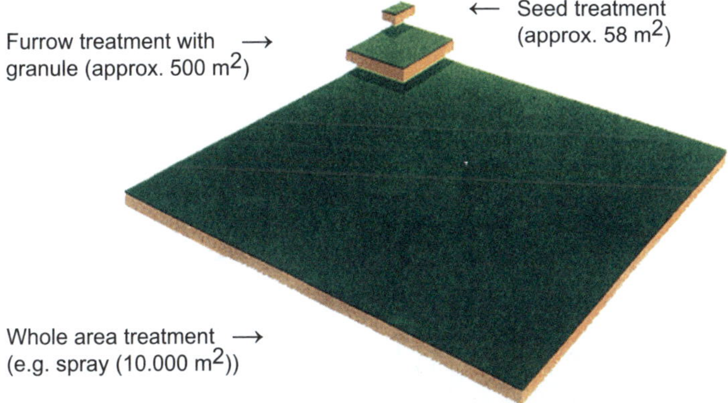

Figure 4.5 Compared to spray application, seed treatment reduces the overall treated area. While spray application of 1 hectare resulted in 10 000 m² of soil surface being in contact with active ingredient (*e.g.* neonicotinoids), an in-furrow treatment with granules covers only 500 m². This is further decreased to less than 60 m², *i.e.* less than 1 percent of one hectar, when choosing seed treatment technology. (Source: Bayer CropScience).

and forms a so-called *protective zone* of ai against insects or fungal pathogens in the soil around the seed grain. Finally, the growing plant transports the ai from the roots to the aerial parts of the plant and uniformly in its tissues of the upper leaves. In this way, the growing plant is protected "from the inside out" against pests and pathogens, which is also of benefit to consumers by making their food safer (see Figure 4.6).

Today, seed treatment is a standard technique used in many crops from high-grade seed, which involve intensive production methods, such as cotton, sugar beet, rice, wheat, barley, maize and market gardening. It offers farmers many advantages such as fewer spray applications, a reduction in agricultural machinery and maintenance cost, and considerable time saving, all of which help to increase income.

New opportunities for an innovative and sustainable agriculture have been opened up by the use of modern seed treatment application with the highly systemic neonicotinoid insecticides.[90] The application amount (g ai ha⁻¹) used per unit area is thereby reduced remarkably, as shown for neonicotinoids in corn (see Figure 4.7; spray > granules > seed treatment).[91]

Today, the three plant systemic neonicotinoids, imidacloprid (Gaucho), thiamethoxam (Cruiser) and clothianidin (Poncho), are widely used in various FS formulations (FS = flowable concentrate) for seed treatment applications in different crops, such as cotton, corn, ccreals, sugar beet, oilseed rape *etc.* (see Table 4.4).

For example, as a corn seed treatment, clothianidin (Poncho) protects the young plants against early-season pests (soil and leaf pests), especially wire-worms (*Agriotes* spp.) and hemipterans.[92] In addition, chlothianidin is very

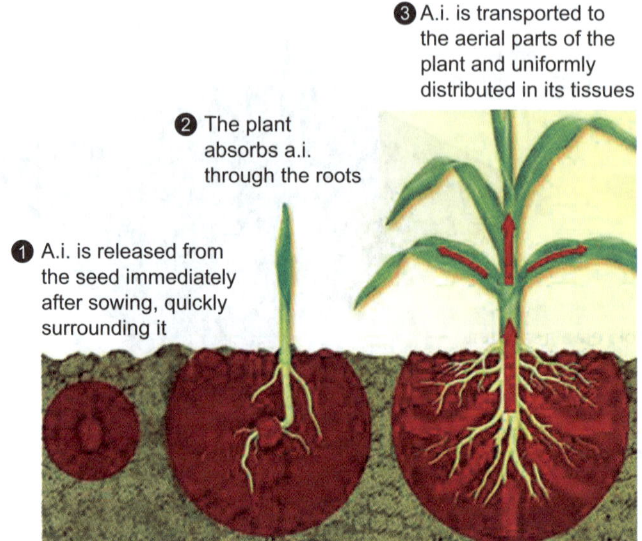

Figure 4.6   Formation of a protected zone in: (1) the soil, (2) root uptake and (3) systemic translocation, after seed treatment. Active ingredient (a neonicotinoid) can be applied in combination with further active substance classes (*e.g.* fungicides, other insecticides) in one product in order to create synergies. (Source: Bayer CropScience).

effective against different species of *Diabrotica* (corn rootworm), such as the western (*Diabrotica virgifera virgifera*), northern (*D. barberi*), southern (*D. undecimpunctata*) and Mexican (*D. virgifera zeae*) corn root worm.

Generally, larvae feed on primary and secondary corn roots. Excessive loss of root tissues from larval feeding can cause instability of the corn plants which results in lodging. Massive lodging reduces the harvest efficiency and, therefore, severe losses in yields. Feeding damage on roots will also reduce water and nutrient uptake, root and plant growth, and ultimately yield, especially under favorable dry soil conditions.[92]

The physico-chemical properties of clothianidin are theoretically very close to ideal, *e.g.* water solubility (0.327 g l^{-1}) and lipophilicity ($\log P_{ow} = 0.7$) (see section 4.3.3; Table 4.2) and are in a moderate range which allow a uniform distribution within the rooting zone of corn establishing a halo around the rhizosphere. In addition, the microbial degradation of clothianidin also remains on a moderate level. Therefore, the ai is continuously available to the plant over an extended period of time at the required dosage. The presence of clothianidin, even in the youngest roots far away from the initial point of application (kernel), was demonstrated in autoradiograms of roots of six-week-old corn plants treated with ^{14}C-labelled clothianidin (see Figure 4.8).

As described earlier, uptake of the ai by insect pests takes place primarily through oral ingestion, and therefore feeding of sufficient amounts of root

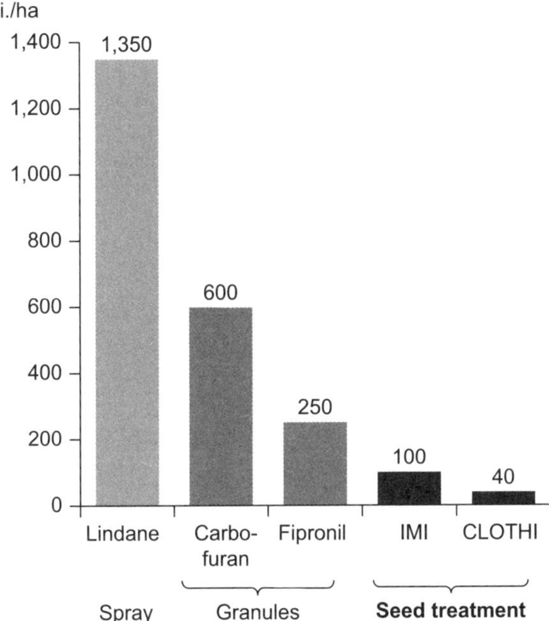

Figure 4.7 Improvements in application methods for crop protection products in maize due to the introduction of neonicotinoids. g a.i./ha (applied grams active ingredient per hectare in maize), IMI (imidacloprid) and CLOTHI (clothianidin). (Source: Bayer CropScience).

Table 4.4 Neonicotinoid insecticides used in seed treatment applications.

Neonicotinoid insecticides	Brand names	Main formulations*	Crops (selected examples)
Imidacloprid	Gaucho	FS 480 FS 600	Canola, cereals, corn, cotton, oil seed Rape, pastures, potatoes, rice, sorghum, sugar beet, sun flowers
Thiamethoxam	Cruiser	FS 350	Cotton, sun flowers, soybeans, maize, sorghum, sweet corn, rice, sugar beet, oil seed rape, cereals, wheat, potatoes
Clothianidin	Poncho	FS 600	Canola, cereals, corn, sunflowers, sugar beet

*FS (Flowable concentrate for seed treatment).

(0.1 to 0.3 μg g^{-1} root mass) or leaf tissue containing clothianidin is essential for pest intoxication.[92]

In addition, synergistic combinations of both imidacloprid (Gaucho) and clothianidin (Poncho) with members of different insecticide classes, such as pyrethroids (*e.g.* Imprimo, Chinook, Poncho Beta) or oxime carbamates like thiodicarb (*e.g.* Aeris, CropStar), have also been developed (see Table 4.5).

Bunt, smut (genus *Tilletia*) and fusarium fungi (*e.g. Fusarium moniliforme, F. proliferatum* in maize) are just a few of the most dangerous diseases of cereal

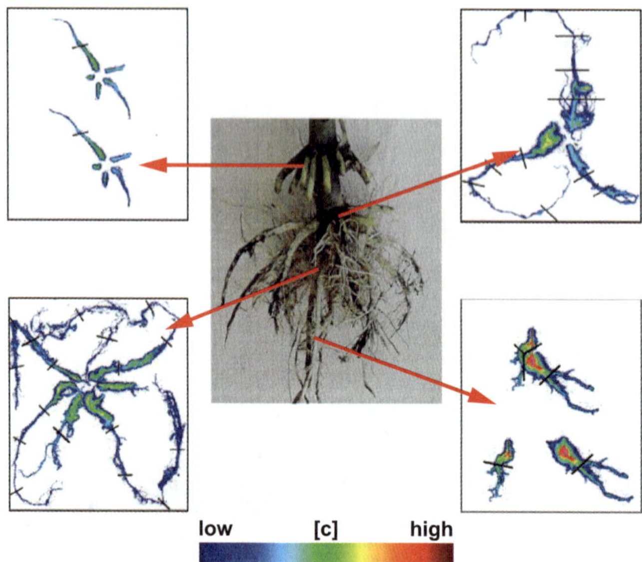

Figure 4.8   Clothianidin ensures long term control of below- and above-ground corn pests. The autoradiograms with ^{14}C-labelled clothianidin equivalents were prepared six weeks after sowing and demonstrate that even the youngest roots far distant from the kernel contain equivalents of the active ingredient. (Source: Bayer CropScience).

Table 4.5   Combination for seed treatments of neonicotinoids with insecticides and fungicides.

Neonicotinoid insecticide	*Brand names*	*Combination partners (insecticide,[a] fungicide[b]) in FS formulation**
Imidacloprid	Aeris; CropStar	Thiodicarb[a]
	Imprimo; Montur	Tefluthrin[a]
	Chinook	*beta*-Cyfluthrin[a]
	Monceren G; Prestige	Pencycuron[b]
Thiamethoxam	Cruiser Extreme 250	Azoxystrobin,[b] fludioxonil,[b] metalaxyl M,[b]
	Cruiser OSR	Metalaxyl M,[b] fludioxonil,[b]
	Helix	Fludioxonil,[b] difenoconazole,[b] metalaxyl M[b]
Clothianidin	Poncho Beta	*beta*-Cyfluthrin[a]
	Prosper	Thiram,[b] carboxin,[b] metalaxyl[b]

*FS = Flowable concentrate for seed treatment.

grains, causing yield losses of up to 60 percent. *Typhula* root pathogens (*Typhula incarnate* Lasch, *T. ishikariensis* Imai) live in the soil and are a serious threat to crops such as winter barley. Powdery mildew (*e.g. Uncinula necator*) is a windborn pathogen which threatens barley and wheat. The three neonicotinoid combination seed treatments have therefore been developed to include up to three different fungicides (four-way mixture = 4 components) as outlined in Table 4.5.

4.5.2 Seedling Box Application with Neonicotinoids

Seedling (or nursery) box application with neonicotinoid insecticides in rice gives an excellent control of rice pests, such as hopper species and rice water weevils.[93,94]

For example, imidacloprid (tradename Admire in Japan) revolutionized the nursery box application (approx. 0.5 to 1.0 g box^{-1}) as a one-shot application, which gives season long control of hoppers and viral infection, so that no field application is necessary. In addition, three nursery box formulations have been developed with the ai clothianidin and one additional component providing control of plant hoppers, water weevils and leaf beetles: Dantotsu Nursery Box Granule (1.5% ai), Windantotsu (4% ai + 4% carpropamid) and Delausdantotsu (1.5% ai + dicyclomet). The latter formulation provides additional control against rice blast disease fungus.

4.6 New Formulation Concepts for Neonicotinoids

With respect to the improved bioavailability of neonicotinoids, their low risk for the user and their reduced environmental impact, considerable efforts have been made in formulation research, which have led to novel and modern application methods.

The distribution of systemic insecticides, such as neonicotinoids, depends largely on conditions during and after application. Often, even when good delivery to the plant surface is ensured after spray application of the insecticide, there are limitations for maximum systemic performance if foliar penetration is low.

On the other hand, several solvents that have been used over the last decades have come under toxicological and subsequently regulatory pressure. As a result, some of the most common solvents are no longer available (*e.g. N*-methylpyrrolidone, *N,N*-dimethylformamide, *N,N*-dimethylacetamide, isopherone, *etc.*). Beneficial effects of adjuvants have been attributed to increased retention, modification of spray deposits or penetration and translocation of an insecticide.[95] Nowadays, new formulation concepts and optimized formulations are seen as important technical steps, which can add significant value to modern crop protection products.

4.6.1 Oil Dispersion Technology

Recently, Bayer CropScience has developed a completely new formulation technology O-TEQ (**o**il **d**ispersion, OD) for foliar application of both imidacloprid (Confidor OD 200) and thiacloprid (Biscaya OD 240).[96–98] This formulation type is defined as "a stable suspension of ai(s) in a water immiscible fluid, which may contain other dissolved ai(s), intended for dilution with water before use".[99] The main characteristics related to this aspect are outlined in Figure 4.9.

O-TEQ can be seen as a vehicle to transport systemic neonicotinoids from the formulation into the plant, from where it can be taken up by the insect. This was achieved by improving leaf penetration, particularly under suboptimal conditions for foliar uptake. Systemicity and rain fastness of neonicotinoids reach a high level not yet observed with other formulation types. In comparison with conventional <u>s</u>uspension <u>c</u>oncentrate (SC) formulation, retention (*e.g.* thiacloprid SC 480), leaf coverage and spreading of the spray deposit on the barley leaf surface are improved remarkably as shown for Biscaya O-TEQ 240 (see Figure 4.10).

Generally, the O-TEQ formulation provides minimized run-off and a better retention on the plant in the event of rainfall. Once applied to the leaf, the neonicotinoid shows a smooth spreading of the oil after the evaporation of the spray water, resulting in even coverage and distribution. Subsequently, the

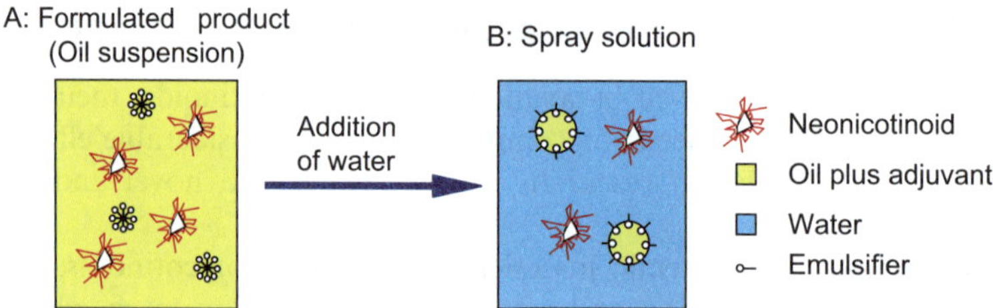

Figure 4.9   Schematic representation of an O-TEQ formulation of neonicotinoids (available for imidacloprid and thiacloprid) as: (A) a concentrate, and (B) after dilution with water. (Source: Bayer CropScience).

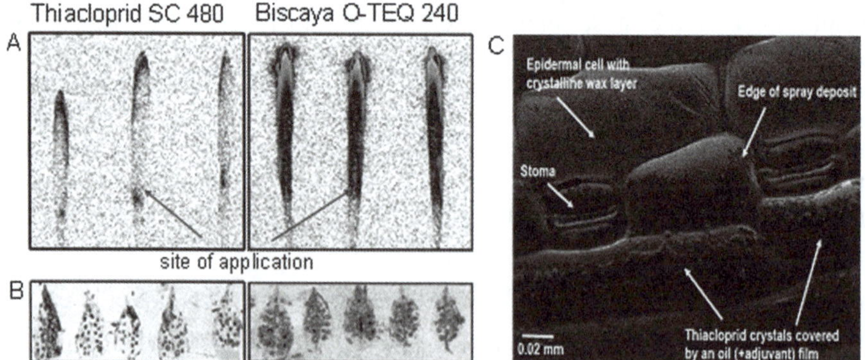

Figure 4.10   Comparison of thiacloprid formulations SC 480 and O-TEQ 240: (A) translocation of thiacloprid in barley leaves 48 h after foliar application of 0.2 g a.i. l^{-1}; (B) Uptake and translocation of thiacloprid into pear leaves after 10 mm rain 3 h after application; and (C) spray deposit of thiacloprid OD on barley leaf (electron micrograph). (Source: Bayer CropScience).

systemic neonicotinoid penetrates through the leaf cuticle and is translocated within the leaf lamina.[100,101]

The presence of the natural oil guarantees such an uptake over an extended period of time, resulting in a better, long-lasting control compared with more conventional formulation types. Furthermore, the excellent efficacy of neonicotinoids formulated with O-TEQ technology and the fact that this performance is related to an increase in the adjuvant concentration in the spray solution rather than the spray volume per hectare, means that the application of O-TEQ formulations using reduced water volumes (lower than $30\,l\,ha^{-1}$) is very reliable. This helps to offer sustainable solutions in agricultural areas where water is scarce. With the application of so-called "in-can" formulations (consumer friendly, ready-to-use), products are safer and more convenient for operators to use.

Overall, such an innovative formulation technology for neonicotinoids can help farmers particularly when unfavorable conditions prevail, allowing more flexibility and reliability of the application by reducing the impact of temperature, humidity, pH value and evaporation.

4.7 Phytotonic Effects of Neonicotinoids

Plant growth and productivity, as well as product quality, are greatly influenced by the environmental stress factors to which plants are continuously exposed.[102] Stress impairs the energy balance of crops, resulting in more energy consumption for cell repair and less energy generation for growth. The optimal growth, development and maximum yield potential are therefore considered far from that achieved in the field or greenhouse (see Figure 4.11).

In recent years, it was found that some members of the systemic neonicotinoid insecticide class, such as imidacloprid, clothianidin and thiamethoxam, show phytotonic or plant stimulating effects.

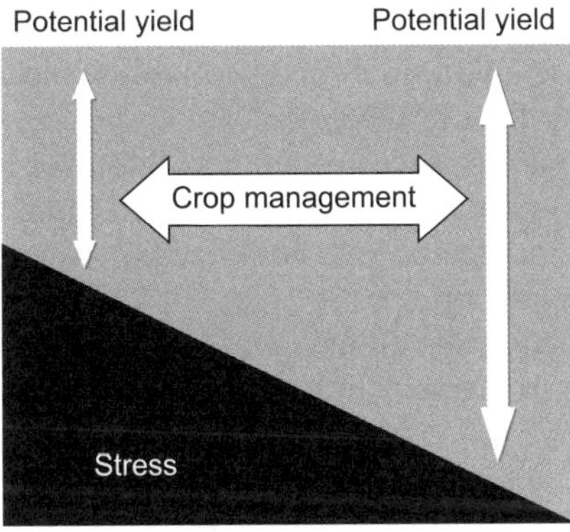

Figure 4.11 Due to multiple stress factors, crops do not reach the maximum yield potential. (Source: Bayer CropScience).

4.7.1 Overcoming Abiotic and Biotic Stress

Stress in plants can be of two types:

(1) Abiotic – arising from fluctuations in their physical or chemical environment (such as cold, heat, excessive light, elevated salt levels in soil, ozone in air, oxygen deficiency, flooding or drought, or mechanical damage); or
(2) Biotic – imposed by insect pests, weed and pathogens (fungi, viruses, bacteria).

Most yield losses, however, are attributed to abiotic stress.

The response of Trimax-treated (Trimax is an optimized imidacloprid formulation) barley to pure abiotic stress stimuli was investigated in detail in order to elucidate the underlying physicochemical and biochemical mechanisms. In this case, a significant leaf-area growth improvement following imidacloprid soil application could be shown after short-term drought stress. Plants from these trials were analysed at different elapsed time intervals using DNA microarrays (barley chip consisting of 400 000 expressed sequence tags with information on 22 840 barley genes), a tool for profiling gene expression in plants. After imidacloprid treatment, results from the DNA microarray experiments showed three different plant reactions to drought stress:

(1) The expression level of drought-stress marker genes in barley was significantly delayed, suggesting a mitigation of drought stress;
(2) Photosynthesis-related genes were simultaneously expressed at a higher level (energy production was ongoing), whereas in untreated plants photosynthesis declined more rapidly; and
(3) In contrast to non-treated plants, numerous pathogenesis-related proteins were found to be overexpressed, explaining field observations of synergistic fungicidal and bactericidal effects.

Imidacloprid treatment led to a considerable reduction in yield losses by drought stress compared with other neonicotinoids, such as thiamethoxam, as established in a pepper field in Georgia.[102]

Since most field-grown cotton is sensitive to abiotic or biotic stress at various times throughout the seasons, the application of neonicotinoids may have a widespread benefit if upcoming stress could be timely detected in field crops to allow for targeted applications.

In 2004, water-deficit field studies confirmed the potential of Trimax to moderate water stress in plants with an average lint yield increase in cotton of 10 percent.[103,104]

Later on, in field trials with broad strip side-by-side aerial applications of Trimax Pro/Provado (imidacloprid) and Centric (thiamethoxam), the abiotic stress (drought, salinity and hypoxia caused by flood irrigation) reducing potential of both neonicotinoids and their impact on lint yield were elucidated

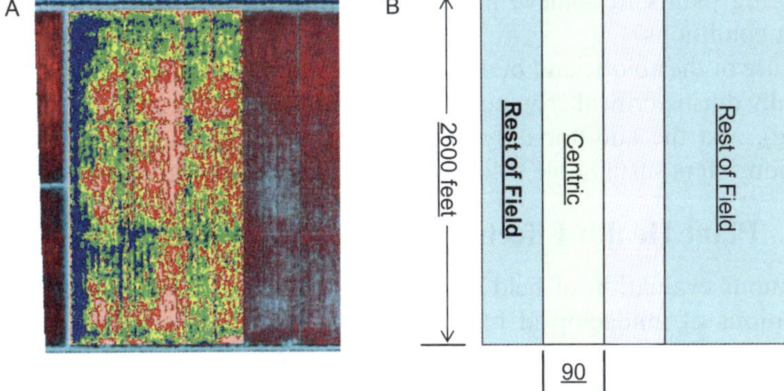

Figure 4.12 Field trials with broad strip side-by-side aerial applications of Trimax Pro/Provado (imidacloprid) and Centric (thiamethoxam); assessment of plant vigor by remote sensing of Near Infrared (NIR) images and Normalized Difference Vegetation Index (NDVI) images analysis: (A) typical NDVI variability (vigor legend: blue = very high, green = high, yellow = medium, red = low, pink = very low); and (B) typical field plot design (adapted from Zelinski and Thielert, 2008).[105]

using remote sensing of reflectance of **N**ear **I**nfra **R**ed (NIR) images and **N**ormalized **D**ifference **V**egetation **I**ndex (NDVI) images analysis (see Figure 4.12).[105]

The NIR (800 nm) and red light (660 nm) can be used to measure plant stress levels. So-called *healthy* cotton plants reflect very little red light due to absorption by their chlorophyll. Therefore, the normalized ratio of NIR to red light; can be a used as a measure of plant stress.

Data suggest a superior stress mitigation performance of Trimax Pro/Provado in cotton over treatments by Centric. This statistical design allows users to investigate plant growth effects across a range of stress levels both quickly and accurately.

All results clearly indicate that, in addition to its insecticidal activity, a *stress shield MoA* (mode of action) of imidacloprid supports plants in moderating the effects of abiotic and biotic stress by its interaction with the NAD-salvage pathway. As a trigger of these effects of imidacloprid, its major metabolite, 6-chloro-nicotinic acid (6-CNA; see section 4.3.1; Figure 4.2), is discussed as a systemic plant inducer which possibly causes physiological changes in the plant that result in stress protection.

On the other hand, thiamethoxam-treated plants are more tolerant toward abiotic stress factors. They can grow more vigorously under suboptimal conditions. However, this vigor effect leading to abiotic stress mitigation is only a sub-set of the whole stress shield (mitigation of abiotic and biotic stress). The bioisosteric metabolite, 5-chloro-1,3-thiazole carboxylic acid (2-CTA; see section 4.3.1; Figure 4.2), could be a trigger of these effects as well.

Finally, the interaction of neonicotinoids like imidacloprid (Trimax) with plants to moderate abiotic and biotic stress leads to the conclusion of a second MoA on top of the well known direct neurotoxic MoA against insect pests, supporting plants to achieve higher yields and better quality under adverse growth conditions.

The use of the abiotic and biotic stress mitigation potential of neonicotinoids, especially during critical crop periods (*e.g.* emergence, transplanting, flowering, fruiting), and the additional synergies with fungicides/bactericides and virus reduction offers sustainable solutions in modern crop protection.

4.7.2 Plant Health Effects

Continuous evaluation of field trials data indicates that multiple foliar spray applications of imidacloprid results in improved health and increased plant growth even in the absence of damaging pest species.

The formation of more of the plant's own substances (see section 4.7.1; pathogenesis-related proteins) associated with its own defense mechanism against fungal diseases could explain the outstanding growth and health of tobacco plants treated with imidacloprid in the nursery box prior to transplanting into soil heavily infested with *Phytophthora nicotianae*. Initial evidence for such an effect by the induction of salicylate-associated plant defense responses has been recently published for *Arabidopsis thaliana*. It was shown that the neonicotinoids, imidacloprid and clothianidin, were similar to the phytohormone, salicylic acid, in inducing systemic acquired resistance in plants.[106]

4.8   Neonicotinoids as Resistance Management Tools

Neonicotinoid insecticides all belong to the same mode of action class, and cross-resistance to the established chemistry of other modes of action such as pyrethroids, organophosphates, carbamates or fiproles is not reported. Neonicotinoids are therefore considered to be useful tools in resistance management strategies. Even though neonicotinoid insecticides have been used for a prolonged period of time (since 1991), there has been little resistance development in some well-known high risk pests, such as *Bemisia tabaci* Gennadius (Homoptera: Aleyrodidae) and *Myzus persicae* Sulzer (Homoptera: Aphididae), and in many cases reported resistance is still manageable and/or geographically localized.[107] In the last five years, however, there have been an increasing number of reports concerning resistance development in some important target pests, albeit still regionally restricted in many cases. The extensive use of neonicotinoids worldwide, unless carefully regulated and coordinated, is likely to increase exposure to this important class of chemistry and enhance conditions favouring resistant phenotypes.[107] After 20 years of use, insect pests such as whiteflies, *Bemisia tabaci* (Gennadius)[107] and *Trialeurodes vaporariorum* (Westwood);[108] the brown planthopper, *Nilaparvata lugens* (Stål);[109] the peach potato aphid, *Myzus persicae* (Sulzer);[110] the Colorado potato beetle, *Leptinotarsa decemlineata* (Say);[111] and a few others like the mango leafhopper, *Idioscopus clypealis* (Lethierry), have developed resistance to neonicotinoids in some parts of the world.[36]

Although resistance reports are increasing in established markets, neonicotinoids still find uses where they are considered as invaluable new chemical options

Figure 4.13 Border application of thiacloprid (Biscaya OD 240) in winter oilseed rape for the control of pyrethroid resistant pollen beetles, *Meligethes aeneus.* The inner part of the field was treated by a pyrethroid. The lack of efficacy is seen by the prevention of flowering in many parts due to pollen beetle feeding on buds. (Source: Bayer CropScience).

for resistance management purposes. One such example is the recent introduction of thiacloprid and acetamiprid in European winter oilseed rape for the control of pyrethroid-resistant pollen beetles, *Meligethes aeneus.*[112] Figure 4.13 demonstrates the efficacy of thiacloprid applied to the margin of a winter oilseed rape field where the rest of the field was treated with a pyrethroid insecticide.

Thiacloprid shows excellent control of resistant beetles, whereas in most parts of the pyrethroid-treated crop, flowering was prevented due to pollen beetle feeding. The example shown demonstrates that neonicotinoids will remain important chemical options for pest control and management in the future.

Acknowledgements

The material covered by this article is based on the product development and scientific support of many colleagues in Bayer CropScience AG, as well as research colleagues in crop protection worldwide. We are extremely grateful to Dr. Christian Maus for his contribution regarding pollinator safety.

References

1. P. Jeschke and R. Nauen, *Pest Manag. Sci.*, 2008, **64**, 1084.
2. K. Matsuda, M. Shimomura, M. Ihara, M. Akamatsu and D. B. Sattelle, *Biosci. Biotechnol. Biochem.*, 2005, **69**, 1442.

3. M. Tomizawa and J. E. Casida, *Annu. Rev. Entomol.*, 2003, **48**, 339.
4. M. Okamoto, T. Kita, H. Okuda, T. Tanaka and T. Nakashima, *Pharmacol Toxicol.*, 1994, **75**, 1.
5. I. B. Karaconji, *Arh. Hig. Toksikol.*, 2005, **56**, 363.
6. K. Matsuda, S. Kanaoka, M. Akamatsu and D. B. Sattelle, *Mol. Pharmacol.*, 2009, **76**, 1.
7. M. Ihara, T. Okajima, A. Yamashita, T. Oda, K. Hirata, H. Nishiwaki, T. Morimoto, M. Akamatsu, Y. Ashikawa, S. Kuroda, R. Mega, S. Kuramitsu, D. B. Sattelle and K. Matsuda, *Invertebrate Neurosci.*, 2008, **8**, 71.
8. T. T. Talley, M. Harel, R. E. Hibbs, Z. Radic, M. Tomizawa, J. E. Casida and P. Taylor, *Proc. Natl. Acad. Sci. U. S. A.*, 2008, **105**, 7606.
9. A. Elbert, B. Becker, J. Hartwig and C. Erdelen, *Pflanzenschutz-Nachrichten Bayer*, 1991, **44**, 113.
10. P. Jeschke, K. Moriya, R. Lantzsch, H. Seifert, W. Lindner, K. Jelich, A. Göhrt, M. E. Beck and W. Etzel, *Pflanzenschutz-Nachrichten Bayer*, 2001, **54**, 147.
11. R. Senn, D. Hofer, T. Hoppe, M. Angst, P. Wyss, F. Brandl, P. Maien-fisch, L. Zang and S. White, *Proc. Brighton Crop Prot. Conf. – Pest and Diseases*, BCPC, Farnham, Surrey, UK, 1998, 27.
12. I. Minamida, K. Iwanaga, T. Tabuchi, I. Aoki, T. Fusaka, H. Ishizuka and T. Okauchi, *J. Pestic. Sci.*, 1993, **18**, 41.
13. H. Takahashi, J. Mitsui, N. Takakusa, M. Matsuda, H. Yoneda, J. Suzuki, K. Ishimitsu and T. Kishimoto, *Proc. Brighton Crop Prot. Conf. – Pest and Diseases*, BCPC, Farnham, Surrey, UK, 1992, 89.
14. Y. Ohkawara, A. Akayama, K. Matsuda and W. Andersch, *Proc. Brighton Crop Prot. Conf. – Pest and Diseases*, BCPC, Farnham, Surrey, UK, 2002, 51.
15. K. Kodaka, K. Kinoshita, T. Wakita, N. Kawahara and N. Yasui, *Proc. Brighton Crop Prot. Conf. – Pest and Diseases*, BCPC, Farnham, Surrey, UK, 1998, 21.
16. R. Nauen, U. Ebbinghaus-Kintscher, A. Elbert, P. Jeschke and K. Tietjen, in *Biochemical Sites of Insecticide Action and Resistance*, ed. I. Ishaaya, Springer, New York, NY, 2001, p. 77.
17. P. Jeschke and R. Nauen, in *Comprehensive Molecular Insect Science*, ed. L. I. Gilbert, L. Iatrou and S. S. Gill, Elsevier, Oxford, UK, 2005, vol. 5, p. 53.
18. P. Jeschke, in *Modern Crop Protection Compounds*, ed. W. Krämer and U. Schirmer, Wiley-VCH, Weinheim, Germany, 2007, p. 958.
19. P. Jeschke, R. Nauen, M. Schindler and A. Elbert, *J. Agric. Food Chem.*, 2010; DOI:10.1021/jf101303g.
20. S. Kagabu, in *Chemistry of Crop Protection: Progress and Prospects in Science and Regulation*, ed. G. Voss and G. Ramos, Wiley-VCH, New York, USA, 2003, p. 193.
21. P. Jeschke, M. Schindler and M. E. Beck, *Proc. Brighton Crop Prot. Conf. – Pest and Diseases*, BCPC, Farnham, Surrey, UK, 2002, 137.

22. R. Nauen, U. Ebbinghaus-Kintscher, V. L. Salgado and M. Kaussmann, *Pestic. Biochem. Physiol.*, 2003, **76**, 55.
23. P. Jeschke and R. Nauen, in *Synthesis and Chemistry of Agrochemicals VII*, ed. J. W. Lyga and G. Theodoridis, ACS, Washington, DC, 2007, p. 51.
24. P. T. Anastas and J. C. Warner, *Green Chemistry: Theory and Practice*, Oxford University Press, New York, USA, 1998, p. 30.
25. S. K. Potluri, A. R. Ramulu and M. Pardhasaradhi, *Synth. Commun.*, 2005, **35**, 971.
26. K. Yamagata and S. Seko, *PCT Int. Appl. WO 2007/105793* A1 (Sumitomo Chem. Comp., Ltd. Japan).
27. H. Uneme, *Abstracts of Papers*, 239th ACS National Meeting, San Francisco, CA, United States, March 21–25, 2010, AGRO-116, Publisher: American Chemical Society, Washington D.C., USA.
28. K. Yasuyuki, H. Uneme and I. Minamida, *Eur. Pat. Appl.* EP 452782 A1, 1991 (Takeda Chemical Industries, Ltd., Japan).
29. S. Kagabu and S. Medej, *Biosci. Biotechnol. Biochem.*, 1995, **59**, 980.
30. S. Kagabu, *Rev. Toxicol.*, 1997, **1**, 75.
31. S. Kagabu and T. Akagi, *J. Pestic. Sci.*, 1997, **22**, 84.
32. S. Kagabu, *J. Pestic. Sci.*, 1996, **21**, 237.
33. G. Briggs, R. H. Bromilow and A. A. Evans, *Pestic. Sci.*, 1982, **13**, 495.
34. A. Buchholz and R. Nauen, *Pest Manag. Sci.*, 2002, **58**, 10.
35. A. Elbert and R. Nauen, in *Insect Pest Management, Field and Protection Crops*, ed. A. R. Horowitz and I. Ishaaya, Springer, Berlin-Heidelberg-New York, 2004, p. 29.
36. A. Elbert, M. Haas, B. Springer, W. Thielert and R. Nauen, *Pest Manag. Sci.*, 2008, **64**, 1099.
37. R. Nauen, B. Koob and A. Elbert, *Entomol. Exp. Appl.*, 1998, **88**, 287.
38. H. J. Knaust and H. M. Poehling, *Pflanzenschutz-Nachrichten Bayer*, 1992, **45**, 381.
39. P. J. Heatherington and R. H. Meredith, *Pflanzenschutz-Nachrichten Bayer* (German edition), 1992, **45**, 491.
40. A. M. Dewar, *Pflanzenschutz-Nachrichten Bayer* (German edition), 1992, **45**, 423.
41. G. M. Tatchell, *Pflanzenschutz-Nachrichten Bayer* (German edition), 1992, **45**, 409.
42. G. M. Tatchell and R. T. Plumb, *Pflanzenschutz-Nachrichten Bayer* (German edition), 1992, **45**, 443.
43. D. J. Bluett and A. Birch, *Pflanzenschutz-Nachrichten Bayer* (German edition), 1992, **45**, 455.
44. R. D. Rudolph and D. W. Rogers, *Pflanzenschutz-Nachrichten Bayer* (English edition), 2001, **54**, 311.
45. K. Epperlein and H. W. Schmidt, *Pflanzenschutz-Nachrichten Bayer* (German edition), 2001, **54**, 369.
46. D. Hernández, V. Mansanét and J. M. Puiggrós Jové, *Pflanzenschutz-Nachrichten Bayer* (German edition), 1999, **52**, 374.

47. R. Schmuck, *Pflanzenschutz-Nachrichten Bayer*, 2001, **54**, 161.

48. T. Iwasa, N. Matoyama, J. T. Ambrose and M. Roe, *Crop Protection*, 2004, **23**, 371.

49. P. Jeschke and R. Nauen, in *Comprehensive Molecular Insect Science*, ed. L. I. Gilbert, K. Iatrou and S. S. Gill, Elsevier BV, Oxford, UK, 2010, p. 61.

50. Ch. Maus, G. Curé and R. Schmuck, *Bull. Insectol.*, 2003, **56**, 51.

51. R. Schmuck and J. Keppler, *Pflanzenschutz-Nachrichten Bayer*, 2003, **56**, 26.

52. R. Schmuck, R. Schöning and R. Sur, in *Das "Bienensterben" im Winter 2002/2003 in Deutschland – Zum Stand der wissenschaftlichen Erkenntnisse*, ed. R. Forster, E. Bode and D. Brasse, Bundesamt für Verbraucherschutz und Lebensmittelsicherheit (BVL), Braunschweig, Germany, 2005, p. 68.

53. D. Ch. Cutler and C. D. Scott-Dupree, *J. Econ. Entomol.*, 2007, **100**, 765.

54. A. Choudhary and D. C. Sharma, *Environ. Monit. Assess.*, 2008, **144**, 143.

55. A. Nikolakis, A. Chapple, R. Friessleben, P. Neumann, T. Schad, R. Schmuck, H. F. Schnier, H. J. Schnorbach and Ch. Maus, *Julius-Kühn-Archiv*, 2009, **423**, 132.

56. Ph. Vermandere, *HiveLights*, February 2003, 24.

57. M. P. Halm, A. Rortais, G. Arnold, J. N. Taséi and S. Rault, *Environ. Sci. Technol.*, 2006, **40**, 2448.

58. S. Maini, P. Medrzycki and C. Porrini, *Bull. Insectol.*, 2010, **63**, 153.

59. K. Wallner, A. Schur and B. Stürz, *Apidologie*, 1999, **30**, 422.

60. R. Schmuck, *Pflanzenschutz-Nachrichten Bayer*, 1999, **52**, 257.

61. J. N. Tasei, G. Ripault and E. Rivault, *J. Econ. Entomol.*, 2001, **94**, 623.

62. G. Curé, H. W. Schmidt and R. Schmuck, Avignon (France), September 7th to 9th, 1999: *Les Colloques INRA (Paris)* 2001, **98**, 49.

63. T. Stadler, D. Martínez Ginés and M. Buteler, *Bull. Insectol.*, 2003, **56**, 77.

64. M. T. Franklin, M. L. Winston and L. A. Morandin, *J. Econ. Entomol.*, 2004, **97**, 369.

65. B. K. Nguyen, C. Saegerman, C. Pirard, J. Mignon, J. Widart, B. Thironet, F. J. Verheggen, D. Berkvens, E. De Pauw and E. Haubruge, *J. Econ. Entomol.*, 2009, **102**, 616.

66. O. Boecking, *ADIZ/db/IF 8/2003*, 2003, 9.

67. Ch. Otten, *ADIZ/db/IF 8/2003*, 2003, 6.

68. M. P. Chauzat, P. Carpentier, A. C. Martel, S. Bourgeard, N. Cougoule, Ph. Porta, J. Lachaize, F. Madec, M. Aubert and J. P. Faucon, *Environ. Entomol.*, 2009, **38**, 514.

69. AFSSA, *Weakening, Collapse and Mortality of Bee Colonies,* 2009, http://www.afssa.fr/Documents/SANT-Ra-MortaliteAbeillesEN.pdf.

70. P. Hendrikx, M. P. Chauzat, M. Debin, P. Neumann, I. Fries, W. Ritter, M. Brown, F. Mutinelli, Y. Le Conte and A. Gregorc, 2009, https://secure.fera.defra.gov.uk/beebase/downloadNews.cfm?id=54.

71. M. P. Chauzat, C. C. Martel, S. Zeggane, P. Drajnudel, F. Schurr, M. C. Clément, M. Ribière-Chabert, M. Aubert and J. P. Faucon, *J. Apicult. Res.*, 2010, **49**, 40.

72. E. Genersch, *Appl. Microbiol Biotech.*, 2010, **87**, 87.

73. F. L. W. Ratnieks and N. L. Carreck, *Science*, 2010, **327**, 151.

74. E. Genersch, W. von der Ohe, H. Kaatz, A. Schroeder, Ch. Otten, R. Büchler, S. Berg, W. Ritter, W. Mühlen, S. Gisder, M. Meixner, G. Liebig and P. Rosenkranz, *Apidologie*, 2010, **41**, 332.

75. D. Van Engelsdorp and M. D. Meixner, *J. Invertebrate Pathol.*, 2010, **103**, 80.

76. P. Neumann and N. L. Carreck, *J. Apicult. Res.*, 2010, **49**, 1.

77. M. P. Chauzat, P. Carpentier, F. Madec, S. Bougeard, N. Cougoule, P. Drajnudel, M. C. Clément, M. Aubert and J. P. Faucon, *J. Apicult. Res.*, 2010, **49**, 31.

78. F. Vejsnaes, S. L. Nielsen and P. Kryger, *J. Apicult. Res.*, 2010, **49**, 109.

79. B. Dahle, *J. Apicult. Res.*, 2010, **49**, 124.

80. J. D. Charrière and P. Neumann, *J. Apicult. Res.*, 2010, **49**, 132.

81. J. Pistorius, G. Bischoff, U. Heimbach and M. Stähler, *Julius-Kühn-Archiv*, 2009, **423**, 118.

82. R. Forster, *Julius-Kühn-Archiv*, 2009, **423**, 126.

83. R. Friessleben, T. Schad, R. Schmuck, H. F. Schnier, R. Schöning and A. Nikolakis, *Aspects Appl. Biol.*, 2010, **99**, 265.

84. P. Balsari, M. Manzone, P. Marucco and M. Tamagnone, *Aspects Appl. Biol.*, 2010, **99**, 297.

85. A. Herbst, D. Rautmann, H. J. Osteroth, H. J. Wehmann and H. Ganzelmeier, *Aspects Appl. Biol.*, 2010, **99**, 265.

86. V. Girolami, L. Mazzon, A. Squartini, N. Mori, M. Marzaro, A. Di Bernardo, M. Greatti, C. Giorio and A. Tapparo, *J. Econ. Entomol.*, 2009, **102**, 1808.

87. J. Keppler, R. Becker, R. Spatz and F. Dechet, *Julius-Kühn-Archiv*, 2010, **428**, 133.

88. R. S. Leal, *Pflanzenschutz-Nachrichten Bayer* (German edition), 2001, **54**, 337.

89. D. Kreutzweiser, K. Good, D. Chartrand, T. Scarr and D. Thompson, *Ecotox. Environm. Safety*, 2007, **68**, 315.

90. R. Altmann, *Pflanzenschutz-Nachrichten Bayer* (German edition), 1991, **44**, 159.

91. R. Altmann, *Pflanzenschutz-Nachrichten Bayer* (English edition), 2003, **56**, 102.

92. W. Andersch and M. Schwarz, *Pflanzenschutz-Nachrichten Bayer* (English edition), 2003, **56**, 147.

93. K. Iwaya and S. Tsuboi, *Pflanzenschutz-Nachrichten Bayer* (German edition), 1992, **45**, 197.

94. A. Elbert, R. Nauen and W. Leicht, in *Insecticides with Novel Modes of Action, Mechanism and Application*, ed. I. Ishaaya and D. Degheele, Springer, New York, NY, USA, 1998, p. 50.

95. J. Green, in *Proceedings of the Sixth International Symposium on Adjuvants for Agrochemicals*, ed. H. de Ruiter, Amsterdam, The Netherlands, 2001, p. 179.

96. R. Vermeer and P. Baur, *Pflanzenschutz-Nachrichten Bayer* (English edition), 2007, **60**, 7.

97. M. Haas and J. Kühnhold, *Pflanzenschutz-Nachrichten Bayer* (English edition), 2007, **60**, 59.

98. W. Thielert and H. Hungenberg, *Pflanzenschutz-Nachrichten Bayer* (English edition), 2007, **60**, 43.

99. Food and Agriculture Organization, *CropLife International Codes for Technical and Formulated Pesticides. Manual on Development and Use of FAO and WHO Specifications for Pesticides*, 1st edn, Appendix E, 2002, p. 150; http://www.fao.org/docrep/007/y4353e/y4353e0i.htm.

100. P. Baur, R. Arnold, S. Giessler, P. Mansour and R. Vermeer, *Pflanzenschutz-Nachrichten Bayer* (English edition), 2007, **60**, 27.

101. J. Van Bömmel and Ch. Nagel, *Pflanzenschutz-Nachrichten Bayer* (English edition), 2007, **60**, 5.

102. W. Thielert, *Pflanzenschutz-Nachrichten Bayer* (English edition), 2006, **59**, 73.

103. D. M. Oosterhuis and R. S. Brown, *Proceedings – Beltwide Cotton Conferences*, 2004, 2225.

104. R. S. Brown, D. M. Oosterhuis and E. Gonias, *Proceedings – Beltwide Cotton Conferences*, 2004, 2231.

105. L. Zelinski and W. Thielert, *Proceedings – Beltwide Cotton Conferences*, 2008, 1.

106. K. A. Ford, J. E. Casida, D. Chandran, A. G. Gulevich, R. A. Okrent, K. A. Durkin, R. Sarpong, E. M. Bunnelle and M. C. Wildermuth, *Proc. Natl. Acad. Sci. U. S. A.*, 2010, **107**, 17527.

107. R. Nauen and I. Denholm, *Arch. Insect Biochem. Physiol.*, 2005, **58**, 200.

108. K. Gorman, G. Devine, J. Bennison, P. Coussons, N. Punchard and I. Denholm, *Pest Manag. Sci.*, 2007, **63**, 555.

109. K. Gorman, Z. Liu, I. Denholm, K. U. Brüggen and R. Nauen, *Pest Manag. Sci.*, 2008, **64**, 1122.

110. A. M. Puinean, S. P. Foster, L. Oliphant, I. Denholm, L. M. Field, N. S. Millar, M. S. Williamson and Ch. Bass, *PloS Genetics*, 2010, **6**, 1.

111. D. Mota-Sanchez, R. M. Hollingworth, E. J. Grafius and D. D. Moyer, *Pest Manag. Sci.*, 2006, **62**, 30.

112. Ch. Zimmer and R. Nauen, *Pest Manag. Sci.*, 2011, *in press*.

CHAPTER 5

The Spinosyn Insecticides

J. E. DRIPPS,[1*] R. E. BOUCHER,[1] A. CHLORIDIS,[2]
C. B. CLEVELAND,[1] C. V. DEAMICIS,[1] L. E. GOMEZ,[1]
D. L. PAROONAGIAN,[1] L. A. PAVAN,[3] T. C. SPARKS[1]
AND G. B. WATSON[1]

[1] Dow AgroSciences LLC, 9330 Zionsville Rd., Indianapolis, IN 46268,
USA; [2] Dow AgroSciences LLC, 2 Kalymnou St., 55133 Thessaloniki, Greece;
[3] Dow AgroSciences LLC, Caixa Postal 226, Mogi Mirim, SP, Brazil,
CEP 13800-970

5.1 Introduction

The spinosyns are a unique family of insecticidal secondary metabolites produced by two relatively rare species of soil-dwelling bacteria, *Sacchar-*
opolyspora spinosa and *Saccharopolyspora* sp. NRRL30141, a new species
tentatively named *Saccharopolyspora pogona* (see Figure 5.1).[1–3] Both species
are aerobic, gram-positive, non-motile, spore-forming, filamentous actinomy-
cetes.[3,4] The term *spinosyns* derives from the species name, "spinosa".

Two insecticide active ingredients, spinosad and spinetoram, have been
developed from the spinosyn family of secondary metabolites. When spinosad
received its first crop registration in the United States in 1997, it was among the
first insecticides recognized as "reduced risk" by the United States Environmental Protection Agency (US EPA).[5] At that time, spinosad provided a
unique combination of high efficacy against a wide range of insect pest species,
low toxicity to mammals and other nontarget organisms, and minimal persistence in the environment. As a result of these attributes, spinosad received a

RSC Green Chemistry No. 11
Green Trends in Insect Control
Edited by Óscar López and José G. Fernández-Bolaños
© Royal Society of Chemistry 2011
Published by the Royal Society of Chemistry, www.rsc.org

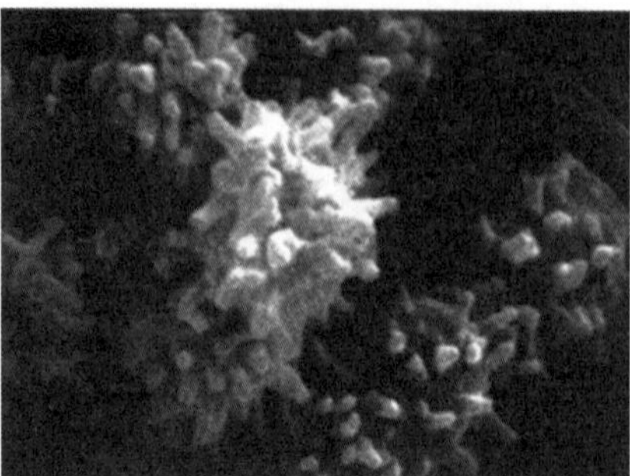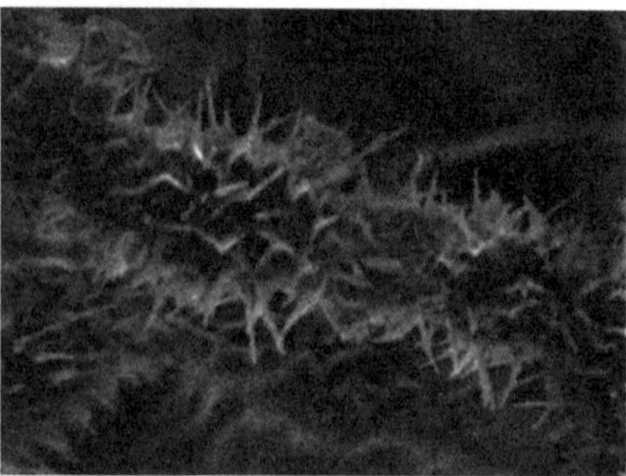

Figure 5.1 Scanning electron micrographs of sporulating *Saccharopolyspora spinosa* (left) and *Saccharopolyspora pogona* (right). (Magnification: 20 000×).

Presidential Green Chemistry Challenge Award in 1999, and has since become a major tool in the management of arthropod pests in agricultural and horticultural crops, animal production and public health.[6] Spinetoram, the second active ingredient in the spinosyn class, received its first registration in 2007. Spinetoram has attributes not possessed by spinosad that enable it to control a wider range of pests, especially in tree fruits, while retaining toxicological and environmental properties similar to spinosad.[7,8] For these reasons, spinetoram received a Presidential Green Chemistry Challenge Award in 2008.

The spinosyns have become a major insecticide group; products containing spinosad or spinetoram are registered in over 80 countries and on more that 250 crops, with several registrations for non-crop uses. In the time since the discovery of the spinosyns in 1991, a large volume of research has been published, including several significant review articles focusing on the chemistry and biochemistry of the spinosyns.[9–13] The purpose of this chapter is to take a broader view and summarize salient information across the diverse disciplines of chemistry, biochemistry, biology, regulatory sciences, manufacturing and commerce to provide a comprehensive understanding of the spinosyns as highly effective green chemicals for insect pest management.

5.2   Spinosyn Chemical Structure

The spinosyns are generally characterized by a unique tetracyclic ring system (a 5,6,5-*cis*-anti-*trans*-tricyclic ring system fused to a 12-membered macrocyclic lactone).[13] In most cases, this tetracycle is substituted with a neutral sugar at the C9 position and an amino sugar at the C17 position (see Figure 5.2).[13]

5.2.1   Naturally Occurring Spinosyns

Saccharopolyspora spinosa produces 23 spinosyns that are designated with letters reflecting the order in which these were identified.[12] For the spinosyns produced by *S. spinosa*, the neutral sugar at C9 is 2,3,4-tri-*O*-methyl-α-ʟ-rhamnose and

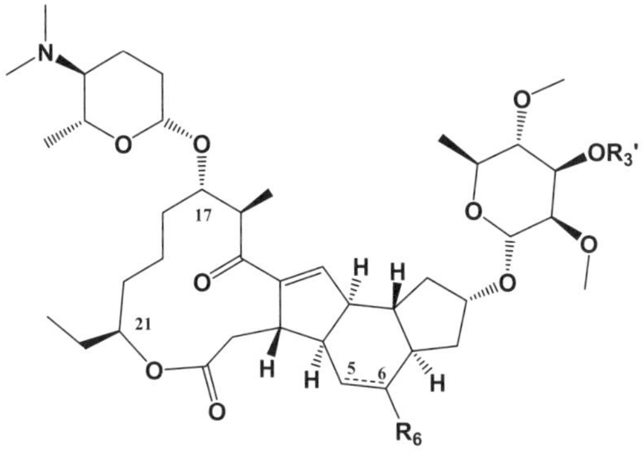

Figure 5.2 Spinosyn chemical structure.

Structure number	Name	$R_6{}^a$	5,6 bond	$R_{3'}{}^a$
5.1	Spinosyn A[b]	H	double	Me
5.2	Spinosyn D[c]	Me	double	Me
5.3	Spinosyn J	H	double	H
5.4	Spinosyn L	Me	double	H
5.5	3'-*O*-ethyl-5,6-dihydro spinosyn J[d]	H	single	Et
5.6	3'-*O*-ethyl spinosyn L[e]	Me	double	Et

[a]H = hydrogen, Me = methyl, Et = ethyl.
[b]Major component of spinosad.
[c]Minor component of spinosad.
[d]Major component of spinetoram, also identified as spinetoram-J and as XDE-175-J.
[e]Minor component of spinetoram, also identified as spinetoram-L and as XDE-175-L.

the amino sugar at C17 is generally β-D-forosamine. The 23 spinosyns produced by *S. spinosa* differ primarily in the presence or absence of methyl groups at eight positions (three positions on the tetracycle, three positions on the 2',3',4'-tri-*O*-methyl rhamnose and two positions on the forosamine). *Spinosad* is the ISO common name for the mixture of spinosyns that is extracted from the fermentation broth of *S. spinosa*. Spinosyn A (see Figure 5.2; structure 5.1) is produced in the greatest quantity by *S. spinosa*, followed by spinosyn D (see Figure 5.2; structure 5.2), which differs from spinosyn A by having a methyl group instead of a hydrogen atom at position C6 on the macrolide. The other 21 spinosyns are produced in very small quantities. Spinosyns A and D have the highest levels of insecticidal activity among the spinosyns produced by *S. spinosa*, the others are generally much less active.[12,14]

Saccharopolyspora pogona produces a similar series of secondary metabolites.[3] These compounds are referred to as butenyl-spinosyns because a majority of the 31 spinosyns produced by *S. pogona* have a but-1-enyl group at the C21 position on the tetracycle, instead of the ethyl group typically found in spinosyns produced by *S. spinosa*. The spinosyns produced by *S. pogona* have greater structural diversity than the spinosyns produced by *S. spinosa*; some have hydroxyl groups instead of methyl groups in several positions, sugars other than forosamine at C17, substituents other than ethyl or but-1-enyl at C21, or a 14-membered macrolide instead of a 12-membered macrolide.[3] Several spinosyns produced by *S. pogona* have high levels of insecticidal activity.[3]

5.2.2   Semi-Synthetic Spinosyns

Spinetoram is the published ISO common name for the semi-synthetic spinosyn mixture derived by making chemical modifications (described in section 5.9.2) to a mixture consisting predominantly of spinosyn J (see Figure 5.2; structure 5.3) and spinosyn L (see Figure 5.2; structure 5.4). There are two small, but significant, structural differences between spinetoram and spinosad. First, the two primary components of spinetoram, 3′-*O*-ethyl-5,6-dihydro spinosyn J (see Figure 5.2; structure 5.5) and 3′-*O*-ethyl spinosyn L (see Figure 5.2; structure 5.6), both have an ethoxy group at the 3′-position on the 2′,3′,4′-tri-*O*-methyl rhamnose instead of the methoxy group present in spinosyns A and D. Second, the 5,6 bond of 3′-*O*-ethyl-5,6-dihydro spinosyn J is a single bond instead of the double bond present in spinosyn A. Like spinosyns A and D in spinosad, 3′-*O*-ethyl-5,6-dihydro spinosyn J is the primary component in spinetoram, followed by 3′-*O*-ethyl spinosyn L. Like spinosad, the balance of spinetoram is also made up of a number of minor spinosyns. The relative abundance of 3′-*O*-ethyl-5,6-dihydro spinosyn J, 3′-*O*-ethyl spinosyn L and other minor spinosyns in spinetoram is comparable to that of spinosyn A, spinosyn D and the minor spinosyns in spinosad.

5.3   Classification of the Spinosyns

Although the spinosyns have previously been identified as *antibiotic insecticides*, this designation is inaccurate because the spinosyns do not have antibiotic effects on bacteria or fungi.[1,13]

The spinosyns, avermectins and milbemycins have been grouped together as *macrocyclic lactones* or *macrolide insecticides*. This designation is not appropriate because the avermectins and milbemycins differ significantly from the spinosyns in both structure and function (mode of action). The avermectins and milbemycins are 16-membered macrocyclic lactones substituted with a spiroketal unit and a hexahydrobenzofuran unit.[15,16] The spinosyns, however, are comprised of a 12-membered macrocycle attached to a unique 5,6,5-*cis*-anti-*trans*-tricyclic ring system.[12,13] Functionally, the avermectins and milbemycins stimulate the release of γ-amino butyric acid (GABA) and enhance GABA binding to insect post-synaptic receptors, which activates glutamate-gated chloride channels and inhibits nerve impulse transmission.[16] In contrast, the spinosyns act in a unique manner at insect nicotinic acetylcholine receptors.[17] For these reasons, most insecticide classification schemes separate the avermectins and milbemycins from the spinosyns.[18–20]

5.4   Synthesis and Design of Spinosyn Insecticides

5.4.1   Spinosyn Biosynthesis

Biosynthesis of the spinosyns begins with sequential incorporation of acetate and propionate into a polyketide synthase (PKS) pathway to form a chain of 23 carbon atoms that is then cyclized.[9,13] The C15-hydroxyl group of this cyclic intermediate is converted to a keto group, which is followed by the formation of three intramolecular carbon-carbon bonds within the macrocycle to form the unique tetracyclic core of the spinosyns.[9,13,21,22] This tetracyclic macrocycle is known as the spinosyn aglycone. Production of the tetracycle (aglycone) is followed by addition of L-rhamnose at position C9 on the tetracycle. The rhamnose is then methylated by a series of methyl transferases to produce the 2′,3′,4′-tri-*O*-methyl L-rhamnosyl derivative, which is known as the spinosyn C17-pseudoaglycone.[9,13,22] The final biosynthetic step is addition of forosamine to the C17 position of the tetracycle to produce the completed spinosyn. The entire biosynthetic gene cluster for the spinosyns and the 21-butenyl spinosyns has been determined.[9,13,23]

5.4.2   Spinosyn Derivatives

The creation of new spinosyns has been achieved through a variety of approaches including genetic modifications to the spinosyn tetracycle PKS and sugar biosynthetic pathways, discovery of new naturally occurring spinosyns (the 21-butenyl spinosyns) and, most commonly, synthetic modification of naturally occurring spinosyns.

5.4.2.1   Modifications at C21 of the Tetracycle

Most of the spinosyns produced by *Saccharopolyspora spinosa* have an ethyl group at position C21 on the tetracycle.[12,22] The C21 position was difficult to access synthetically so other approaches were taken to provide access to this region of the molecule. Modification of the starter unit for the PKS pathway provided a means to incorporate new substituents at position C21, including *n*-propyl, isopropyl, cyclopropyl, cyclobutyl, *etc.* (see Table 5.1).[24] Some of these modifications at the C21 position (*n*-propyl, cyclobutyl) provided activity against lepidopterous pests such as *Heliothis virescens* (tobacco budworm) and *Spodoptera exigua* (beet armyworm) that was similar, or superior to, the naturally occurring spinosyns from *Saccharopolyspora spinosa* (see Table 5.1). Some of these same modifications also significantly increased toxicity to aphids and whiteflies (see Table 5.1).[24]

The 21-butenyl spinosyns produced by *Saccharopolyspora pogona* provided an opportunity to further understanding of the structure–activity relationships (SAR) within the spinosyns. Like the PKS-based C21 modifications, presence of a but-1-enyl moiety at C21 of the tetracycle (see Table 5.1; structures 5.18 and 5.19) imparts activity against *Heliothis virescens* larvae that is equal to C21-ethyl

Table 5.1 Effect of modifications of the C21 position of the spinosyn tetracycle on insecticidal activity.

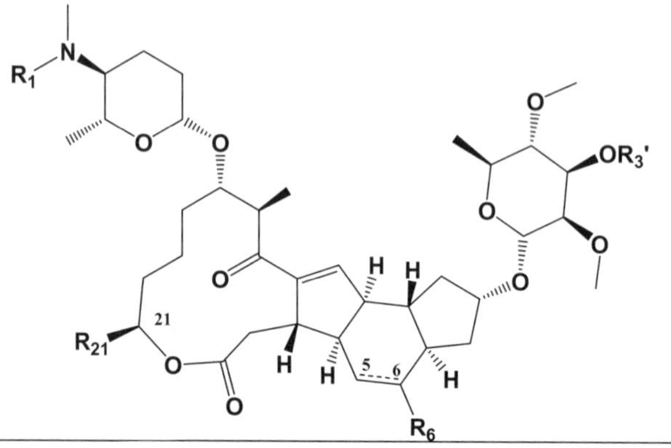

Structure number	R_{21}	R_1	R_6	$R_{3'}$	$5,6^a$	$TBWn^b$	$TBWt^c$	$BAWo^d$	$BAWt^e$	CA^f	SWF^g
5.1	ethyl[h]	Me	H	Me	D	0.31	0.03	0.079	0.63	18–55	5–23
5.2	ethyl[i]	Me	Me	Me	D	0.8	–	–	–	65	1.7
5.7	methyl[j]	Me	H	Me	D	4.6	–	–	–	>50	3.9
5.8	ethyl	Me	H	Et	D	0.03	–	0.012	0.039	8.1	5.3
5.9	ethyl	Me	H	Et	S	0.05	–	–	0.04	15	–
5.10	vinyl	H	H	Et	D	0.2	–	–	–	–	–
5.11	vinyl	Me	H	Et	D	0.5	–	0.018	–	–	–
5.12	*n*-propyl	Me	H	Me	D	0.16	0.025	–	0.031	27	–
5.13	isopropyl	Me	H	Me	D	–	0.02	–	0.123	–	–
5.14	cyclopropyl	Me	H	Me	D	–	0.125	–	≈1	≈46	>50
5.15	cyclobutyl	Me	H	Me	D	–	0.013	–	>1	≈3	55
5.16	cyclobutyl	Me	Me	Me	D	–	0.016	–	0.39	15	13.5
5.17	cyclobutyl	Me	H	Me	S	–	0.02	–	0.13	1.7	0.7
5.18	but-1-enyl	Me	H	Me	D	0.29	–	–	0.035	5–11	2
5.19	but-1-enyl	Me	Me	Me	D	0.3	–	–	–	2	–
5.20	but-1-enyl	Me	H	Et	D	–	–	–	0.004	0.95	1.65
5.21	hex-1-enyl	Me	H	Et	D	–	–	<0.012	–	15.2	–
5.22	oct-1-enyl	Me	H	Et	D	–	–	0.022	–	26	–
5.23	styryl	Me	H	Et	D	–	–	0.013	–	16.9	–

(Data in table adapted from ref. 9, 12, 14, 24 and 25).
[a] 5,6 bond: S = single bond, D = double bond.
[b] TBW*n* = tobacco budworm (*Heliothis virescens*) neonate drench bioassay (LC_{50} ppm).
[c] TBW*t* = tobacco budworm (*Heliothis virescens*) topical bioassay (LD_{50} µg larva^{-1}).
[d] BAW*o* = beet armyworm (*Spodoptera exigua*) oral (diet) bioassay (LC_{50} µg cm^{-2}).
[e] BAW*t* = beet armyworm (*Spodoptera exigua*) topical bioassay (LD_{50} µg larva^{-1}).
[f] CA = cotton aphid (*Aphis gossypii*) on plant bioassay (LC_{50} ppm).
[g] SWF = sweet potato whitefly (*Bemisia tabaci*) on plant bioassay (LC_{50} ppm).
[h] Spinosyn A.
[i] Spinosyn D.
[j] Spinosyn E.

homologs such as spinosyn A and improved activity against *Spodoptera exigua* larvae (see Table 5.1).[3] Activity against *Aphis gossypii* (cotton aphid) and, to a lesser extent, *Bemisia tabaci* (sweetpotato whitefly) is also improved (see

Table 5.1).[3] The C21-but-1-enyl spinosyns also enabled further synthetic modifications of the C21 position by means of cross-metathesis reactions.[25] The resulting spinosyn analogs had position C21 substituted with a variety of moieties including monounsaturated 2-, 6- and 8-carbon chains and styrene. These substitutions had little effect on activity (see Table 5.1).[25]

5.4.2.2 Other Tetracycle Modifications

In addition to modifications at the periphery of the tetracycle, the internal structure of the tetracycle has been investigated. Reduction of the 5,6 double bond imparts greater photolytic stability, but further internal unsaturation of the tetracycle (*e.g.* 7,11-dihydro, 7,8-dihydro) has little effect on lepidopteran activity.[10,26] Aromatization of the six-membered ring of the tetracycle appears to eliminate activity.[11,26] Simplification of the spinosyn tetracycle by replacing the cyclopentane ring that links to the 2′,3′,4′-tri-*O*-methyl rhamnose in spinosyn A with a benzene ring recently has been investigated and may provide synthetic access to a new range of spinosyn analogs.[27]

5.4.2.3 Sugar Modifications

Numerous studies have explored the effect of modifications of the two sugars attached to the spinosyn tetracycle.[10,11,22,26,28–32] All of these studies point to changes at the 3′-position of the rhamnose sugar as producing the largest increases in activity relative to the naturally occurring spinosyns.

5.4.3 Spinosyn Quantitative Structure–Activity Relationships (QSAR)

5.4.3.1 Multiple Linear Regression Approaches

Hansch-style multiple linear regression techniques are a well-established approach to the quantification of structure–activity relationships.[33] Although the spinosyns are large, complex molecules, relatively simple multiple linear regression equations have been developed to define relationships between insecticidal activity and the physicochemical properties of spinosyns from *Saccharopolyspora spinosa* and semi-synthetic derivatives of these.[29] The activity of 34 spinosyns against neonates of *Heliothis virescens* is well defined by an equation consisting of three whole molecule properties: CLogP (calculated log P), Mopac dipole moment and HOMO (highest occupied molecular orbital).[29] These trends also appear to extend to spinosyns modified at C21 of the tetracycle. The topical toxicity of 18 semi-synthetic 21-ethyl and 21-but-1-enyl spinosyns to *H. virescens*, is well explained by an equation combining CLogP and SPAM, (the average SPAN (R), which is the radius of a sphere encompassing all of the atoms in the molecule).[34,35] In a like manner, the *Aphis gossypii* activity of 50 spinosyns, 21-butenyl spinosyns and semi-synthetic spinosyn derivatives is also well explained simply based on CLogP.[35]

5.4.3.2 *Neural Network Approach*

While Hansch-style multiple linear regression techniques proved useful for quantifying spinosyn structure–activity relationships on the basis of molecular properties, other QSAR approaches provided the means to predict specific directions for synthesis of analogs with improved insecticidal activity. The use of artificial neural network (ANN) algorithms, a form of artificial intelligence, is a relatively new approach for solving QSAR problems.[36,37] An ANN-based approach was used to analyze the patterns of activity within the existing spinosyns and predict new, and potentially more active, semi-synthetic derivatives. The results of these studies led to the synthesis of modified-rhamnose spinosyn derivatives more active against *Heliothis virescens* larvae.[28,29]

As mentioned earlier, spinosyn analogs in which the 5,6 double bond has been reduced exhibited longer residual activity than those with the 5,6 double bond intact. Studies on the photolytic degradation of spinosad identified the 5,6 double bond as being involved in an early step in spinosad photolysis. This knowledge, combined with the insight on rhamnose modifications gained from the ANN QSAR studies, led to the synthesis and discovery of spinetoram.[8,11,30]

5.5   Spinosyn Mode of Action

Spinosad and spinetoram share a common insecticidal mode of action that involves a novel site on the nicotinic acetylcholine (nACh) receptor. Although many of the studies conducted to determine the spinosyn mode of action employed only spinosyn A, the actions of spinosyn D, spinosad and spinetoram have been equivalent to those of spinosyn A in all studies where comparisons were made. Also, insect strains resistant to spinosad exhibit resistance to spinetoram, providing further evidence for a common mode of action.[17]

5.5.1   Evidence of a Novel Mode of Action for the Spinosyns

Initial studies demonstrated that treatment of susceptible insects with spinosyn A resulted in a number of excitatory symptoms such as involuntary muscle contractions and tremors.[38] Subsequent studies showed that these symptoms resulted from widespread neuronal excitation caused by an interaction of spinosyn A with the insect central nervous system.[38,39] Symptoms induced by spinosyn A appeared to be different from those produced by other known insecticides, and it was hypothesized that the mode of action was likely different from that of known insecticides. Subsequent studies on the mode of action failed to demonstrate a meaningful interaction with known insecticidal target sites, such as the target sites for the neonicotinoids and the avermectins.[40]

5.5.2   Interaction with Ligand-Gated Ion Channels

Neuronal excitation can result from interactions with a number of ligand- or voltage-gated ion channels, or a perturbation within the signal transduction pathway of a G-protein-coupled receptor. The available evidence supports a unique interaction of the spinosyns with ligand-gated ion channels, with compelling evidence for an insecticidally relevant interaction with nACh receptors in some insects. Early studies on *Periplaneta americana* (American cockroach) neurons demonstrated that spinosyn A induced an α-bungarotoxin-sensitive current and prolonged acetylcholine-induced currents.[41] These effects strongly suggested that spinosyns interact with insect nACh receptors. Subsequently, it was demonstrated that spinosyn A interacts with a non-desensitizing subtype of nACh receptor in *P. americana* neurons, and it is likely that spinosyns generate the above mentioned acetylcholine-induced inward current through this receptor subtype.[42]

5.5.3 Molecular Target Site in *Drosophila melanogaster*

Subsequent studies demonstrated that in *Drosophila melanogaster* (common fruit fly), a null mutation of the Dα6 nACh receptor subunit underlies pronounced resistance to spinosad and spinetoram.[17] Similarly, a Dα6 knockout strain of *D. melanogaster* was also shown to have high levels of spinosad resistance.[43] These studies strongly suggest a role for the Dα6 subunit in the nACh-receptor-mediated toxicity of the spinosyns to *D. melanogaster*, and open the possibility that Dα6-like subunits might underlie the insecticidal actions of spinosyns in other insect species. It was recently shown that in *Plutella xylostella* (diamond-back moth), field-selected spinosad resistance is likely due to a mis-spliced transcript of a Dα6 homolog, Pxα6, resulting in the expression of truncated, and presumably nonfunctional nACh receptor proteins.[44] Therefore, it appears that Dα6-like nACh receptors are the insecticidal target site for the spinosyns in these two species, and most likely other insect species as well. The Insecticide Resistance Action Committee (IRAC) has assigned spinosad and spinetoram as the sole members of insecticide mode of action Group 5: *Nicotinic acetylcholine receptor (nAChR) allosteric modulators*.[20]

Spinosad resistance in *Musca domestica* (house fly) also appears to be target-site related, but has not been found to involve the homologs of Dα6 in *M. domestica* (Mdβ3, Mdα5, or Mdα6).[45] In addition, an interaction of insecticidal spinosyns with certain types of insect GABA receptors has been demonstrated.[46] These findings leave open the possibility of alternative or secondary target sites for spinosyns in some susceptible insect species. However, interactions with nACh receptors appear to underlie spinosyn toxicity in most insect species studied to date.

5.5.4 Expression and Characterization of a Spinosyn-Sensitive Nicotinic Acetylchloline Receptor

The coexpression of the Dα6 and Dα5 nACh receptor subunits in oocytes of *Xenopus laevis* (African clawed frog) resulted in a nACh receptor with pharmacological properties consistent with a native nACh receptor.[17] Further, the coexpressed Dα6/Dα5 nACh receptor was sensitive to spinosyn A, spinetoram and acetylcholine, but insensitive to the neonicotinoid insecticide imidacloprid.[17] This strongly supports the earlier conclusion that the spinosyn target site is different from the target sites for other insecticides that act at the nACh receptor, such as the neonicotinoids.[40] Additionally, spinosyn A does not directly activate other *Drosophila melanogaster* neonicotinoid-sensitive nACh receptors (Dα2 coexpressed with chicken β2) or chicken α7 nACh receptors.[47] This suggests that spinosyns are highly specific for certain nACh receptor subtypes. It is likely that this specificity contributes to the high degree of selectivity for arthropod species, rather than vertebrate species, demonstrated by the spinosyns (see section 5.8.1).

5.6 Insect Resistance and Cross-Resistance to the Spinosyns

5.6.1 Resistance Mechanisms, Inheritance and Fitness Costs

Many of the studies that elucidated the spinosyn mode of action also demonstrated target-site resistance to be the primary resistance mechanism. As mentioned in section 5.5.3, recent studies have shown that spinosad resistance in *Plutella xylostella* is associated with a genetic mutation related to the Pxα6 subunit in the nACh receptor, and spinosad resistance in *Drosophila melanogaster* is associated with a genetic mutation related to the Dα6 subunit of the nACh receptor.[17,43,44,48] Studies with laboratory-selected, spinosad-resistant strains of *Musca domestica* and *Heliothis virescens* suggested the presence of target-site resistance.[45,49–51] A study involving a field-selected, spinosad-resistant strain of *Frankliniella occidentalis* (western flower thrips), also points to target-site resistance.[52] In contrast, a recent study of spinosad resistance in a spinosad-selected strain of *P. xylostella* was associated with the presence of enhanced monooxygenases.[53] However, with few exceptions, target-site-based mechanisms predominate as the basis for resistance to the spinosyns.

Most studies that have explored the inheritance of spinosad resistance indicate that it involves a single, recessive locus.[17,54,55] However, there are a few examples of spinosad resistance that may involve multiple genes. A strain of *Frankliniella occidentalis* was found to inherit spinosad resistance in an incompletely-dominate manner, possibly involving more than one gene.[56] In this instance, spinosad resistance appeared to be related to the target site; however, more than one resistance mechanism may have been involved.[56] In another study, spinosad resistance in a multi-resistant strain of *Plutella xylostella* (that was first exposed to abamectin and then spinosad), was associated with a single locus inherited as a codominant trait.[57]

As noted above, spinosad resistance in most studies is autosomal, recessive or incompletely recessive, and associated with a single locus. These genetic characteristics are consistent with a target-site resistance mechanism, and have been observed in spinosad-resistant strains of *Frankliniella occidentalis*, *Plutella xylostella*, *Drosophila melanogaster*, *Heliothis virescens* and *Musca domestica*.[17,43,49,52,54,55] In *P. xylostella* and *D. melanogaster*, these genetic characteristics were directly associated with target-site resistance.[52,55] Thus, the observed genetic characteristics of spinosad resistance also support target-site resistance as the most common resistance mechanism for the spinosyns.

There is evidence of significant fitness costs associated with spinosyn resistance in lepidopterous pest species. Increased development times and reduced survival rates for various life stages, reduced pupal weight, reduced number of eggs produced, and reduced rate of egg hatch have been observed in spinosad-resistant strains of *Plutella xylostella*, *Heliothis virescens* and *Helicoverpa armigera* (old world bollworm).[53,58,59] In *Frankliniella occidentalis*, however, no decrease in fitness was found in spinosad-resistant strains.[60] The stability of spinosyn resistance in laboratory populations has been variable, occurring in some cases but not in others.[53,58,61,62] When susceptible individuals were introduced into colonies of spinosad-resistant *H. virescens* or *F. occidentalis*, reversion to susceptibility occurred within a few generations.[58,61]

5.6.2 Cross-Resistance between Spinosad and Other Insecticides

Insect strains that have become resistant to spinosad are usually susceptible to other insecticides.[22,63–65] In instances where spinosyn target-site resistance is the likely resistance mechanism, there is a uniform lack of cross-resistance to other insecticides (abamectin, indoxacarb, imidacloprid, acetamiprid, pyrethroids and organophosphates).[17,49,51,55] There are a few reports of significant cross-resistance between spinosad and other insecticides, but in some of these reports the cross-resistance is associated with prior exposure to other insecticides. For example, a strain of *Plutella xylostella* which had been extensively treated in the field with a variety of insecticides, including abamectin and later spinosad, displayed a high level of resistance to both insecticides, likely because multiple resistance mechanisms were present.[57]

Insect strains selected for resistance to other insecticides have shown little cross-resistance to spinosad.[22] For example, strains selected for, or found to be highly resistant to abamectin or emamectin benzoate show little cross-resistance to spinosad.[53,62,66–69] Likewise, strains highly resistant to indoxacarb, imidacloprid, pyrethroids, or other insecticides typically show no cross-resistance to spinosad.[64,66,70–77] One study with a *Plutella xylostella* strain that was resistant to fipronil, indoxacarb and spinosad did find evidence of possible cross-resistance among these three insecticides, however, the presence of multiple resistance mechanisms was also considered likely.[78]

5.6.3 Spinosyn Resistance Management

Resistance management guidelines for spinosad were developed before its first commercial use in 1997.[55,79] Despite these guidelines, reduced susceptibility to spinosad was detected in *Spodoptera exigua* populations in Thailand and Arizona, and in *Plutella xylostella* populations in Hawaii, within a few years of being introduced.[55,80] These two pest species have a history of rapidly developing resistance to a wide range of insecticides.[81] In Hawaii, *P. xylostella* populations developed high levels of resistance to spinosad after more than two years of nearly continuous exposure.[55] Temporary suspension of spinosad use against *P. xylostella* and focused resistance management efforts by University of Hawaii extension advisers resulted in restoring spinosad efficacy in most brassica production areas in Hawaii.[82]

During the 13 years following the first comercial use of spinosad, Dow AgroSciences has investigated 21 incidents of reduced field efficacy that proved to be a result of resistance development. All of these incidents occurred within small geographic areas, widespread resistance to spinosyns has not been detected. Product overuse was involved in two-thirds of the incidents. Half of the incidents occurred in greenhouse production settings, where pest populations may be isolated from susceptible immigrants. Thrips species are at greatest risk for resistance development; about half of the incidents involved *Frankliniella occidentalis* and an additional 10% involved *Thrips palmi* (palm thrips). About 20% of the incidents involved *Plutella xylostella* and 10% involved *Liriomyza* spp. (dipterous leafminers). This handful of pest species accounts for more than 90% of the resistance incidents investigated; most pest species targeted by spinosyn insecticide applications have not demonstrated a great risk of developing resistance.

The biological attributes of spinosyn resistance (lack of genetic dominance, genetic instability with immigration of susceptible individuals and fitness cost in some species) coupled with lack of significant cross-resistance, have made it possible to successfully manage spinosyn resistance globally with the use of basic insecticide resistance management (IRM) tactics. Where spinosyn resistance has been detected, the actions taken to reduce spinosyn selection pressure have varied with the severity of each situation. These actions have been as simple as educating growers on IRM tactics or as severe as temporary withdrawal of spinosyn products from a specific area. When these actions have been taken quickly, susceptibility has usually returned.

5.7 Biological Properties of the Spinosyns

5.7.1 Speed of Action

The spinosyns are fast acting. Spinosyn-intoxicated insects exhibit lack of coordinated movement and cessation of feeding that can occur within minutes of first exposure. These effects are followed by mortality that can occur within one to 24 hours after first exposure (see Figures 5.3 and 5.4). A lethal dose in lepidopterous larvae can be accumulated in exposure periods of one to 30 minutes.[83] In laboratory bioassays, mortality may continue to increase up to three days after initial exposure, but in the field, full control is normally observed within one to two days after a crop is treated.[84] Spinetoram demonstrates faster action than spinosad when injected into lepidopterous larvae (see Figure 5.3), or when exposed to surface residues (see Figure 5.4).[8]

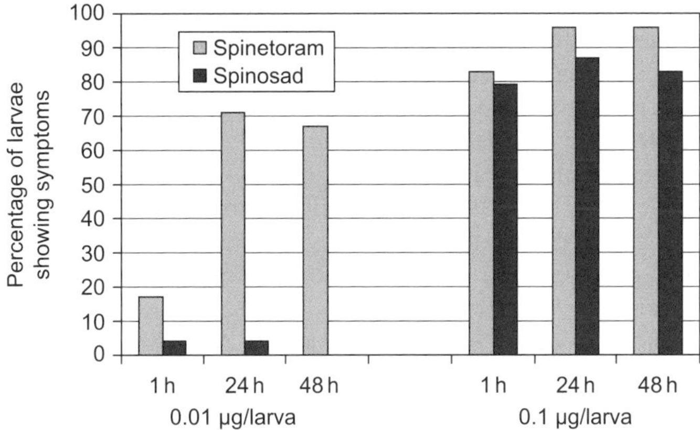

Figure 5.3 Injection toxicity of spinosad and spinetoram to fourth instar *Spodoptera exigua* (beet armyworm) larvae.

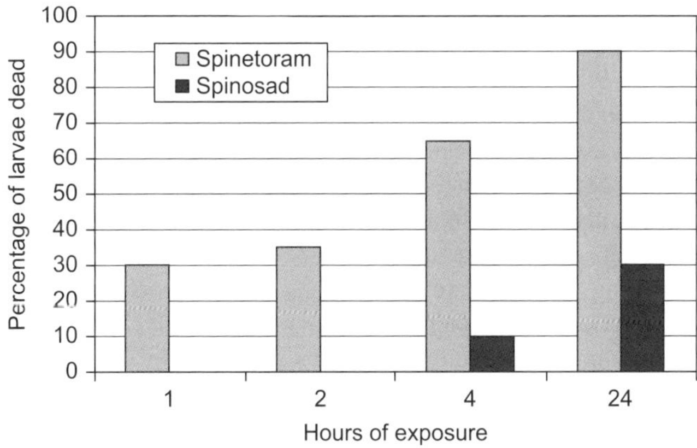

Figure 5.4 Contact toxicity of spinosad and spinetoram to second instar *Spodoptera exigua* (beet armyworm) larvae. (Exposure to 250 ppm spray residues on glass Petri plates).

5.7.2 Uptake and Metabolism in Insects

The spinosyns enter insects either by ingestion of treated food materials or by penetrating the cuticle.[83,85,86] In practice, the spinosyns are considered to be more active when ingested by pest species (particularly lepidopterous larvae) than when exposed by contact.[84]

In cuticular penetration studies, spinosyns A and D were found to penetrate into *Heliothis virescens* or *Trichoplusia ni* (cabbage looper) larvae more slowly than most pyrethroid insecticides (*e.g.* permethrin, cypermethrin) and organophosphorus insecticides (*e.g.* chlorpyrifos, methyl parathion).[87,88]

These observations are supported by bioassay results. When injected into *H. virescens* larvae, the insecticidal activities of spinosyns A and D were nearly equivalent to cypermethrin. However, when applied topically, spinosyns A and D were five-fold less active than cypermethrin.[87,89] When spinosad and spinetoram were applied topically to *Spodoptera exigua* larvae, both exhibited rates of cuticular penetration similar to earlier results with spinosyns A and D; no statistically significant differences between spinosad and spinetoram were observed.[90]

The slower rate of penetration by spinosyns compared with other insecticides appears to be offset by the relative stability of these molecules to insect metabolic enzymes. In *Heliothis virescens* larvae, no metabolism of spinosyn A was observed over a period of nine to 24 hours.[22,51,87] When injected into *Spodoptera exigua* larvae, spinosad and spinetoram both exhibited modest rates of metabolism, with no statistically significant differences between spinosad and spinetoram observed.[90]

Since spinosad and spinetoram appear to have similar rates of penetration and metabolic stability, the difference in efficacy between spinosad and spinetoram evident in Figures 5.3 and 5.4 is most likely based on differences in the potency of spinosad and spinetoram at the insect nACh receptor target site.

5.7.3 Stage-Specific and Sublethal Effects

The spinosyns can cause mortality to all stages of susceptible arthropod pests that consume or contact a residue. Given the relatively greater activity by ingestion than by contact described in section 5.7.2, the spinosyns are generally most effective against those life stages that consume treated food materials; in many insect orders this is the larval stage. However, toxicity to adult lepidopterans resulting from contact with surface residues can be significant.[91,92] The spinosyns do not appear to have full ovicidal activity (penetrating into the insect egg and killing the developing embryo). The spinosyns do demonstrate significant ovi-larvicidal activity in lepidopterous species; larvae that consume spinosyn residues on the chorion (eggshell) are killed during or shortly after eclosion (hatching).[83]

A number of insect-specific sublethal effects have been observed for lepidopterous larvae that have survived exposure to spinosyns, including prolonged larval, prepupal and pupal development times, reduced larval and pupal weights, reduced proportion of larval pupation, reduced proportion of pupae surviving to the adult stage and decreased adult longevity and fecundity.[92–95] Reduced egg size, reduced pupation rate, prolonged larval development times and reduced adult female longevity have been observed in the offspring of adults that had been exposed to sublethal doses of spinosad as larvae.[95]

5.7.4 Spectrum of Arthropod Pest Activity

The spinosyns have demonstrated toxicity to a wide range of arthropod pest species. Table 5.2 lists the orders and families of arthropod pest species that

Table 5.2 Taxonomic distribution of arthropod pests susceptible to the spinosyns.

Arthropod pest order	*Arthropod pest families*
Acari: mites, ticks	Argasidae, Ixodidae, Tetranychidae
Anoplura: sucking lice	Pediculidae
Blattodea: cockroaches	Blattellidae
Coleoptera: beetles, weevils	Anobiidae, Bostrichidae, Chrysomelidae, Curculionidae, Dermestidae, Elateridae, Laemophloeidae, Scarabaeidae, Silvanidae, Tenebrionidae
Dermaptera: earwigs	Forficulidae
Diptera: flies, gnats, midges, mosquitoes	Agromyzidae, Anthomyiidae, Calliphoridae, Cecidomyiidae, Culicidae, Drosophilidae, Ephydridae, Glossinidae, Muscidae, Phoridae, Sciaridae, Tephritidae
Hemiptera: true bugs, aphids, psyllids, whiteflies, leafhoppers, planthoppers, scales	Aphididae, Aleyrodidae, Cicadellidae, Delphacidae, Diaspididae, Miridae, Pseudococcidae, Psyllidae
Hymenoptera: ants, sawflies	Diprionidae, Formicidae, Tenthredinidae
Isoptera: termites	Kalotermitidae
Lepidoptera: moths, butterflies	Arctiidae, Crambidae, Gelechiidae, Geometridae, Gracillariidae, Lasiocampidae, Lymantriidae, Lyonetiidae, Noctuidae, Pieridae, Plutellidae, Psychidae, Pyralidae, Sphingidae, Thaumetopoeidae, Tortricidae, Zygaenidae
Mallophaga: chewing lice	Trichodectidae
Orthoptera: grasshoppers, crickets	Acrididae, Gryllotalpidae, Tettigoniidae
Psocoptera: booklice	Liposcelididae, Trogiidae
Siphonaptera: fleas	Pulicidae
Thysanoptera: thrips	Phlaeothripidae, Thripidae

have shown some level of susceptibility to the spinosyns, based on a review of spinosad and spinetoram product labels around the world, Dow AgroSciences field development trials and laboratory screens, and selected literature references. As indicated by the many families listed for the Lepidoptera, Coleoptera and Diptera in Table 5.2, the spinosyns demonstrate a high level of activity against many species in these important insect orders. A fairly unique attribute of the spinosyns is exceptional activity against thrips (Thysanoptera). Activity against sap-feeding insects (Hemiptera) is variable; the spinosyns (particularly spinetoram) are highly active against psyllids (Psyllidae) but are much less active against most aphids (Aphididae) and true bugs (Heteroptera). The spinosyns have demonstrated innate toxicity to whiteflies (Aleyrodidae), certain scale insects (Diaspididae), leafhoppers (Cicadellidae), planthoppers (Delphacidae) and mealybugs (Pseudococcidae); however, efficacy in field applications has been inconsistent. The spinosyns have also demonstrated activity against mites and ticks (Acari). As new arthropod pests are tested, it is possible that the spinosyns will be found to be active against additional taxonomic groups.

Table 5.3   Toxicity of spinosad, spinetoram, indoxacarb, and cypermethrin to larvae of four lepidopterous pest species (diet assay).

	LC$_{90}$ (ppm, applied to diet)			
Pest species	*Spinosad*	*Spinetoram*	*Indoxacarb*	*Cypermethrin*
Spodoptera exigua beet armyworm	0.58	0.053	0.59	0.14
Helicovepa zea corn earworm	0.19	0.032	0.67	0.079
Plutella xylostella diamondback moth	0.20	0.020	0.20	0.12
Cydia pomonella codling moth	0.46	0.010	0.080	–

The intrinsic activity spectrum of spinosad and spinetoram is similar; however, the greater potency of spinetoram enables it to control a greater number of pest species than spinosad at cost-effective use rates. For example, in laboratory bioassays of four important lepidopterous pest species (see Table 5.3), spinetoram was six to 11 times more potent than spinosad to the vegetable pests *Spodoptera exigua*, *Helicoverpa zea* (corn earworm) and *Plutella xylostella*, but was 46 times more potent than spinosad to *Cydia pomonella* (codling moth), a major insect pest of apples, pears and walnuts.

5.7.5   Effects on Beneficial Arthropods

5.7.5.1   Natural Enemies

The spinosyns demonstrate innate toxicity to several arthropod natural enemy species in laboratory bioassays, however, this innate toxicity is offset by the relatively short persistence of the spinosyns in the environment. Lasting disruptions of field or greenhouse populations of natural enemy species generally are not observed.[96,97] The spinosyns, in particular spinosad, have low to moderate toxicity to species of predatory insects and mites, particularly ladybird beetles (Coccinellidae), lacewings (Chrysopidae), damsel bugs (Nabidae), big-eyed bugs (Geocoridae) and pirate bugs (Anthocoridae); and sublethal effects are generally minimal.[97,98] Reduced fecundity has been observed in females of predatory phytoseiid mites exposed to spinosad.[99,100] Hymenopterous parasitoids (parasitic wasps) are more susceptible to the spinosyns than predatory species and the impact of sublethal effects may be more significant.[98] Release of parasitic wasps to control certain insect pests is practiced in a number of production systems, particularly greenhouses. Due to the relatively short duration of spinosad residues on treated crops, parasitic wasps can be safely released one to two weeks after a spinosad application.[101] Spinetoram is more potent than spinosad and the implications of this for natural enemies are not fully understood, but there is evidence that spinetoram can be equally compatible with the use of biological control.[102]

5.7.5.2 Pollinators

The spinosyns are acutely toxic to *Apis mellifera* (honey bee) when administered in oral and contact bioassays. The acute oral LD_{50} (48 hours) values for spinosad and spinetoram are 0.057 and 0.11 µg per bee, respectively, and the acute contact LD_{50} (48 hours) values for spinosad and spinetoram are 0.0036 µg per bee and 0.024 µg per bee, respectively.[103,104] The contact toxicity of spinosad to other pollinators such as *Bombus impatiens* (common eastern bumble bee), *B. terrestris* (large earth bumble bee), *Nomia melanderi* (alkali bee), *Megachile rotundata* (alfalfa leafcutting bee) and *Osmia lignaria* (blue orchard bee) is equal to or less than that of *A. mellifera*.[105–107] Although contact with liquid spray residues of spinosad (either directly or on treated crops) may cause significant mortality to pollinators, dried residues are not acutely toxic and the risk posed by spinosad to adult bees is negligible after residues have dried.[104,105,108,109]

Pollen and nectar from sprayed plants may have transient effects on brood development, but residues in food sources do not overtly affect hive viability of *Apis mellifera* or *Bombus* spp.[105,106,108] Relatively few sublethal effects on pollinator species have been observed. A decrease in foraging efficiency was observed when *Bombus impatiens* larvae were exposed to a 0.8 mg spinosad per kg pollen continuously for the entire duration of larval development.[110] However, the researchers who conducted this study acknowledge that *B. impatiens* larvae are unlikely to encounter this level of spinosad exposure when it is applied according to label recommendations, and they observed no adverse effects on the foraging efficiency of *B. impatiens* workers continuously exposed to 0.2 mg spinosad per kg pollen as larvae.[110] The overwhelming majority of field studies in which typical application methods were used to apply spinosad at recommended use rates have demonstrated that spinosad poses a low risk to adult bees and has little or no effect on hive activity and brood development.[105]

5.7.6 Movement in Plant Tissues

The spinosyns have demonstrated significant translaminar and systemic mobility in plants. Spinosad and spinetoram control larvae of *Liriomyza* spp. and the early larval instars of some lepidopterous larvae by moving through the upper epidermis into the mesophyll tissue within treated leaves.[111,112] Spinosad movement through the leaf epidermis can be increased with the addition of penetrating surfactants to the spray mixture.[112] Once in the mesophyll, spinosyn residues do not appear to move out of treated leaves. The spinosyns can also be taken up by plant roots and move acropetally in vascular tissue. When administered to the roots of hydroponically-grown plants, spinosad can control *Spodoptera littoralis* (Egyptian cotton leafworm) larvae, *Trialeurodes vaporariorum* (greenhouse whitefly) nymphs and *Tetranychus urticae* (twospotted spider mite) motile stages.[113,114] Spinosad applied as a soil drench has significantly reduced emergence of adult *Liriomyza huidobrensis* (pea leafminer) from infested leaves.[115]

5.8 Overview of Regulatory Studies for Spinosad and Spinetoram

Dow AgroSciences has conducted hundreds of proprietary regulatory studies on spinosad and spinetoram. The majority of these studies were performed according to national or international regulatory guidelines.[116–118] The studies encompass mammalian toxicity, ecotoxicology, environmental fate (lab and field), crop metabolism and field residue trials. Information on environmental fate is used to predict potential off-target movement and resultant environmental concentrations. Crop residue studies are used to understand potential human dietary exposure following use in crops. Results from toxicity studies, *e.g.* NOELs (No Observed Effect Levels) or NOAELs (No Observed Adverse Effect Levels) are used to set protective reference dose (RfD) levels for dietary (acute and chronic), worker (applicator) and bystander exposures. Results from ecotoxicity studies on multiple species are used to set protective endpoints for several environmental compartments. The selected reference doses and endpoints serve as the safety thresholds by which predicted exposures can be assessed. Risk/safety assessments and applied margins of safety then drive the specific use recommendations on the product label, including maximum use rates, minimum preharvest intervals, buffer zones, appropriate personal protective equipment, *etc.*

The details of individual regulatory studies are too extensive to relate here. Fortunately, key study results are distilled by regulatory agencies during their review process and reported in registration decision documents. Most of these documents are in the public domain, and the regulatory overview for spinosad and spinetoram presented in this section draws on these publicly available documents (see Table 5.4).

In the major regulatory submissions for spinetoram, 3′-*O*-ethyl-5,6-dihydro spinosyn J (see Figure 5.2; structure 5.5) and 3′-*O*-ethyl spinosyn L (see Figure 5.2; structure 5.6) were referred to as *XDE-175-J* and *XDE-175-L*, respectively. These designations were derived from *XDE-175*, the Dow AgroSciences internal code for spinetoram during its development. Following registration, US EPA and some other regulatory agencies adopted *spinetoram-J* and *spinetoram-L* to refer to the two primary components of spinetoram. To be consistent with the terminology in the regulatory summaries used to prepare this section, we refer to 3′-*O*-ethyl-5,6-dihydro spinosyn J as *XDE-175-J* and 3′-*O*-ethyl spinosyn L as *XDE-175-L* in this section.

5.8.1 Mammalian Toxicity

A full battery of acute, subchronic, chronic, reproduction, developmental, neurotoxicity and mutagenicity/genotoxicity tests exists for spinosad and spinetoram. The structural similarity of spinosad and spinetoram results in highly similar toxicity profiles in mammalian systems. The majority of mammalian toxicological studies are summarized in Table 5.5. Key points for the mammalian toxicity of spinosad and spinetoram are: (1) low acute toxicity for

Table 5.4 Publically available regulatory summaries for spinosad and spinetoram.

Agency	Regulatory summary document
US EPA[a]	EPA Pesticide Fact Sheet – Spinosad[119]
US EPA	EPA Pesticide Fact Sheet – Spinetoram[120]
US EPA	Spinetoram and Spinosad – Chronic Dietary Exposure and Risk Assessments[121]
US EPA	Environmental Fate and Ecological Risk Assessment for Spinetoram New Registration[122]
US EPA	Human Health Water Assessment for Spinosad (Naturalyte GF-1592 and GF-1593) Mosquito Larvicide Use[123]
US EPA	Spinetoram. Residue Chemistry Summary Concerning the Application of Spinetoram to Numerous Crops[124]
Canada PMRA[b]	Regulatory Note: Spinosad - Success™ 480SC Naturalyte Insect Control Product and Conserve™ 480SC Naturalyte Insect Control Product[125]
Canada PMRA	Evaluation Report: Spinetoram (XDE-175)[104]
Australia PVMA[c]	Public Release Summary on Evaluation of the New Active Spinosad in the Products Laser Naturalyte Insect Control, Tracer Naturalyte Insect Control[126]
Australia PVMA	Public Release Summary on Evaluation of the New Active Spinetoram in the Product Delegate[109]
European Union[d]	Review Report for the Active Substance Spinosad[103]
EFSA[e]	Reasoned Opinion: Setting of an Import Tolerance for Spinetoram in Peaches (Including Nectarines) and Apricots[127]
JMPR[f]	Pesticide Residues in Food, Toxicological Evaluations, Spinosad[128]
JMPR	Pesticide Residues in Food, Report of the 2001 Joint FAO/WHO Meeting of Experts[129]
JMPR	Pesticide Residues in Food 2008, Joint FAO/WHO Meeting on Pesticide Residues, 5.20 Spinetoram[130]
WHOPES[g]	WHO Specifications and Evaluations for Public Heath Pesticides, Spinosad[131]

[a]United States Environmental Protection Agency.
[b]Health Canada Pest Management Regulatory Agency.
[c]Australian Pesticide and Veterinary Medicine Authority.
[d]European Commission, Health and Consumer Protection Directorate.
[e]European Food Safety Authority.
[f]United Nations, Joint Food and Agriculture Organization/World Health Organization Meeting on Pesticide Residues.
[g]United Nations, World Health Organization Pesticide Evaluation Scheme.

both technical product and end-use formulations; (2) no target organ is identified; (3) not teratogenic; (4) protective reproductive/developmental NOELs are established; (5) weight of evidence indicates no endocrine alterations; (6) negative in multiple *in vitro* and *in vivo* mutagenicity/genotoxicity tests; (7) no neurotoxicity effects noted; (8) not carcinogenic; and (9) dogs appear to be the most toxicologically sensitive species.

No target organ is identified for either molecule; instead toxicological profiles for spinosad and spinetoram indicate generalized systemic toxicity with effects of cell vacuolation (lysosomal lamellar bodies) and histiocyte/macrophage aggregation noted as the primary effects and at similar doses.

Table 5.5 Mammalian toxicological endpoints for spinosad and spinetoram.

Category	Study/Endpoint	Spinosad[103,126]	Spinetoram[109,120]
Acute	Oral LD$_{50}$	>3738 mg/kg	>5000 mg/kg
	Dermal LD$_{50}$	>5000 mg/kg	>5000 mg/kg
	Inhalation LD$_{50}$	>5.18 mg/kg	>5.50 mg/kg
	Eye irritation	mild, transient irritation	mild, transient irritation
	Dermal irritation	no irritation	no irritation
	Dermal sensitization	negative[a]	moderate sensitizer[b]
Subchronic	28-d dietary (Rat) NOAEL	250 ppm (21 mg/kg/d)	500 ppm (48 mg/kg/d)
	28-d dietary (Mouse) NOEL	–	24.5 mg/kg/d
	28-d dietary (Dog) NOAEL	200 ppm (7 mg/kg/d)	200 ppm (5.9–8.1 mg/kg/d)
	90-d dietary (Rat) NOEL	120 ppm (8.6 mg/kg/d)	120 ppm (10.1 mg/kg/d)
	90-d dietary (Mouse) NOEL	50 ppm (6 mg/kg/d)	50 ppm (7.5 mg/kg/d)
	90-d dietary (Dog) NOAEL	150 ppm (4.89 mg/kg/d)	150 ppm (5.7 mg/kg/d)
Long-Term	2-year dietary (Rat) NOEL	50 ppm (2.4 mg/kg/d)	250 ppm (10.8 mg/kg/d), not carcinogenic
	18-month dietary (Mouse) NOEL	80 ppm (11.4 mg/kg/d) not carcinogenic	150 ppm (19 mg/kg/d), not carcinogenic
	1-year dietary (Dog) NOEL	100/120 ppm (2.68 mg/kg/d)	100 ppm (2.5 mg/kg/d), not carcinogenic
Developmental/ Reproductive	Maternal (Rat) NOEL	50 mg/kg/d	100 mg/kg/d
	Developmental (Rat) NOEL	200 mg/kg/d	300 mg/kg/d
	Maternal (Rabbit) NOEL	10 mg/kg/d	10 mg/kg/d
	Developmental (Rabbit) NOEL	50 mg/kg/d	60 mg/kg/d
	2-generation reproductive (Rat) NOEL maternal	10 mg/kg/d	10 mg/kg/d
	2-generation reproductive (Rat) NOEL fetal/neonatal	10 mg/kg/d	10 mg/kg/d
Mutagenicity	Ames	Negative	Negative
	Chromosomal aberration	Negative	Negative
	Forward mutation	Negative	Negative
	Mouse micronucleous	Negative	Negative
Neurotoxicity	Acute NOEL	2000 mg/kg	>2000 mg/kg
	Chronic NOEL	1000 ppm (51 mg/kg/d)[c] 1-year study	>750 ppm (36.7 mg/kg/d)[c] 2-year study

[a]Magnusson and Kligman Guinea Pig Maximisation Test (GPMT).
[b]Local Lymph Node Assay conducted in the mouse.
[c]Highest dose tested.

Teratogenicity was not observed or reported for spinosad or spinetoram in rabbits or rats. Maternal NOEL values were consistently lower than developmental NOEL values for the offspring; maternal toxicity was not accompanied by embryo toxicity, fetal toxicity, or teratogenicity. In the two-generation reproduction study in rats, parental toxicity was observed in both males and females at the highest dose tested ($100\,mg\,kg^{-1}\,d^{-1}$ for spinosad and $75\,mg\,kg^{-1}\,d^{-1}$ for spinetoram). Effects in the offspring (decreased litter size and pup weight) at high dose were attributed to maternal toxicity. The systemic and reproductive NOELs were both $10\,mg\,kg^{-1}\,d^{-1}$. Dystocia (difficulty in labor observed as protracted delivery of pups) was noted in the two-generation reproductive toxicity studies within the highest dose groups of $100\,mg\,kg^{-1}\,d^{-1}$ or $75\,mg\,kg^{-1}\,d^{-1}$ for spinosad and spinetoram, respectively. US EPA reduced the Food Quality Protection Act (FQPA) safety factor from 10X to 1X for both molecules, in large part as a result of these studies.[121]

Subchronic (28-day and 90-day) oral feeding studies in rats, mice and dogs have been conducted for spinosad. Effects at the LOEL (Lowest Observed Effect Level) were typically vacuolation, aggregates of macrophages, or increased liver weight. For dogs only, the LOEL included observations of arteritis (inflammation of the walls of arteries). Beagle dogs are genetically predisposed to developing arteritis, and this effect is an exacerbation of this predisposition. US EPA established joint short-term exposure endpoints (inhalation and incidental oral) for spinosad and spinetoram based on the oral NOAEL of $4.9\,mg\,kg^{-1}\,d^{-1}$ from the spinosad subchronic feeding study in dogs and a required Margin of Exposure (MOE) of 100. A NOAEL of $0.249\,mg\,kg^{-1}\,d^{-1}$ from the spinetoram chronic toxicity study in dogs is representative of the long-term oral toxicity. Spinosad and spinetoram are considered "Not likely to be Carcinogenic to Humans", based on the lack of evidence for carcinogenicity in mice and rats.

5.8.2 Animal Metabolism

ADME (absorption, distribution, metabolism and elimination) studies in rats indicate spinosad and spinetoram are rapidly absorbed, extensively distributed among tissues and extensively metabolized. Highest spinetoram concentrations were found in the gastrointestinal tract, fat, carcass and liver.[130] For spinosad, key tissues were perirenal fat, liver, kidneys and thyroid.[103,127] Fecal excretion was the major route of elimination, while urine was a minor route. Rat metabolism studies for spinosad showed no major differences between the bioavailability, routes or rates of excretion, or metabolism of spinosyn A or spinosyn D. Low (<2%) dermal absorption has been documented for spinosad.[103]

For both spinosad and spinetoram, the primary metabolic pathway in rats was glutathione conjugation, either the parent molecules, or the products of *N*-demethylation or *O*-dealkylation; cysteine conjugation was noted as well.[103,130]

For spinetoram, deglycosylation of each parent factor, as well as hydroxylation of parent XDE-175-J, was also noted, and the aglycone of XDE-175-L was subject to sulfate and glucuronide conjugation.[130] Livestock metabolism studies conducted with ^{14}C-labeled materials confirmed that overall dealkylation is a primary component of the metabolic pathway for both spinosad and spinetoram in animals. *O*-dealkylation in the 2′,3′,4′-tri-*O*-methyl rhamnose and demethylation in the forosamine were observed for both molecules in the goat and hen. In addition, hydroxylation of the macrolide ring was noted for spinosad in goats and in hens, more extensive metabolism and eventual loss of the forosamine sugar was noted. For spinetoram and spinosad, unchanged parent compound was the primary residue component in milk, eggs and animal tissues, which is a consideration for potential human consumption. Generalized animal metabolism pathways are shown in Figures 5.5 and 5.6, with metabolites that could arise from rats, hens and/or goats. Pathways include both major (>10%) and minor compounds as well as both tentative and confirmed structures.

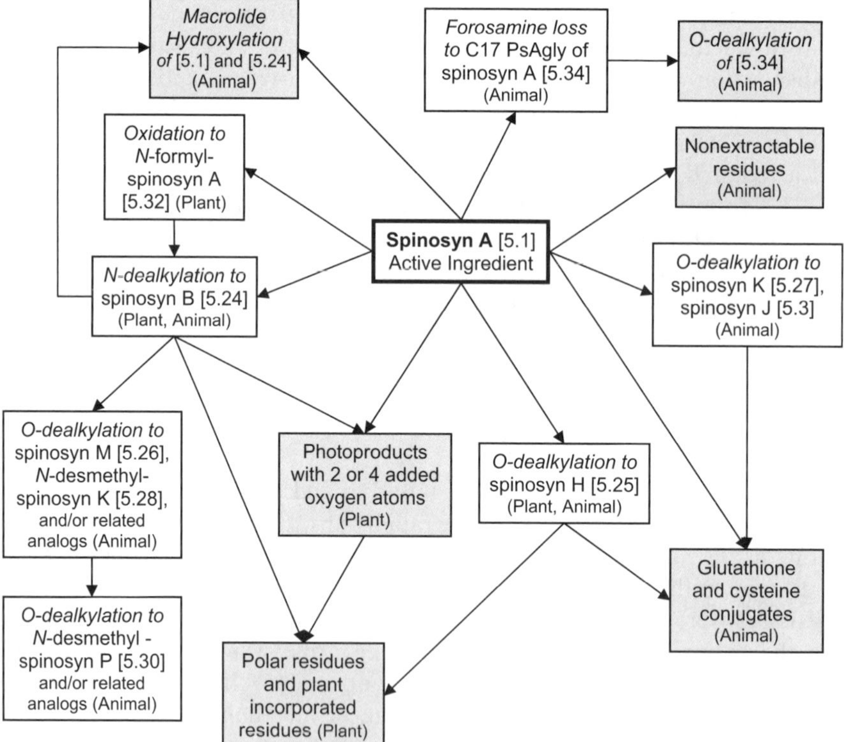

Figure 5.5 General plant and animal metabolism pathways for spinosyn A, the major component of spinosad. (Plant = representative crops [apple, lettuce, and turnip]; Animal = rat or hen, there was minimal metabolism in goat). See Figures 5.2 and 5.7 for structures.

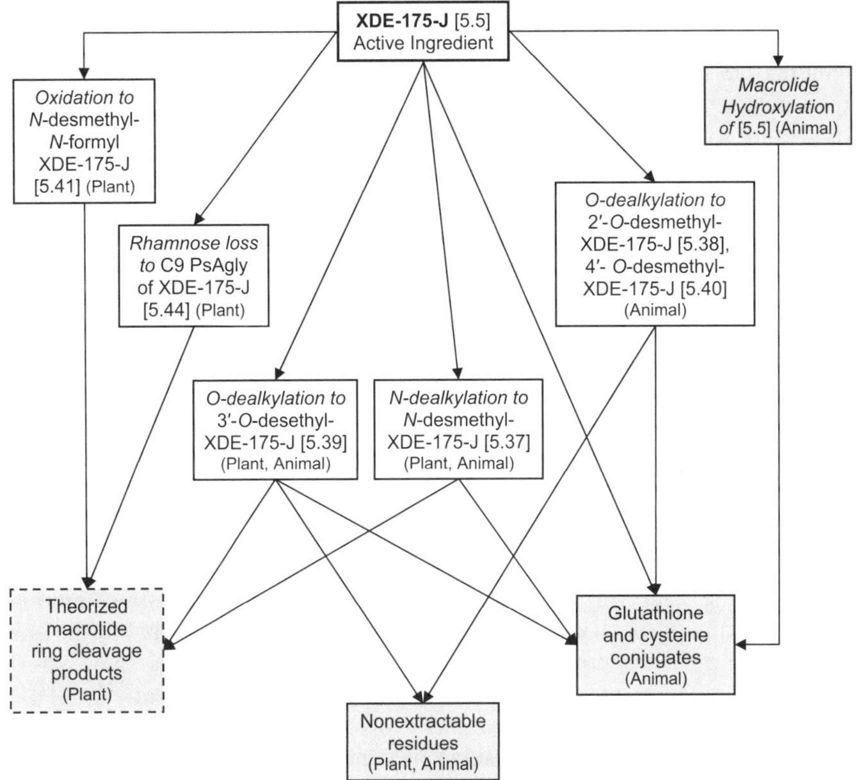

Figure 5.6   General plant and animal metabolism pathways for XDE-175-J, the major component of spinetoram. (Plant = representative crops [apple, lettuce, and turnip]; Animal = rat or hen, there was minimal metabolism in goat). See Figures 5.2 and 5.7 for structures.

5.8.3   Plant Metabolism and Crop Residues

Several metabolic pathways are involved in the breakdown of spinosad and spinetoram in plants. Fundamental understanding of these degradation paths is gained from Nature of Residue (NOR) studies in plants conducted with active ingredient uniformly ^{14}C-labeled in the macrolide (^{14}C-spinosyn A, ^{14}C-spinosad D, ^{14}C-XDE-175-J or ^{14}C-XDE-175-L).

Studies with grapes, apple fruit and apple leaves have compared exposed and control (covered) plant tissues and concluded that photolysis plays a dominant role in the degradation of spinosyns on plants. For example, during the first seven days following spinosad application to apple leaves, minimal degradation of spinosyns A and D occurred on covered leaves whereas more than 90% degradation occurred on leaves exposed to sunlight.

Spinosad metabolism studies on cotton seed, apple fruit and leaves, turnip leaves and roots, cabbage, tomato and grapes demonstrate that spinosyns A and D are the primary residues and that these rapidly degrade by photolysis to give numerous, highly polar breakdown products.[129] The nature of spinetoram residues in turnip, apple and lettuce has been reviewed by multiple agencies including US EPA and FAO/WHO, and a general consistency in residues was noted between the crops tested.[124,130] The metabolic pathways for both spinosad and spinetoram include rapid photolysis and plant metabolism that involve: (1) simple dealkylation on either sugar; (2) modification of the forosamine; and (3) complete loss of the 2,3,4-tri-*O*-methyl rhamnose. The ultimate fate of spinosyns in plants is extensive modification (including macrolide cleavage) to give small fragments that are incorporated or encapsulated into various plant components. Generalized plant metabolism pathways are shown in Figures 5.5 and 5.6, with metabolites that could arise from several crops. Pathways include both major ($>10\%$) and minor compounds, as well as tentative and confirmed structures.

5.8.4   Residue Definitions and Global Maximum Residue Limits

Nature of Residue (NOR) studies are used to establish which residues are monitored in the field Magnitude of Residues (MOR) studies, which in turn are used to set the legal Maximum Residue Limits (MRLs) for the trace amount allowed in human dietary exposure in food. For spinosad, global authorities have agreed on a harmonized "parent-only" (spinosyn A plus spinosyn D) residue definition for monitoring in crops and MRL enforcement.[121,126,129] A divergence in the residue definition has evolved globally for spinetoram. The terminology used for the US "residue of concern" in plants for the purpose of tolerance enforcement is spinetoram (XDE-175-J and XDE-175-L) plus the two metabolites: *N*-desmethyl-XDE-175-J (see Figure 5.7; structure 5.37) and *N*-desmethyl-*N*-formyl-XDE-175-J (see structure 5.41).[121] The proposed EU residue definition from EFSA for risk assessment also includes the two metabolites.[127] In contrast, Australia and FAO/WHO Codex Alimentarius (CODEX) have established a "parent-only" residue definition (XDE-175-J and XDE-175-L).[109,130]

5.8.5   Environmental Fate

Existing lab studies for ^{14}C-labeled spinosyns in prototype environmental compartments include aerobic and anaerobic soil metabolism, aqueous photolysis, hydrolysis, aqueous aerobic and anaerobic degradation, and adsorption/desorption to soils. Field studies have focused on field degradation rates in soil and studies with formulated products and aquatic systems. Key points for the environmental fate of spinosad and spinetoram are: (1) photolysis is the major route of degradation for spinosad and spinetoram in water and on soil and plant surfaces; (2) microbial degradation is an important secondary dissipation pathway; (3) primary environmental degradates arise from initial *N*-demethylation of parent, followed at times by loss of sugars, with ultimate breakdown to many fragments with increased polarity; (4) sorption

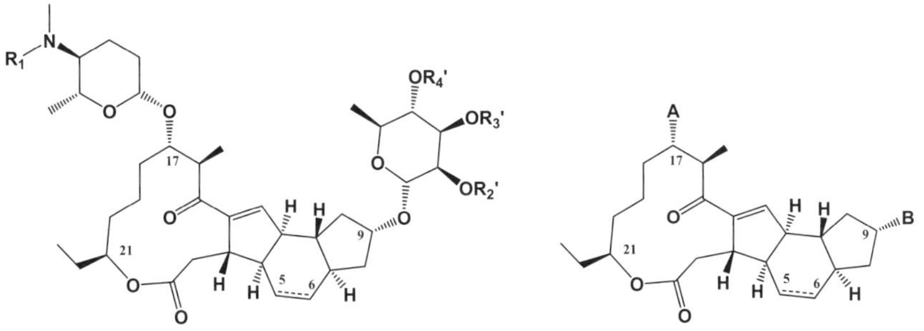

Form I

Form II

Structure number	Name	Form	A	B	R_1	5,6 bond	$R_{2'}$	$R_{3'}$	$R_{4'}$
Structures related to spinosyn A metabolism and degradation pathways									
5.24	Spinosyn B	I	–	–	H	double	Me	Me	Me
5.25	Spinosyn H	I	–	–	Me	double	H	Me	Me
5.26	Spinosyn M	I	–	–	H	double	Me	H	Me
5.27	Spinosyn K	I	–	–	Me	double	Me	Me	H
5.28	*N*-desmethyl spinosyn K	I	–	–	H	double	Me	Me	H
5.29	Spinosyn P	I	–	–	Me	double	Me	H	H
5.30	*N*-desmethyl spinosyn P	I	–	–	H	double	Me	H	H
5.31	Spinosyn T	I	–	–	Me	double	H	H	Me
5.32	*N*-formyl spinosyn A	I	–	–	formyl	double	Me	Me	Me
5.33	C9 pseudoaglycone of spinosyn A	II	forosamine	OH	Me	double	–	–	–
5.34	C17 pseudoaglycone of spinosyn A	II	OH	triMeRh[a]	–	double	Me	Me	Me
5.35	C9 ketone of spinosyn A pseudoaglycone	II	forosamine	keto	Me	double	Me	Me	Me
5.36	C9,17 diketone of spinosyn A aglycone	II	keto	keto	–	double	–	–	–
Structures related to XDE-175-J metabolism and degradation pathways									
5.37	*N*-desmethyl-XDE-175-J	I	–	–	H	single	Me	Et	Me
5.38	2′-desmethyl-XDE-175-J	I	–	–	Me	single	H	Et	Me
5.39	3′-desethyl-XDE-175-J	I	–	–	Me	single	Me	H	Me
5.40	4′-desmethyl-XDE-175-J	I	–	–	Me	single	Me	Et	H
5.41	*N*-desmethyl,*N*-formyl-XDE-175-J	I	–	–	formyl	single	Me	Et	Me
5.42	*N*-desmethyl,*N*-succinyl-XDE-175-J	I	–	–	succinyl	single	Me	Et	Me
5.43	*N*-desmethyl,*N*-nitroso-XDE-175-J	I	–	–	nitroso	single	Me	Et	Me
5.44	C9 pseudoaglycone of XDE-175-J	II	forosamine	OH	–	single	–	–	–

[a]triMeRh = 2′,3′,4′-tri-*O*-methyl rhamnose

Figure 5.7 Structures related to the general metabolic and environmental degradation pathways for spinosyn A (Figures 5.5 and 5.8) and XDE-175-J (Figures 5.6 and 5.9).

partitioning coefficients indicate spinosyns are unlikely to move with the water phase; (5) sorption to sediment will be a primary dissipation route only when photolysis is not available in an aquatic system; (6) minimal ground or surface

Table 5.6   Environmental fate parameters for spinosad and spinetoram.

Study	*Type*	*Spinosad*[103,119,126]	*Spinetoram*[122]
Aqueous photolysis	Lab	$t_{\frac{1}{2}} = <1\,d$	$t_{\frac{1}{2}} = <0.5\,d$
Soil photolysis	Lab	$t_{\frac{1}{2}} = 9\text{--}10\,d$	$t_{\frac{1}{2}} = 10.5\text{--}24\,d^a$
Aerobic soil metabolism[b]	Lab	$t_{\frac{1}{2}} = 9\text{--}17\,d$	$t_{\frac{1}{2}} = 8\text{--}29\,d$ (XDE-175-J) $t_{\frac{1}{2}} = 3\text{--}17\,d$ (XDE-175-L)
Aerobic aquatic metabolism[b]	Lab	–	$DT_{50} = 117\,d$ (XDE-175-J) $DT_{50} = 124\,d$ (XDE-175-L)
Anaerobic aquatic metabolism[b]	Lab	$t_{\frac{1}{2}} = 161\text{--}250\,d$ (Spinosyn A) $t_{\frac{1}{2}} = 250\text{--}495\,d$ (Spinosyn D)	$DT_{50} = 385\,d$ (XDE-175-J) $DT_{50} = 1386\,d$ (XDE-175-L)
Hydrolysis	Lab	Stable at pH 5 and 7. At pH 9, $DT_{50} = 200\text{--}259\,d$	Stable at pH 5 and 7. At pH 9, $DT_{50} = 154\,d$ (XDE-175-L) Slight hydrolysis (XDE-175-J)
Field dissipation	Field	$DT_{50} <1\,d$	$t_{\frac{1}{2}} = 3\text{--}5\,d$ (XDE-175-J) $t_{\frac{1}{2}} = 1\text{--}2\,d$ (XDE-175-L)
Aquatic field study	Field	$t_{\frac{1}{2}} = 2.7\,d^c$	$t_{\frac{1}{2}} = <0.05\,d$

[a]Adjusted to 40°N latitude in summer sunlight.
[b]Dark.
[c]Total measured spinosyns.

water concerns are predicted (depending on use rates and models); and (7) hydrolysis is not a major route of degradation at environmental pH levels. Table 5.6 summarizes available information on degradation or dissipation in terms of a first-order half-life ($t_{1/2}$) or time required for dissipation of 50% of the applied compound (DT_{50}).

5.8.5.1   Fate in Soil

Fate studies in soil demonstrate that the spinosyns dissipate rapidly in the field and that the *N*-demethylated degradates of spinosad and spinetoram are not persistent. For spinosad, radiolabeled test material allowed further investigation of the degradation pathway and revealed classes of degradates that display significantly reduced biological activity. Differences in soils studied and national calculation methods result in slight differences in absolute sorption coefficient values. For spinosad, K_{oc} values range from >2800 to at least $22\,000\ 1\ kg^{-1}$.[103,123] For spinetoram, US EPA reports K_{oc} values of 1800 to $43\,873\ 1\ kg^{-1}$ and states spinetoram is expected to be generally immobile in soil.[120]

5.8.5.2   Aquatic Fate

A challenge in interpreting aquatic fate of spinosyns is the numerous degradates that can be generated in low quantities. A further challenge is that the pattern of degradates shifts depending on the dominant degradation process associated with the test system or conditions. For spinosad, the traditional approach of using radiolabeled test substance in dark laboratory studies to identify metabolites for development of methods for nonradiolabeled field studies was generally followed. This approach was augmented by: (1) development of an immunoassay kit used in the aquatic microcosm study; (2) insecticidal screening of numerous spinosyns and several degradates; (3) ecotoxicology testing of degradates; and (4) the use of radiolabeled test material in an outdoor study. This approach provided a comprehensive characterization of the degradate profile and potential environmental impact, but required integration of multiple studies. For spinetoram, an aquatic field study at two sites in the USA resulted in reported half-lives of less than one day. The respective *N*-demethylated metabolites were formed rapidly following application, and these dissipated with a first order half-life of one to two days. Although spinetoram has high sorption potential for soil and sediment, no partitioning to sediment was observed in these shallow water bodies because of the rapid degradation in the water column.

Collectively, the studies indicate that photolysis, rather than adsorption, is the major route of dissipation of spinosyn residues in water. Dissipation of the active ingredient occurs very rapidly and the *N*-demethylated products of

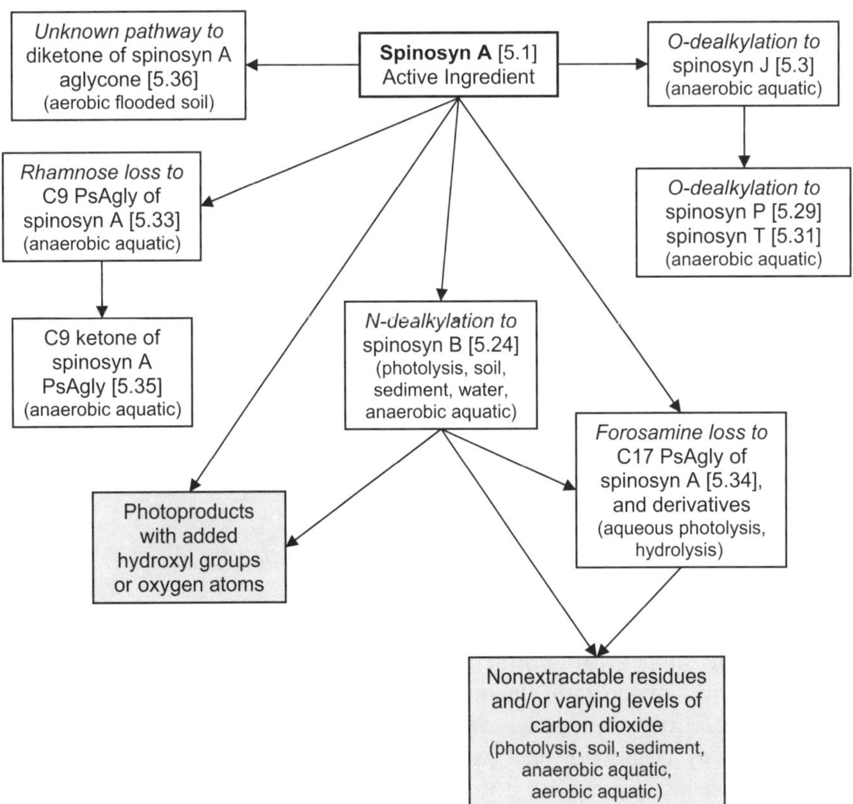

Figure 5.8 General environmental degradation pathways for spinsosyn A, the major component of spinosad. See Figures 5.2 and 5.7 for structures.

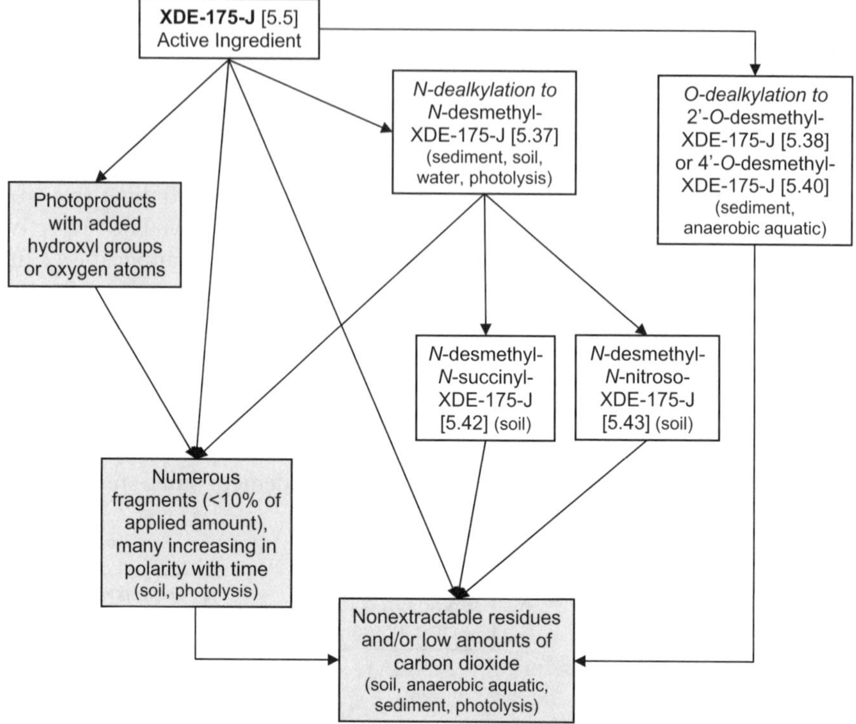

Figure 5.9 General environmental degradation pathways for XDE-175-J, the major component of spinetoram. See Figures 5.2 and 5.7 for structures.

spinosad and spinetoram are not persistent. For spinosad, it has been demonstrated that secondary pseudoaglycone metabolites are not persistent and have little biological activity. Generalized environmental degradation pathways based on studies in several environmental compartments (*e.g.* soil metabolism or aqueous photolysis, *etc.*) are shown in Figures 5.8 and 5.9. Pathways include both major (>10%) and minor compounds as well as both tentative and confirmed structures.

5.8.6   Ecotoxicity

A full set of studies for sentinel test organisms has been conducted with spinosad and spinetoram including birds, fish, aquatic invertebrates, aquatic algae, earthworms and bees. Results of these studies are summarized in Table 5.7. Key points for the ecotoxicity of spinosad and spinetoram are: (1) low acute and reproductive toxicity for birds; (2) toxicity of dealkylated metabolites is similar to parent spinosyn compounds; (3) substantially reduced bioactivity for pseudoaglycone metabolites with either sugar removed; (4) no bioaccumulation in fish is predicted; (5) adequate margins of safety for aquatics

Table 5.7 Nontarget organism toxicological endpoints for spinosad and spinetoram.

Toxicological endpoint	Spinosad[103,126]	Spinetoram[122]
Acute toxicity to birds	$LD_{50} > 2000$ mg/kg bw[a] (mallard duck and bobwhite quail)	$LD_{50} > 2250$ mg/kg bw (mallard duck and bobwhite quail)
Dietary toxicity to birds	$LC_{50} > 5253$ mg/kg feed (bobwhite quail) $LC_{50} > 5156$ mg/kg feed (mallard duck)	$LC_{50} > 5620$ mg/kg feed (bobwhite quail) $LC_{50} > 5640$ mg/kg feed (mallard duck)
Reproductive toxicity to birds	NOEC = 550 mg/kg (mallard duck and bobwhite quail)	NOEC = 1000 mg/kg (mallard duck and bobwhite quail)
Acute toxicity to fish	96-h $LC_{50} = 4.0$ mg/l (carp) 96-h $LC_{50} = 5.94$ mg/l (bluegill sunfish) 96-h $LC_{50} = 30$ mg/l (rainbow trout) 96-h $LC_{50} = 7.87$ mg/l (sheepshead minnow)	96-h $LC_{50} = 2.69$ mg/L; NOEC 2.09 mg/l (bluegill sunfish) 96-h $LC_{50} > 3.49$; NOEC = 3.49 mg/l (rainbow trout) 96-h $LC_{50} > 2.05$; NOEC = 3.49 mg/l (sheepshead minnow)
Acute metabolite toxicity to fish	– 96-h $LC_{50} = 1.55$ mg/l (bluegill sunfish) *N*-desmethyl-XDE-175-L	96-h $LC_{50} = 2.98$ mg/l (bluegill sunfish) *N*-desmethyl-XDE-175-J
Long term toxicity to fish	NOEC (ELS) = 0.5 mg/l (trout)	37-d NOAEC = 1.73 mg/l (sheepshead minnow) 32-d NOAEC = 0.186 mg/l (fathead minnow)
Acute toxicity to invertebrates	48-h $EC_{50} > 1.0$ mg/l (*Daphnia magna*)	48-h $EC_{50} = 3.4$ mg/l (*D. magna.*) 48-h $LC_{50} = 0.355$ mg/l (mysid shrimp)
Acute metabolite toxicity to invertebrates (*Daphnia* sp.)	48-h $EC_{50} = 6.5$ mg/l for spinosyn B 48-h $EC_{50} = 3.8$ mg/l for *N*-desmethyl spinosyn D 48-h $EC_{50} = 65.8$ to > 197 mg/L for spinosad pseudoaglycones	Not available
Chronic toxicity to invertebrates	21-d NOEC (flow-through) = 0.0012 mg/l (*D. magna*) 21-d NOEC (semi-static) = 0.008 mg/l (*D. magna*) NOEC = 0.0842 (mysid shrimp)	21-d NOAEC = 0.0000624 mg ai/l (*D. magna*) 28-d NOAEC = < 0.0196 mg/l (mysid shrimp)
Chronic toxicity to sediment dwellers (*Chironomus riparius*)	25-d NOEC = 0.0016 mg/L 28-d NOEC = 0.080 mg/kg dry sediment (equivalent pore water NOEC = 0.00228 mg/l)	28-d NOEC = 0.00075 mg/l 28-d NOEC = 0.097 mg/kg dry sediment (equivalent pore water NOEC = 0.00160 mg/l)
Toxicity to earthworms	$EC_{50} > 970$ mg/kg dry soil (acute) NOEC ≥3.6 mg/kg dry soil (chronic)	$EC_{50} > 1000$ mg/kg soil (acute) NOAEC = 18.65 mg/kg soil (chronic)
Bioconcentration factor (BCF)	114 l/kg for spinosyn A 115 l/kg for spinosyn D	46 mL/g for XDE-175-J (whole fish) 344 mL/g for XDE-175-L (whole fish)

[a] mg/kg body weight.

under approved use patterns; and (6) bluegill is a sensitive and protective test species for fish.

5.8.7 Biological Relevance of Metabolites

Dow AgroSciences has developed information on the biological relevance of spinosyn metabolites from several disciplines. Ecotoxicity studies for the key environmental degradates (see Table 5.7) demonstrate that the toxicity of the initial *N*-demethylated degradates to aquatic invertebrates is similar to the parent materials; however, once a sugar is lost, the pseudoaglycone metabolites are several orders of magnitude less toxic than the parent materials. These results are consistent with the Dow AgroSciences insecticidal screening results for these compounds.[12] Acute oral LD_{50} values in rat for spinosad metabolites spinosyn B (see Figure 5.7; structure 5.24) and spinosyn K (see Figure 5.7; structure 5.27) are $> 2000\,\mathrm{mg\,kg^{-1}}$ body weight. Leachate samples from a spinosad lysimeter study resulted in negative findings in genotoxicity studies.[103] Collectively, these findings indicate that the biological activity of spinosad is lost with any of the following structural changes: (1) addition of a hydroxyl group almost anywhere on the spinosyn molecule; (2) loss of either sugar; or (3) the addition of hydrogen in the enone region of the macrolide even if both sugars remain.

5.9 Spinosyn Manufacturing

The manufacturing processes for spinosad and spinetoram utilize several green chemistry principles.[132] Both spinosad and spinetoram are derived primarily from fermentation processes which use proteins, carbohydrates, oils and minerals from renewable agricultural sources as feedstocks. The solvents used in the extraction and precipitation steps that isolate the desired materials from the fermentation broth are recycled. Spinosad is produced entirely through fermentation, whereas spinetoram requires additional chemical steps. These synthetic steps are atom economical and significant amounts of the solvents used in these synthetic steps are recycled.

5.9.1 Production of Spinosad

Spinosad is produced directly from the fermentation of a strain of *Saccharopolyspora spinosa*. Production strains of *S. spinosa* have been selected for increased titers of spinosyns A and D, however, no genetic engineering techniques have been used in this process and no genetically-modified organisms are used in the production process. After fermentation, the spinosyn A and D mixture is extracted from the fermentation broth, precipitated and dried to create technical spinosad, which is then formulated into end-use products. Spinosad technical material is also produced under pharmaceutical manufacturing guidelines to be used as a flea control agent in companion animals.

5.9.2 Production of Spinetoram

Production of spinetoram begins with the fermentation of a mutant strain of *Saccharopolyspora spinosa* that produces primarily spinosyns J and L, rather than spinosyns A and D. This strain was generated through mutagenesis of *S. spinosa*. However, like the spinosad-producing strains, no genetic engineering techniques were used in this process and no genetically-modified organisms are used in the production process. After fermentation, the spinosyn J and L mixture is extracted from the fermentation broth and precipitated in preparation for the two chemical synthesis steps required to produce spinetoram. The solvents used in extracting and precipitating the spinosyn J and L mixture are recycled.

Spinosyns J and L, unlike spinosyns A and D, have a free hydroxyl group at the 3′-position on the rhamnose sugar, which allows for chemical manipulation of this site (see Figure 5.10). In the first synthetic step, the free hydroxyl at the 3′-position in spinosyn J and spinosyn L is ethylated to yield a mixture of 3′-*O*-ethyl spinosyn J and 3′-*O*-ethyl spinosyn L. This material is then hydrogenated to yield a mixture of spinetoram-J (3′-*O*-ethyl-5,6-dihydro spinosyn J; see Figure 5.2, structure 5.5) and spinetoram-L (3′-*O*-ethyl spinosyn L; see Figure 5.2, structure 5.6). The hydrogenation conditions are selective and reduce only the disubstituted double bond between C5 and C6 in the 3′-*O*-ethyl spinosyn J intermediate, leaving the 3′-*O*-ethyl spinosyn L unchanged. The material is crystallized from the reaction mixture and dried to create technical spinetoram, which is then formulated into end-use products.

5.9.3 Formulation Attributes of the Spinosyns

To meet a variety of market needs, spinosad and spinetoram products span a very wide range of formulation types (see Table 5.8).

The range of possible formulations for any pesticide is determined by the physical and chemical properties of the active ingredient. Three primary properties determine the formulation characteristics of the spinosyns: (1) both

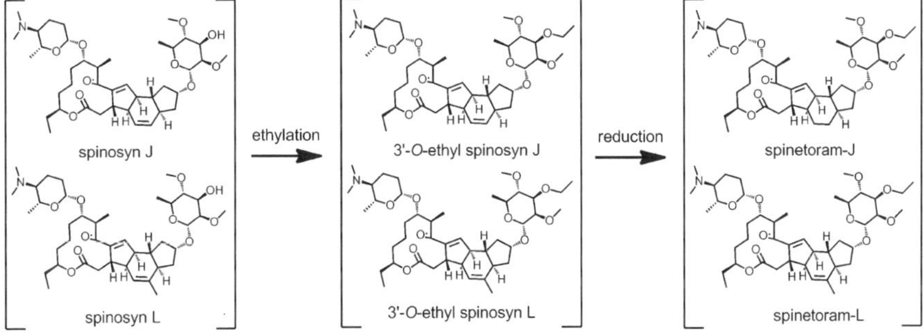

Figure 5.10 Chemical synthesis steps in spinetoram manufacturing.

Table 5.8 Spinosyn product formulation types and associated uses.

Formulation type	Use pattern
Suspension concentrate	Crops, ornamentals, forestry, stored grain, animal health, public health, turf, home and garden
Emulsifiable concentrate	Public health
Wettable granule	Crops
Wettable powder	Crops, ornamentals, seed treatment
Dustable powder	Stored grain, crops
Sprayable bait	Crops
Granular bait	Crops, animal health, urban pests
Bait stations	Urban pests
Granules	Public health
Tablets	Public health
Chewable tablets	Animal health
Gel, paste	Urban pests
Crème rinse	Public health

are fermentation-derived mixtures; (2) both are weak bases; and (3) both have significant solubility in organic solvents.

As fermentation-derived products, spinosad and spinetoram are mixtures composed primarily of two similar, but not identical molecules. In terms of physical properties, a significant difference between the major and minor components of both spinosad and spinetoram is the presence or absence of a methyl group at C6 on the tetracycle (see Table 5.9). With regard to components of spinosad, spinosyn D (methyl group at C6) has a melting point 71 °C higher than that of spinosyn A (hydrogen at C6), and the water solubility of spinosyn D (at pH 7) is almost 1000-fold lower than that of spinosyn A. With regard to the components of spinetoram, spinetoram-L (methyl group at C6) has a melting point 72 °C lower than that of spinetoram-J (hydrogen at C6), and the water solubility of spinetoram-L (at pH 7) is four-fold higher than that of spinetoram-J. The melting points and water solubilities of the mixtures that constitute technical spinosad and technical spinetoram are determined by the relative ratios of the major and minor components.

The predominant components of both spinosad and spinetoram all have pK_a values of about 8 (see Table 5.9). As a weak base, the solubility of spinosyns in water increases as the pH is reduced. From a formulation perspective, at pH level above 5, the spinosyns behave like high-melting solids with little water solubility, which results in the predominant agricultural formulations being suspension concentrates and wettable granule formulations composed of milled crystalline particles. Acid salts of spinosyns can be produced and are used in animal health formulations. The basic nature of the spinosyns is also a consideration when combining multiple active ingredients into the same formulation.

The spinosyns have significant solubility in organic solvents (see Table 5.9). This property is relatively rare in high-melting solids with limited water solubility, and has proven to be useful in a number of formulations for

Table 5.9 Selected physical properties of spinosyn A, spinosyn D, spinetoram-J, and spinetoram-L.

Property	Spinosyn A[133]	Spinosyn D[133]	Spinetoram-J[134]	Spinetoram-L[134]
Melting point, °C	84–99.5[a]	161.6–170[a]	143.4[b]	70.8[b]
Water solubility, mg/l[c, d, e]	235	0.332	11.3	46.7
pK_a[f]	8.10[e]	7.87[e]	7.86[g]	7.59[g]
Solubility in organic solvents, mg/L[c]				
Acetone	168 000	10 100	>250 000	>250 000
Ethyl acetate	194 000	19 000	>250 000	>250 000
n-Heptane	12 400	300	23 900	>250 000
Methanol	190 000	2520	163 000	>250 000
Xylene	>250 000	64 000	>250 000	>250 000

[a]Visual determination.
[b]Diffential scanning calorimetry.
[c]Shake flask.
[d]Buffered to pH 7.
[e]At 20 °C.
[f]Capillary zone electrophoresis.
[g]At 25 °C.

non-agricultural markets, such as mosquito control and animal health. It is also a consideration when combining the spinosyns with other active ingredients.

5.10 Use of Spinosad and Spinetoram for Arthropod Pest Management

The first spinosad registration for agricultural crop use was in Korea in 1996, followed by the US registration for use on cotton in 1997. Spinosad is now widely registered around the world to manage a broad spectrum of arthropod pests in a variety of agricultural, horticultural, forestry, animal health, public health, stored grain and residential environments. The first spinetoram registrations for agricultural crop use were granted in New Zealand, USA and Canada in 2007. Although spinetoram registrations are currently focused on use in agricultural crops, development in other use patterns is anticipated.

5.10.1 Spray Applications to Plants

The most common application method to crop plants is the use of pressurized spray equipment to distribute spinosad or spinetoram in aqueous suspension over the above-ground surfaces of plants. Depending on the crop and local practices, these applications are typically made with hand-operated or powered backpack sprayers, truck mounted low-volume sprayers, tractor-powered boom sprayers and airblast sprayers, or aircraft. Formulated spinosad products applied as foliar sprays are sold in various countries under brand names including Laser™, Tracer®, Conserve®, SpinTor®, Success®, SpinoAce™ and

Entrust®. Formulated spinetoram products for this use pattern are sold under brand names including Delegate®, Radiant®, Exalt™, Endure™ and Palgus™.

Spinosad and spinetoram are well-suited for managing insect pests on crops. Foliar applications of both spinosad and spinetoram have short re-entry intervals (REIs) and preharvest intervals (PHIs), which gives growers flexibility in scheduling harvest and other crop production practices following applications. As described in section 5.7.5.1, spinosad and spinetoram have minimal or short-lived effects on populations of most biological control agents (natural enemies) in most crop protection use patterns. This makes the spinosyns suitable components of IPM programs that employ biological control agents. Spinosad and spinetoram do not cause phytotoxic effects (leaf burning or discoloration, russetting or other fruit finish effects) on treated plants. No effects on photosynthetic rate or dry weight gain were found following spinosad application to lettuce seedlings.[135]

5.10.1.1 *Perennial Horticultural Crops*

Spinosad or spinetoram are used on more than 50 tree-fruit crops, including pome fruits, stone fruits and citrus. The spinosyns are also widely used in tree nuts and other perennial crops such as blueberry, currant, blackberry, raspberry, cranberry, table and wine grapes, and tea. The major insect pests controlled by spinosad and spinetoram in perennial crops are shown in Table 5.10. The spinosyns are used primarily to control tortricid fruit moths and leafrollers, lepidopterous leafminers, thrips, psyllids and tephritid fruit flies. The duration of control demonstrated by spinetoram in tree crops makes it well suited for this use, and it is an effective alternative to azinphos-methyl, an organophosphorous insecticide that has long dominated the control of tortricid fruit moths (especially *Cydia* spp.) in tree fruits.[8]

5.10.1.2 *Annual Horticultural and Agronomic Crops*

The most significant use pattern for spinosad and spinetoram is the control of insect pests attacking vegetable crops grown outdoors and in greenhouses. Spinosad or spinetoram are registered for use in more than 100 annual horticultural crops including bulb vegetables, brassica vegetables, sweet corn, cucurbits, solanaceous fruiting vegetables, leafy vegetables, potato, sweetpotato, root vegetables, strawberry and pineapple. The key insect pests controlled by spinosad and spinetoram in several major vegetable crop groups are listed in Table 5.11. The spinosyns are used primarily to control a number of important pest lepidopterans, thrips, (primarily *Frankliniella* spp. and *Thrips* spp.) and dipterous leafminers (*Liriomyza* spp.).

The spinosyns are also used to manage insect pests in a number of major agronomic crops. In cotton, spinosyns are used to control thrips (primarily

Table 5.10 Key pest species controlled by spinosad or spinetoram in major perennial crops.

Crop group	Key pests controlled	Use rate (g ai ha^{-1})
Pome fruits	*Cydia pomonella* (codling moth); *Cydia molesta* (oriental fruit moth); *Archips* spp., *Adoxophyes* spp., *Argyrotaenia* spp., *Choristoneura roseceana, Pandemis* spp. (tortricid leafroller moths); *Platynota idaeusalis* (tufted apple bud moth); *Epiphyas postvittana* (light brown apple moth); *Leucoptera scitella* (pear leaf blister moth); *Phyllonorycter* spp. (blotch leafminers); *Cacopsylla* spp. (psyllids); *Frankliniella occidentalis* (western flower thrips); *Caliroa cerasi* (pear sawfly)	Spinetoram 32.5–123 Spinosad 96–216
Stone fruits	*Cydia molesta* (oriental fruit moth); *Cydia funebrana* (plum fruit moth); *Orthosia hibisci* (green fruitworm); *Argyrotaenia* spp., *Adoxophyes* spp., *Pandemis* spp., *Choristoneura roseceana, Platynota flavedana* (tortricid leafroller moths); *Anarsia lineatella* (peach twig borer); *Phyllonorycter* spp. (blotch leafminers); *Frankliniella occidentalis* (western flower thrips); *Thrips major* (rubus thrips); *Thrips meridionalis* (peach thrips); *Rhagoletis* spp. (cherry fruit flies)	Spinetoram 32.5–123 Spinosad 96–216
Citrus fruits	*Phyllocnistis citrella* (citrus leafminer); *Papilio* spp., *Diaphorina citri* (Asian citrus psyllid); *Scirtothrips citri* (citrus thrips); *Pezothrips kellyanus* (Kelly's citrus thrips); *Scudderia furcata, Microcentrum retinerve* (katydids); *Gracillaria perseae* (avocado leafroller); *Marmara gulosa* (citrus peelminer); *Egira curialis* (citrus cutworm); *Archips argyrospila* (fruit tree leafholler); *Epiphyas postvittana* (light brown apple moth); *Argyrotaenia franciscana* (orange tortrix); *Orgyia vetusta* (western tussock moth); *Cryptophlebia leucotreta* (false codling moth)	Spinetoram 53–105 Spinosad 70–175
Tree nuts	*Cydia pomonella* (codling moth); *Cydia caryana* (hickory shuckworm); *Cydia latiferreana* (filbertworm moth); *Hyphantria cunea* (fall webworm); *Amyelois transitella* (navel orangeworm); *Choristoneura roseceana* (oblique banded leafroller); *Anarsia lineatella* (peach twig borer); *Acrobasis nuxvorella* (pecan nut casebearer); *Rhagoletis completa* (walnut husk fly)	Spinetoram 32.5–123 Spinosad 144–216
Grapes	*Lobesia botrana* (European grapevine moth); *Clysia ambiguella* (European grape berry moth); *Endopiza viteana* (grape berry moth); *Sparganothis pilleriana* (grape leafroller); *Platynota stultana* (omnivorous leafroller); *Epiphyas postvittana* (light brown apple moth); *Desmia funeralis* (grape leaf folder); *Harrisina americana* (grape leaf skeletonizer); *Argyrotaenia franciscana* (orange tortrix); *Frankliniella occidentalis* (western flower thrips); *Drepanothrips reuteri* (grape thrips); *Otiorrhynchus* spp. (grape weevils)	Spinetoram 24–87.5 Spinosad 48–120

Table 5.11 Key pest species controlled by spinosad or spinetoram in major vegetable crops.

Crop group	Key pests controlled	Use rate (g ai ha^{-1})
Bulb vegetables	*Spodoptera exigua* (beet armyworm); *Liriomyza* spp. (dipterous leafminers); *Ostrinia nubilalis* (European corn borer); *Epitrix* spp. (flea beetles); *Trichoplusia ni* (cabbage looper); *Thrips tabaci* (onion thrips); *Frankliniella fusca* (tobacco thrips)	Spinetoram 9–88 Spinosad 36–140
Brassica vegetables	*Plutella xylostella* (diamondback moth); *Trichoplusia ni* (cabbage looper); *Pieris rapae* (imported cabbageworm); *Pieris brassicae* (large white cabbage butterfly); *Spodoptera exigua* (beet armyworm); *Spodoptera littoralis* (Egyptian cotton leafworm); *Liriomyza* spp. (dipterous leafminers); *Thrips tabaci* (onion thrips)	Spinetoram 9–88 Spinosad 24–175
Cucurbit vegetables	*Spodoptera exigua, S. frugiperda, S. littoralis* (armyworms); *Trichoplusia ni* (cabbage looper); *Diaphania hyalinata* (melonworm); *Diaphania nitidalis* (pickleworm); *Diabrotica undecimpunctata howardi, Acalymma vittata* (cucumber beetles); *Liriomyza* spp. (dipterous leafminers); *Frankliniella occidentalis* (western flower thrips); *Thrips palmi* (palm thrips)	Spinetoram 15–88 Spinosad 36–140
Fruiting vegetables	*Leptinotarsa decemlineata* (Colorado potato beetle); *Ostrinia nubilalis* (European corn borer); *Manduca* spp. (hornworms); *Trichoplusia ni, Chrysodeixis chalcites* (loopers); *Helicoverpa zea, Heliothis virescens* (tomato fruitworms); *Keiferia lycopersicella* (tomato pinworm); *Tuta absoluta* (tomato moth); *Spodoptera exigua, S. frugiperda* (armyworms); *Frankliniella occidentalis* (western flower thrips); *Thrips palmi* (palm thrips); *Liriomyza* spp. (dipterous leafminers)	Spinetoram 9–88 Spinosad 25–175
Leafy vegetables	*Trichoplusia ni* (cabbage looper); *Plutella xylostella* (diamondback moth); *Pieris rapae* (imported cabbageworm); *Spodoptera exigua* (beet armyworm); *Spodoptera litura* (common cutworm); *Liriomyza* spp. (dipterous leafminers); *Thrips tabaci* (onion thrips)	Spinetoram 9–88 Spinosad 25–175
Root vegetables	*Spodoptera exigua* (beet armyworm); *Liriomyza* spp. (dipterous leafminers); *Ostrinia nubilalis* (European corn borer); *Epitrix cucumeris* (potato flea beetle); *Trichoplusia ni* (cabbage looper); *Thrips tabaci* (onion thrips)	Spinetoram 40–70 Spinosad 53–175

Frankliniella spp.) and lepidopterous larvae including *Helicoverpa* spp., *Heliothis virescens*, (tobacco budworm), *Alabama argillacea* (cotton leafworm), *Earias* spp. (spotted/spiny bollworms) and *Spodoptera* spp. In field corn, the primary pest targets of the spinosyns are lepidopterous larvae such as *Helicoverpa* spp., *Ostrinia nubilalis*, (European corn borer), *Sesamia nonagriodes* (corn stalk borer) and *Spodoptera frugiperda* (fall armyworm). In soybean, spinosyns are also used to control lepidopterous larvae such as *Pseudoplusia includens* (soybean looper) and *Anticarsia gemmatalis* (velvetbean caterpillar). In rice, foliar sprays of spinosad are applied to control *Cnaphalocrocis medinalis* (grass leaf roller) and *Scirpophaga incertulas* (yellow stem borer).

5.10.1.3 Floriculture, Turfgrass, Ornamental Plants and Forestry

Spinosad is registered for use on many noncrop plants. A major use is to control thrips species including *Frankliniella occidentalis*, *Scirtothrips dorsalis* (yellow tea thrips/chilli thrips) and *Thrips* spp. on floricultural crops grown in greenhouses, plastic tunnels, shadehouses or outdoors. Spinosad is also used to control several species of lepidopterous larvae, *Liriomyza* spp. and ephydrid flies (shore flies).

In turfgrass, spinosad is used primarily to control lepidopterous larvae, including *Crambus* spp., *Fissicrambus* spp. and related genera (sod webworms), *Spodoptera frugiperda*, *Agrotis ipsilon* and *Peridroma saucia* (variegated cutworm). It is also used to control coleopterous turf pests such as *Listronotus maculicollis* (annual bluegrass weevil) and *Ataenius spretulus* (black turfgrass ataenius). *Solenopsis* spp. (fire ant) colonies in turf may be controlled by drench applications of spinosad directly to mounds.

Spinosad is used to control a range of insect pests attacking woody ornamental shrubs and trees grown in nurseries, residences, public parks and in forest environments. These pests include several species of chrysomelid beetles, such as *Pyrrhalta luteola* (elm leaf beetle); a number of sawfly species, such as *Neodiprion* spp. (pine sawflies); several thrips species; gall midges (Cecidomyidae), and a wide variety of lepidopterous pests including *Choristoneura* spp. (conifer budworms), *Malacosoma* spp. (tent caterpillars), *Lymantria dispar* (gypsy moth), bagworms (Psychidae), leaf blister miners (Lyonetiidae) and leaf blotch miners (Gracillaridae).

5.10.2 Treatment of Seeds and Seedlings

In Japan, it is standard practice to apply granular formulations of insecticides and fungicides to rice nursery boxes, the shallow trays where rice seedlings are grown before being transplanted into flooded paddies. Spinosad, either alone or in combination with other insecticides and/or fungicides, is registered in Japan for this purpose. Immediately before transplanting, spinosad granules are applied at a rate of 50 g of formulated product per nursery box (which delivers 75 g of spinosad per ha). Spinosad applied in this manner controls *Cnaphalocrocis medinalis* (grass leaf roller), *Chilo suppressalis* (Asiatic rice borer), *Parnara guttata* (paddy skipper) and *Naranga aenescens* (green rice caterpillar) for up to five weeks after transplanting.

In Europe, brassica seedlings are typically grown in modular trays (composed of many individual cells) until being transplanted into the field. Immediately before transplanting, spinosad is applied to these trays as a water drench treatment at the rate of 6 g ai (active ingredient) per 1000 plants. This treatment protects the seedlings from *Delia radicum* (cabbage maggot) and lepidopterous larvae for up to six weeks after transplanting.

Spinosad is being developed as a seed treatment to control *Delia platura* (seedcorn maggot) and *Delia antiqua* (onion maggot) in several crops including bulb vegetables, beans and peas and cucurbit vegetables. Spinosad is applied in commercial seed coatings at rates between 0.1 and 0.75 mg ai per seed.

5.10.3 Insecticidal Baits

5.10.3.1 Sprayable Baits

Sprayable bait formulations of spinosad have been developed to control several economically important and invasive fruit fly species. GF-120® NF Naturalyte® Fruit Fly Bait (also sold as Spintor® Cebo, Spintor® Fly, Success®, Success® Appat, Naturalure™ and Flipper™) is a pre-mixed fruit fly bait concentrate. The food-based attractants in GF-120® (sugars, plant proteins and extracts) have been optimized to attract and control male and female fruit flies in the families Tephritidae, Drosophilidae and Lonchaeidae. GF-120® is currently registered in 33 countries to control fruit flies in multiple crops.

Another innovation in fruit fly control is SPLAT-MAT™ Spinosad ME, a novel sprayable formulation containing spinosad and methyl eugenol (ME), a naturally occurring attractant for male fruit flies in the genus *Bactrocera*. The basis of the formulation is a proprietary matrix that protects spinosad and methyl eugenol from degradation and extends the effectiveness of each for a longer period of time. Control is achieved by means of the *male annihilation technique* (MAT), in which male fruit flies are attracted and killed. With no males available for mating, female flies are unable to produce offspring. The level of fruit protection achieved matches or exceeds existing pyrethroid and organophosphorous insecticide options. Male fruit flies are attracted from long distances, making SPLAT-MAT™ Spinosad ME suitable for off-crop application, which eliminates pesticide and attractant residues on the fruit.

5.10.3.2 Granular Baits

Granular bait formulations of spinosad (Conserve® Fire Ant Bait and Justice™ Fire Ant Bait) have been developed to control *Solenopsis invicta* (red imported fire ant) and *Solenopsis richteri* (black imported fire ant). These baits are applied primarily as mound treatments to control fire ants in agricultural, ornamental plant production and residential settings.

Spinosad bait formulations have also been developed for control of adult *Musca domestica*. Elector® Bait Granular Fly Control, produced by Elanco Animal

Health, employs (Z)-9-tricosene and fluorescent yellow pigment to attract the adult flies. It is used around animal housing, handling and feeding facilities, such as poultry houses, dairies and dairy barns, hog farms and pens, cattle barns and pens, feed lots, and horse barns and stables. Elector® bait granules may be scattered in areas where the adult flies congregate, placed in trays or bait stations, adhered to hanging cards, or dissolved with water and applied to surfaces.

5.10.4 Organic Agriculture

Spinosad has become a major component in organic agricultural production; few other insect pest management agents recognized as suitable for organic agriculture provide a comparable level of control and reliability. To date, spinosad as an active ingredient and certain formulations containing spinosad, have received organic certification from a variety of government and private organizations in 18 countries, including the USA, Canada, Mexico, Brazil, Argentina, Chile, Australia and Spain and from the European Union. In the US, organic certifications for spinosad include technical spinosad (USDA National Organic Standards Board), and several formulated products containing spinosad, including Entrust®, GF-120® NF Fruit Fly Bait, Conserve®/ Justice™ Fire Ant Bait and a 5% suspension concentrate formulation for home and garden use (Organic Materials Review Institute, OMRI).[136] In 2009, spinosad was approved for the European Union Positive Organic List by the EU Organic Standing Committee.

5.10.5 Stored Grain Protection

Spinosad has been developed to control insect pests in stored grain. The spectrum of stored grain pests controlled by spinosad under commercial conditions is still under study, but it is clear from research trials that adult and/or immature *Rhyzopertha dominica* (lesser grain borer), *Prostephanus truncatus* (larger grain borer), *Cryptolestes ferrugineus* (rusty grain beetle), *Tribolium castaneum* (red flour beetle), *Tribolium confusum* (confused flour beetle), *Oryzaephilus surinamensis* (sawtoothed grain beetle), *Oryzaephilus mercator* (merchant grain beetle), *Sitophilus oryzae* (rice weevil), *Sitophilus zeamais* (maize weevil), *Plodia interpunctella* (Indianmeal moth), *Sitotroga cerealella* (Angoumois grain moth) and the psocid pests, *Liposcelis entomophila* (grain psocid) and *Lepinotus reticulatus* (reticulate-winged trogiid), are susceptible to spinosad.[137–142] Spinosad breaks the life cycles of stored grain pest species by controlling at least one life stage. Depending on conditions and pest species, spinosad provides protection ranging from six months to two years.

The maximum application rate for spinosad as a grain protectant is 1 mg ai kg^{-1} grain (1 ppm ai). Spinosad is registered for grain protection in a number of countries, but widespread commercial use awaits final MRL approvals in all grain-importing countries.

5.10.6 Animal Health

Spinosad has been developed by Elanco Animal Health for a number of significant animal health uses that involve either direct application to animals, application to animal production premises, or bait applications (as described in section 5.10.3.2) to control adult *Musca domestica*.

For sheep, spinosad is marketed under the Extinosad® brand and is applied as a plunge dip, shower, or handgun spray to control *Bovicola ovis* (sheep biting louse) infestations. Spinosad is also used as handgun spray and wound dressing for the treatment and prevention of "blow fly strike", wounds caused by the larvae of *Lucilia cuprina* (Australian sheep blow fly). Control of *B. ovis* lasts up to 20 weeks and protection from blow fly strike lasts two to four weeks. Due to minimal risks associated with worker exposure, spinosad may be used right up to shearing. There is no withholding period for wool and meat from spinosad-treated sheep.

For cattle, spinosad (Elector® insecticide) is applied as a pour-on or on-animal spray to control *Bovicola bovis* (cattle biting louse), *Linognathus vituli* (long-nosed cattle louse), *Solenopotes capillatus* (little blue cattle louse), *Haematopinus eurysternus* (short-nosed cattle louse) and *Haematobia irritans* (horn fly) on lactating and non-lactating dairy and beef cattle. Elector® may also be used as a premise spray to control *Musca domestica* and *Stomoxys calcitrans* (stable fly) in and around livestock and poultry facilities. Spinosad is also active against *Rhipicephalus* (=*Boophilus*) spp. (cattle ticks) and is registered for this use in several South American countries.[143]

In poultry production, spinosad (Elector® PSP) is used as a premise spray in poultry houses to control *Alphitobius diaperinus* (lesser mealworm/darkling beetle), *Dermestes* spp. (hide beetles), *Musca domestica*, *Stomoxys calcitrans* and *Fannia canicularis* (little house fly). Recently, spinosad has been shown to control *Dermanyssus gallinae* (chicken mite) infesting poultry premises in the United Kingdom.[144] Elector® PSP is also used as a premise spray in livestock housing to control the fly species listed earlier.

For companion animals, spinosad (Comfortis® insecticide) is administered orally to dogs in a chewable, beef-flavored tablet to control *Ctenocephalides felis* (cat flea). Flea mortality begins within 30 minutes of administration, reaches 100% within four hours and control lasts for up to 30 days. Oral administration of spinosad to dogs has also been shown to control *Rhipicephalus sanguineus* (brown dog tick).[145]

5.10.7 Public Health

5.10.7.1 Mosquito Control

Spinosad controls the larvae of many important species in the mosquito genera *Aedes*, *Culex*, *Ochlerotatus* and *Anopheles*.[146] Several new spinosad formulations have been developed by Clarke Mosquito Control Products or Dow AgroSciences to control mosquito larvae in diverse aquatic environments, such as artificial and natural containers, temporary flooded areas, swampy woodlands, marshes, sewage lagoons, storm water and drainage canals, catch basins and coastal impoundments. These formulations can be divided into single-brood and multi-brood formulations. Single-brood formulations, such as Mozkill™ 120SC, Natular™ 2EC and Natular™ G, provide immediate release of spinosad and control one mosquito generation. Multi-brood formulations, such as Natular™ T30 (30-day release tablet), Natular™ XRG (30-day extended release granule) and Natular™ XRT (180-day extended release tablet), release spinosad over time to control multiple generations. Natular™ DT is a specially designed bi-layered tablet for multi-brood mosquito control in cisterns and other potable water containers. The outer layer of the tablet effervesces and releases spinosad immediately, while the inner layer provides controlled, sustained release of spinosad over 60 days. Clarke Mosquito Control Products received a Presidential Green Chemistry Challenge Award in 2010 for its innovative Natular™ formulations containing spinosad. Natular™ 2EC, Natular™ T30, Natular™ XRG, Natular™ XRT are certified by the Organic Material Research Institute (OMRI).

Spinosad has been tested by WHOPES (World Health Organization Pesticide Evaluation Scheme) and has been recommended as an effective and safe mosquito larvicide.[131] WHOPES approved two single-brood formulations of spinosad, the 120 g ai 1^{-1} suspension concentrate (Mozkill™ 120SC) and the 0.5% granule (Natular™ G) in 2007, and the multi-brood 7.48% spinosad tablet (Natular™ DT) in 2010 for direct application to potable water containers.

5.10.7.2 Treatment of Human Head Lice

Spinosad is being developed by ParaPRO LLC as a promising new treatment for *Pediculus humanus capitis* (human head lice). One application of a crème rinse containing 0.9% spinosad was highly effective in controlling *P. humanus capitis* in two Phase 3 clinical trials.[147] The U.S. Food and Drug Administration has given regulatory approval to a proprietary formulation of spinosad as a prescription treatment, branded as Natroba™.

5.11 Conclusions

Green chemistry, as it relates to the manufacture and use of insecticides, implies minimizing waste, using non-hazardous and renewable raw materials, efficient chemical synthesis, and creating products that have minimal environmental and nontarget organism effects.[148–150]

The spinosyn insecticides minimize waste and increase efficiency in agricultural production by reducing the losses caused by many important insect pests in almost all food and fiber crops. Because of the microbial origin of the spinosyns, fermentation using renewable agricultural feedstocks is the primary component of spinosad and spinetoram manufacturing. Spinosad and spinetoram have relatively short residence times in the environment and are degraded by physical and microbial processes into simple fragments. This rapid environmental degradation minimizes exposure to nontarget organisms and residue levels on treated crops. The low toxicity of the spinosyns to mammals provides a wide margin of safety for pesticide applicators and other agricultural workers. Likewise, the low toxicity of the spinosyns to other nontarget vertebrates, such birds and fish, minimizes risks for these species. The spinosyns generally have greater effects on pest insect populations than beneficial arthropod populations, facilitating use in IPM programs. Compatibility with targeted delivery systems, such as insecticidal baits, further facilitates spinosyn use in IPM programs.

The natural origin of spinosad has allowed it to be certified for use in organic agriculture, a crop production system gaining interest among consumers. In its most recently developed use patterns, namely fruit fly control, stored grain protection, mosquito control and human head lice treatment, spinosad is the first truly new active ingredient in decades. Spinosad has significantly better human and environmental safety profiles than currently used products and provides equal or superior efficacy against the target insect pest species, many of which have developed resistance to current products.

Moving beyond spinosad, innovative approaches to understand spinosyn structure–activity relationships led to the discovery of spinetoram, a semi-synthetic spinosyn that has broader utility than spinosad, but retains the favorable toxicological and environmental attributes of spinosad. Potential new uses for the spinosyns are being identified through the continuing efforts of Dow AgroSciences scientists, academic and government researchers, and entrepreneurial technology companies. Advancements in formulation science and targeted delivery systems, as well as changes in regulatory requirements and crop production trends around the world, will provide the opportunity to find new and useful applications for the spinosyns.

Acknowledgements

The authors thank J. Bailey, M. Douglas, L. Laughlin and S. Rontondaro for review and content contributions to this manuscript; G. Deboer for providing results of unpublished insect metabolism studies with spinosad and spinetoram; M. Miles for assistance with the section on non-target arthropod effects; D. Hahn for providing the scanning electron micrographs in Figure 1; N. Simmons, J. Babcock and D. Juberg for review and helpful comments; and finally, J. Tolliver and R. Krell for editorial assistance.

This manuscript draws upon the contributions made by many Dow AgroSciences and Dow Chemical scientists during the last 25 years. The authors also acknowledge the many university, government, contract and corporate researchers around the world who have contributed to our understanding of the spinosyns and the development of spinosad and spinetoram.

Laser™, Tracer®, Conserve®, SpinTor®, Success®, SpinoAce™, Entrust®, Delegate®, Radiant®, Exalt™, Endure™, Palgus™, GF-120®, Naturalyte®, Spintor®, Naturalure™, Flipper™, SPLAT-MAT™, Justice™ and Mozkill™ are trademarks of Dow AgroSciences LLC. Elector®, Extinosad®, and Comfortis® are trademarks of Elanco. Natular™ is a trademark of Clarke Mosquito Control Products, Inc. NatrOVA™ is a trademark of ParaPRO LLC.

References

1. H. A. Kirst, K. H. Michel, J. W. Martin, L. C. Creemer, E. H. Chio, R. C. Yao, W. M. Nakatsukasa, L. D. Boeck, J. L Occolowitz, J. W. Paschal, J. B. Deeter, N. D. Jones and G. D. Thompson, *Tetrahedron Lett.*, 1991, **32**, 4839.
2. US Pat. 6 455 504, 2002.
3. P. Lewer, D. R. Hahn, L. L. Karr, D. O. Duebelbeis, J. R. Gilbert, G. D. Crouse, T. Worden, T. C. Sparks, P. McKamey, R. Edwards and P.R. Graupner, *Bioorg. Med. Chem.*, 2009, **17**, 4185.
4. F. P. Mertz and R. C. Yao, *Int. J. Syst. Bacteriol.*, 1990, **40**, 34.
5. http://www.epa.gov/opprd001/workplan/completionsportrait.pdf (last accessed August 2010).
6. K. Smith, D. A. Evans and G. A. El-Hiti, *Philos. Trans. R. Soc. London, Ser. B*, 2008, **363**, 623.
7. A. Chloridis, P. Downard, J. E. Dripps, K. Kaneshi, L. C. Lee, Y. K. Min and L. A. Pavan in *XVI International Plant Protection Congress, Congress Proceedings*, British Crop Production Council, Alton, UK, 2007, p. 68.
8. J. Dripps, B. Olson, T. C. Sparks and G. Crouse, *Plant Health Prog.*, 2008; doi:10.1094/PHP-2008-0822-01-PS.
9. K. X. Huang, L. Xia, Y. Zhang, X. Ding and J. A. Zahn, *Appl. Microbiol. Biotechnol.*, 2009, **82**, 13.
10. G. D. Crouse, T. C. Sparks, J. Schoonover, J. Gifford, J. Dripps, T. Bruce, L. L. Larson, J. Garlich, C. Hatton, R. L. Hill, T. V. Worden and J. G. Martynow, *Pest Manag. Sci.*, 2001, **57**, 177.
11. G. D. Crouse, J. E. Dripps, N. Orr, T. C. Sparks and C. Waldron in *Modern Crop Protection Compounds,* ed. W. Krämer and U. Schirmer, Wiley-VCH Verlag GmbH & Co. KGaA, Weinheim, Germany, 2007, p. 1013.
12. T. C. Sparks, G. D. Thompson, H. A. Kirst, M. B. Hertlein, J. S. Mynderse, J. R. Turner and T. V. Worden in *Biopesticides: Use and Delivery*, ed. F. R. Hall and J. J. Menn, Humana Press, Totowa, NJ, USA, 1999, p. 171.
13. H. A. Kirst, *J. Antibiotics,* 2010, **63**, 101.
14. C. V. DeAmicis, J. E. Dripps, C. J. Hatton and L. L. Karr in *Phytochemicals for Pest Control*, ed. P. A. Hedin, R. M. Holligworth, E. P. Masler, J. Miyamoto and D. G. Thompson, American Chemical Society, Washington, DC, USA, 1997, p. 144.
15. M. H. Fisher and H. Mrozik in *Ivermectin and Abamectin*, ed. W. C. Campbell, Springer-Verlag, New York, USA, 1989, p. 1.

16. D. Rugg, S. D. Buckingham, D. B. Sattelle and R. K. Jansson in *Comprehensive Insect Molecular Science*, ed. L. I. Gilbert, K. Iatrou and S. S. Gill, Elsevier, New York, USA, 2005, p. 25.

17. G. B. Watson, S. W. Chouinard, K. R. Cook, C. Geng, J. M. Gifford, G. D. Gustafson, J. M. Hasler, I. M. Larrinua, T. J. Letherer, J. C. Mitchell, W. L. Pak, V. L. Salgado, T. C. Sparks and G. E. Stillwell, *Insect Biochem. Mol. Biol.*, 2010, **40**, 376.

18. G. W. Ware and D. W. Whitacre, *The Pesticide Book*, Meister Publishing, Willoughby, OH, USA, 2003.

19. A. Elbert, R. Nauen, A. McCaffery in *Modern Crop Protection Compounds*, ed. W. Krämer and U. Schrimer, Wiley-VCH, Weinheim, Germany, 2007, p. 753.

20. http://eclassification.irac-online.org (last accessed August 2010).

21. C. Waldron, P. Matsushima, P. R. Rosteck Jr., M. C. Broughton, J. Turner, K. Madduri, K. P. Crawford, D. J. Merlo and R. H. Baltz, *Chem. Biol.*, 2001, **8**, 487.

22. V. L. Salgado and T. C. Sparks in *Comprehensive Insect Molecular Science*, ed. L. I. Gilbert, K. Iatrou and S. S. Gill, Elsevier, New York, USA, 2005, p. 137.

23. D. R. Hahn, G. Gustafson, C. Waldron, R. Bullard, J. D. Jackson and J. Mitchell, *J. Ind. Microbiol. Biotechnol.*, 2006, **33**, 94.

24. L. S. Sheehan, R. E. Lill, B. Wilkinson, R. M. Sheridan, W. A. Vousden, A. L. Kaja, G. D. Crouse, J. Gifford, P.R. Graupner, L. Karr, P. Lewer, T. C. Sparks, P. F. Leadlay, C. Waldron and C. J. Martin, *J. Nat. Prod.*, 2006, **69**, 1702.

25. J. Daeuble, T. C. Sparks, P. Johnson, P. R. Graupner, *Bioorg. Med. Chem.* 2009, **17**, 4197.

26. G. D. Crouse, T. C. Sparks, C. V. DeAmicis, H. A. Kirst, J. G. Martynow, L. C. Creemer, T. V. Worden and P. B. Anzeveno in *Pesticide Chemistry and Bioscience: The Food-Environmental Challenge*, ed. G. T. Brooks and T. R. Roberts, Royal Society of Chemistry, Cambridge, UK, 1999, p. 155.

27. L. F. Tietze, G. Brasche, A. Grube, N. Bohnke and C. Stadler, *Chem. Eur. J.*, 2007, **13**, 8543.

28. T. C. Sparks, P. B. Anzeveno, J. G. Martynow, J. M. Gifford, M. B. Hertlein, T. V. Worden and H. A. Kirst, *Pestic. Biochem. Physiol.*, 2000, **67**, 187.

29. T. C. Sparks, G. D. Crouse and G. Durst, *Pest Manag. Sci.*, 2001, **57**, 896.

30. T. C. Sparks, G. D. Crouse, J. E. Dripps, P. Anzeveno, J. Martynow, C. V. DeAmicis and J. Gifford, *J. Computer-Aided Mol. Des.*, 2008, **22**, 393.

31. L. C. Creemer, H. A. Kirst, J. W. Paschal and T. V. Worden, *J. Antibiotics*, 2000, **53**, 171.

32. P. B. Anzeveno and F. R. Green III in *Synthesis and Chemistry of Agrochemicals VI*, ed. D. R. Baker, J. G. Fenyes, G. P. Lahm, T. P. Selby and T. M. Stevenson, American Chemical Society, Washington DC, USA, 2002, p. 262.

33. H. Kubinyi, *QSAR: Hansch Analysis and Related Approaches*, VCH Publishers, New York, USA, 1993.
34. R. Todeschini and V. Consonni, *Handbook of Molecular Descriptors*, Wiley-VCH, Weinheim, Germany, 2000.
35. T. C. Sparks, G. D. Crouse, J. Dripps, J. Daeuble, M. P. Oliver and J. Gifford. Abstract I-1-i-36C, 11th IUPAC International Congress of Pesticide Chemistry, August 2006.
36. D. W. Salt, N. Yildiz, D. J. Livingstone and C. J. Tinsley, *Pestic. Sci.,* 1992, **36**, 161.
37. J. DeVillers, ed. *Neural Networks in QSAR and Drug Discovery*, Academic Press, New York, USA, 1996.
38. V. L. Salgado, *Pestic. Biochem. Physiol.*, 1998, **60**, 91.
39. V. L. Salgado, J. J. Sheets, G. B. Watson and A. L. Schmidt, *Pestic. Biochem. Physiol.*, 1998, **60**, 103.
40. N. Orr, A. J. Shaffner, K. Richey and G. D. Crouse, *Pestic. Biochem. Physiol.*, 2009, **95**, 1.
41. V. L. Salgado, G. B. Watson and J. J. Sheets in *Proceedings of the 1997 Beltwide Cotton Production Conference,* National Cotton Council, Memphis, TN, USA, 1997, p. 1082.
42. V. L. Salgado and R. Saar, *J. Insect Physiol.*, 2004, **50**, 867.
43. T. Perry, J. A. McKenzie and P. Batterham, *Insect Biochem. Mol. Biol.*, 2007, **37**, 184.
44. S. W. Baxter, M. Chen, A. Dawson, J.-Z. Zhao, H. Vogel, A. M. Shelton, D. G. Heckel and C. D. Jiggins, *PLoS Genet.*, 2010, **6**, e1000802. doi:10.1371/journal.pgen.1000802.
45. J. G. Scott, *J. Pestic. Sci.*, 2008, **33**, 221.
46. G. B. Watson, *Pestic. Biochem. Physiol.*, 2001, **71**, 20.
47. G. B. Watson, unpublished work.
48. F. D. Rinkevich, M. Chen, A. M. Shelton and J. G. Scott. *Invert. Neurosci.*, 2010, **10**, 25.
49. T. Shono and J. G. Scott, *Pestic. Biochem. Physiol.*, 2003, **75**, 1.
50. H. P. Young, W. B. Bailey, R. M. Roe, T. Iwasa, T. C. Sparks, G. D. Thompson and G. B. Watson in *Proceedings of the 2001 Beltwide Cotton Production Conference,* National Cotton Council, Memphis, TN, USA, 2001, p. 1167.
51. R. M. Roe, H. P. Young, T. Iwasa, C. F. Wyss, C. F. Stumpf, T. C. Sparks, G. B. Watson, J. J. Sheets and G. D. Thompson, *Pestic. Biochem. Physiol.*, 2010, **96**, 8.
52. P. Bielza, Q. E. Fernández, C. Grávalos and J. Contreras, *J. Econ. Entomol.*, 2007, **100**, 916.
53. A. H. Sayyed, S. Saeed, M. Noor-ul-ane and N. Crickmore, *J. Econ. Entomol.*, 2008, **101**, 1658.
54. C. F. Wyss, H. P. Young, J. Shukla and R. M. Roe in *Proceedings of the 2001 Beltwide Cotton Production Conference,* National Cotton Council, Memphis, TN, USA, 2001, p. 1163.

55. J.-Z. Zhao, Y.-X. Li, H. L. Collins, L. Gusukuma-Minuto, R. F. L. Mau, G. D. Thompson and A. M. Shelton, *J. Econ. Entomol.*, 2002, **95**, 430.
56. S.-Y. Zhang, S. Kono, T. Murai and T. Miyata, *Insect Sci.*, 2008, **15**, 125.
57. A. H. Sayyed, D. Omar and D. J. Wright, *Pest Manag. Sci.*, 2004, **60**, 827.
58. C. F. Wyss, H. P. Young, J. Shukla and R. M. Roe, *Crop Protect.*, 2003, **22**, 307.
59. D. Wang, X. H. Qiu, H. Y. Wang, K. Qiao and K. Y. Wang, *Phytoparasitica*, 2010, **38**, 103.
60. P. Bielza, V. Quinto, C. Grávalos, J. Abellán and E. Fernández, *J. Econ. Entomol.*, 2008, **101**, 499.
61. P. Bielza, V. Quinto, C. Grávalos, E. Fernández, J. Abellán and J. Contreras, *Bull. Entomol. Res.*, 2008, **98**, 355.
62. J. S. Ferguson, *J. Econ. Entomol.*, 2004, **97**, 112.
63. W. Wang, J. Mo, J. Cheng, P. Zhung and Z. Tang, *Pestic. Biochem. Physiol.*, 2006, **84**, 180.
64. D. Wang, X. Aiu, X. Ren, F. Niu and K. Wang, *Pestic. Biochem. Physiol.*, 2009, **95**, 90.
65. J.-Z. Zhao, H. L. Collins, Y.-X. Li, R. F. L. Mao, G. D. Thompson, M. Hertlein, J. T. Andaloro, R. Boykin and A. M. Shelton, *J. Econ. Entomol.*, 2006, **99**, 176.
66. T. C. Sparks, G. D. Thompson, L. L. Larson, H. A. Kirst, O. K. Jantz, T. V. Worden, M. B. Hertlein and J. D. Busacca in *Proceedings of the 1995 Beltwide Cotton Production Conference*, National Cotton Council, Memphis, TN, USA, 1995, p. 903.
67. J. G. Scott, *J. Pestic. Sci.*, 1998, **54**, 131.
68. C.-H. Kao and E. Y. Cheng, *J. Agric. Res. China*, 2001, **50**, 80.
69. M. N. R. Attique, A. Khaliq and A. H. Sayyed, *J. Appl. Entomol.*, 2006, **130**, 122.
70. T. Shono, L. Zhang and J. G. Scott. *Pestic. Biochem. Physiol.*, 2004, **80**, 106.
71. J. E. Dunley, J. F. Brunner, M. D. Doerr and E. H. Beers, *J. Insect. Sci.*, 2006, **6**, 14.
72. D. Mota-Sanchez, R. M. Hollingworth, E. J. Grafius and D. D. Moyer, *Pest Manag. Sci.*, 2006, **62**, 30.
73. A. Khaliq, M. N. R. Attique and A. H. Sayyed, *Bull. Entomol. Res.*, 2007, **97**, 191.
74. R. Nauen, S. Konanz, T. Hirooka, T. Nishimatsu and H. Kodama, *Pflanzenschutz-Nachrichten Bayer*, 2007, **60**, 247.
75. M. Kristensen and J. B. Jepersen, *J. Econ. Entomol.*, 2004, **97**, 1042.
76. G. E. Abo-Eighar, Z. A. Elbermaway, A. G. Yousef, H. K. A. Elhady, *J. Asia-Pac. Entomol.*, 2005, **8**, 397.
77. C. Y. Kang, G. Wu and T. Miyata, *J. Appl. Entomol.*, 2006, **130**, 377.
78. A. H. Sayyed and D. J. Wright, *J. Econ. Entomol.*, 2004, **97**, 2043.
79. G. D. Thompson, R. Dutton and T. C. Sparks, *Pest Manag. Sci.*, 2000, **56**, 696.

80. J. K. Moulton, D. A. Pepper and T. J. Dennehy, *Pest Manag. Sci.*, 2000, **56**, 842.

81. M. E. Whalon, D. Mota-Sanchez and R. M. Hollingworth in *Global Pesticide Resistance in Arthropods*, ed. M. E. Whalon, D. Mota-Sanchez and R. M. Hollingworth, CABI, Wallingford, UK, 2008, p. 5.

82. R. F. Mau and L. Gusukuma-Minuto in *The Management of Diamondback Moth and Other Crucifer Pests: Proceedings of the 4th International Workshop*, ed. N. M. Endersby and P. M. Ridland, The Regional Institute, Melbourne, Australia, 2004, p. 307.

83. G. Boiteau and C. Noronha, *Pest Manag. Sci.*, 2007, **63**, 1230.

84. L. D. Godfrey, E. E. Grafton-Cardwell, H. K. Kaya and W. E. Chaney, *California Agri.*, 2005, **59**, 35.

85. S. Pineda, G. Smagghe, M.-I. Schneider, P. Del Estal, E. Viñuela, A.-M. Martínez and F. Budia, *Environ. Entomol.*, 2006, **35**, 856.

86. K. W. Wanner, B. V. Helson and B. J. Harris, *Pest Manag. Sci.*, 2002, **58**, 817.

87. T. C. Sparks, J. J. Sheets, J. R. Skomp, T. V. Worden, M. B. Hertlein, L. L. Larson, D. Bellows, S. Thibault and L. Wally in *Proceedings of the 1997 Beltwide Cotton Production Conference*, National Cotton Council, Memphis, TN, USA, 1997, p. 1259.

88. G. D. Crouse and T. C. Sparks, *Rev. Toxicol.*, 1998, **2**, 133.

89. T. C. Sparks, G. D. Thompson, H. A. Kirst, M. B. Hertlein, L. L. Larson, T. V. Worden and S. A. Thibault, *J. Econ. Entomol.*, 1998, **91**, 1277.

90. G. DeBoer, unpublished work.

91. T. A. Hill and R. E. Foster, *J. Econ. Entomol.*, 2000, **93**, 763.

92. A. L. Knight, *Pest Manag. Sci.*, 2010, **66**, 709.

93. S. Pineda, M.-I. Schneider, G. Smagghe, A.-M. Martínez, P. Del Estal, E. Viñucla, J. Valle and F. Budia, 2007, *J. Econ. Entomol.*, 2007, **100**, 773.

94. D. Wang, P. Gong, M. Li, X. Qiu and K. Wang, *Pest Manag. Sci.*, 2009, **65**, 223.

95. X.-H. Yin, Q.-J. Wu, X.-F. Li, Y.-J. Zhang and B.-Y. Xu, *Crop Protect.*, 2008, **27**, 1385.

96. M. Miles and R. Dutton, *Pesticides and Beneficial Organisms, IOBC/wprs Bulletin*, 2003, **26**, 9.

97. M. Miles and H. Eelen, *Comm. Agric. Appl. Biol. Sci.*, 2007, **71**, 275.

98. T. Williams, J. Valle and E. Viñuela, *Biocontrol Sci. Technol.*, 2003, **13**, 459.

99. R. T. Villanueva and J. F. Walgenbach, *J. Econ. Entomol.*, 2005, **98**, 2114.

100. C. Duso, V. Malagnini, A. Pozzebon, F. M. Buzzetti and P. Tirello, *Biocontrol Sci. Technol.*, 2008, **18**, 1027.

101. M. Miles, *Pesticides and Beneficial Organisms, IOBC/wprs Bulletin*, 2006, **29**, 53.

102. M. Srivastava, L. Bosco, J. Funderburk, S. Olson and A. Weiss, *Plant Health Prog.*, 2008; doi:10.1094/PHP-2008-0118-02-RS.

103. http://ec.europa.eu/sanco_pesticides/public/index.cfm (last accessed August 2010).
104. http://www.hc-sc.gc.ca/cps-spc/pubs/pest/_decisions/erc2008-01/index-eng.php (last accessed August 2010).
105. M. A. Mayes, G. D. Thompson, B. Husband and M. M. Miles, *Rev. Environ. Contam. Toxicol.*, 2003, **179**, 37.
106. J. Bailey, C. Scott-Dupree, R. Harris, J. Tollman and B. Harris, *Apidologie*, 2005, **36**, 623.
107. C. D. Scott-Dupree, L. Conroy and C. R. Harris, *J. Econ. Entomol.*, 2009, **102**, 177.
108. M. Miles, *Bull. Insectol.*, 2003, **56**, 119.
109. http://permits.nra.gov.au/registration/assessment/public/index.php#ag (last accessed August 2010).
110. L. A. Morandin, M. L. Wilson, M. T. Franklin and V. A. Abbott, *Pest Manag. Sci.*, 2005, **61**, 619.
111. http://ag.arizona.edu/pubs/crops/az1143 (last accessed August 2010).
112. http://ag.arizona.edu/pubs/crops/az1292 (last accessed August 2010).
113. T. Van Leeuwen, W. Dermauw, M. Van de Veire and L. Tirry, *Exp. Appl. Acarol.*, 2005, **37**, 93.
114. T. Van Leeuwen, M. Van de Veire, W. Dermauw and L. Tirry, *Phytoparasitica*, 2006, **34**, 102.
115. P. G. Weintraub and N. Mujica, *Phytoparasitica*, 2006, **34**, 21.
116. http://www.epa.gov/ocspp/pubs/frs/home/guidelin.htm (last accessed August 2010).
117. http://www.oecd.org/document/40/0,3343,en_2649_34377_37051368_1_1_1_1,00.html (last accessed August 2010).
118. http://ec.europa.eu/food/plant/protection/resources/publications_en.htm#council (last accessed August 2010).
119. http://www.epa.gov/opprd001/factsheets (last accessed August 2010).
120. http://www.epa.gov/opprd001/factsheets/spinetoram.html (last accessed August 2010).
121. http://www.regulations.gov/search/Regs/home.html#documentDetail?R=09000064802fe7a5 (last accessed August 2010).
122. http://www.regulations.gov/search/Regs/home.html#documentDetail?R=0900006480a3956a (last accessed August 2010).
123. http://www.regulations.gov/search/Regs/home.html#documentDetail?R=09000064802fe7bd (last accessed August 2010).
124. T. Bloem, *Memorandum: Spinetoram. Residue Chemistry Summary Concerning the Application of Spinetoram to Numerous Crops, 12-September-2007*, Document number D325387, US Environmental Protection Agency, Washington, DC, USA.
125. http://www.hc-sc.gc.ca/cps-spc/pubs/pest/_decisions/reg2001-10/index-eng.php (last accessed August 2010).
126. http://permits.nra.gov.au/registration/assessment/docs/prs_spinosad.pdf (last accessed August 2010).

127. http://www.efsa.europa.eu/en/scdocs/scdoc/1312.htm (last accessed August 2010).
128. http://www.inchem.org/documents/jmpr/jmpmono/2001pr12.htm#2.0 (last accessed August 2010).
129. http://www.fao.org/agriculture/crops/core-themes/theme/pests/pm/jmpr/jmpr-rep/en/ (last accessed August 2010).
130. http://www.fao.org/docrep/011/i0450e/i0450e00.htm (last accessed August 2010).
131. http://www.who.int/entity/whopes/quality/Spinosad_eval_only_March_2008.pdf (last accessed August 2010).
132. http://www.epa.gov/gcc/pubs/prinicples.html (last accessed August 2010).
133. http://www.dowagro.com/PublishedLiterature/dh_0064/0901b803800647cc.pdf?filepath=/PublishToInternet/InternetDOWAGRO/usag/pdfs/noreg/010-80032&fromPage=BasicSearch (last accessed August 2010).
134. http://www.dowagro.com/PublishedLiterature/dh_0072/0901b8038007298a.pdf?filepath=/PublishToInternet/InternetDOWAGRO/usag/pdfs/noreg/010-80088&fromPage=BasicSearch (last accessed August 2010).
135. F. J. Haile, D. L. Kerns, J. M. Richardson and L. G. Higley, *J. Econ. Entomol.*, 2000, **93**, 788.
136. K. D. Racke in *Crop Protection Products for Organic Agriculture: Environmental, Health and Efficacy Assessment*, ed. A. S. Felsot and K. D. Racke, American Chemical Society, Washington, DC, USA, 2007, p. 92.
137. M. D. Toews, B. Subramanyam and J. M. Rowan, *J. Econ. Entomol.*, 2003, **96**, 1967.
138. F.-N. Huang, B. Subramanyam and M. D. Toews, *J. Econ. Entomol.*, 2004, **97**, 2154.
139. M. K. Nayak, G. J. Daglish and V. S. Byrne, *J. Stored Prod. Res.*, 2005, **41**, 455.
140. F. Huang, B. H. Subramnyam and X. Hou, *Biopestic. Int.*, 2007, **3**, 117.
141. C. G. Athanassiou, N. G. Kavallieratos, A. E. Yiatilis, B. J. Vayias, C. S. Mavrotas and Ž. Tomanović, *J. Insect Sci.* 2008, **8**, 60.
142. C. G. Athanassiou, F. H. Arthur and J. E. Throne, *J. Stored Prod. Res.*, 2009, **45**, 236.
143. R. B. Davey, J. A. Miller, J. E. George and D. E. Snyder, *Southwest. Entomol.*, 2005, **30**, 245.
144. D. R. George, R. S. Shiel, W. G. C. Appleby, A. Knox and J. H. Guy, *Vet. Parasitol.*, 2010, **173**, 307.
145. D. E. Snyder, L. R. Cruthers and R. L. Slone, *Vet. Parasitol.*, 2009, **166**, 131.
146. M. B. Hertlein, C. Mavrotas, C. Jousseaume, M. Lysandrou, G. D. Thompson, W. Jany and S. A. Ritchie, *J. Am. Mosq. Control Assoc.*, 2010, **26**, 67.
147. D. Stough, S. Shellabarger, J. Quiring and A. A. Gabrielson, *Pediatrics,* 2009, **124**, e389.

148. G. D. Thompson and T. C. Sparks in *Advancing Sustainability through Green Chemistry and Engineering*, ed. R. L. Lankey and P. T. Anastas, American Chemical Society, Washington, DC, USA, 2002, p. 61.
149. Ó. López, J. G. Fernández-Bolaños and M. V. Gil, *Green Chem.*, 2005, **7**, 431.
150. G. Centi and S. Perathoner in *Sustainable Industrial Processes*, ed. F. Cavani, G. Centi, S. Perathoner and F. Trifiro, Wiley-VCH Verlag GmbH & Co. KgaA, Weinheim, Germany, 2009, p. 1.

The Bisacylhydrazine Insecticides

LUIS E. GOMEZ,[1*] KERRY HASTINGS,[1] HARVEY
A. YOSHIDA,[2] JAMES E. DRIPPS,[1] JASON BAILEY,[1]
SANDRA ROTONDARO,[1] STEVE KNOWLES,[3]
DORIS L. PAROONAGIAN,[1] TARLOCHAN SINGH
DHADIALLA[1] AND RAYMOND BOUCHER[1]

[1] Dow AgroSciences LLC, 9330 Zionsville Rd., Indianapolis, IN 46268,
USA; [2] Dow AgroSciences LLC, 432 Aimee Dr., Richland, WA 99352,
USA; [3] Dow Chemical Services UK Ltd., European Development Centre,
2nd Floor, 3 Milton Park, Milton Park EN, OX14 4RN, UK

6.1 Introduction

The two principle hormones that regulate insect growth, development and
reproduction are the steroidal molting hormone, 20-hydroxyecdysone (20E),
and the sesquiterpenoid juvenile hormone. Any interference with the titers of
these hormones or the mechanisms by which they manifest their action would
result in detrimental growth, development and reproduction. Over the past four
decades a number of juvenile hormone analogs have been discovered and
commercialized. In addition, a new class of non-steroidal insecticides that
mimic the action of 20E, the bisacylhydrazines (BAH), was discovered in the
early 1990s. Since its discovery, several BAH analogs have been developed and
commercialized.

Another important physiological event for insect growth and development is
the coordinated synthesis of the new cuticle. Interference with the deposition of
the cuticle or its components also results in a disruption of growth and
development. This has been achieved by the use of inhibitors of chitin synthesis

RSC Green Chemistry No. 11
Green Trends in Insect Control
Edited by Óscar López and José G. Fernández-Bolaños

Published by the Royal Society of Chemistry, www.rsc.org

that represent a third group of insecticides; this group along with the BAH and the juvenile hormone analogs are categorized as *insect growth regulators*.

This book chapter will focus on reviewing the BAH insecticides, which have been the subject of over two thousand refereed journal articles and other research reports since their discovery by Hsu *et al.* in 1991.[1] Most reviews of this non-steroidal class of chemical insecticides have focused on their mode of action as 20E agonists, their selectivity to target pests and their use in noninsecti-cidal applications such as gene switches in plant and animal systems.[2–5] How-ever, their utility as a class of green insecticides has not been reported extensively.

There are currently five registered BAH insecticides, three of which (tebufe-nozide, methoxyfenozide and halofenozide) were discovered by Rohm and Haas Company (R&H), one by Nippon Kayaku/Sankyo Companies (chro-mafenozide) and one by Jiangsu Pesticide Research Institute Company Limited (fufenozide). Of the three BAH insecticides discovered and originally com-mercialized by R&H, two of them are currently owned by Dow AgroSciences LLC (DAS) (methoxyfenozide and halofenozide) and one was purchased by Nippon Soda Co. Ltd. (tebufenozide) from DAS in the year 2010. Four of the BAH compounds (methoxyfenozide, tebufenozide, chromafenozide and fufenozide) have a spectrum of control specific for lepidopteran larvae, while halofenozide expresses a wider spectrum of control for coleopteran and lepidopteran larvae showing special utility as a soil insecticide.

Methoxyfenozide is the most widely registered and used BAH insecticide, having registrations in more than 50 countries in a variety of crops ranging from vegetables to specialty uses such as in forestry and tea production. This chapter focuses on reviewing specific properties of methoxyfenozide, such as its toxicological profile and spectrum of control within the order Lepidoptera, which make it a good candidate for inclusion in insecticide programs to manage key global pests. Where appropriate, other BAH insecticides are also mentioned throughout the chapter to exemplify the utility of these compounds for insect control.

6.2   Synthesis and Design

The discovery of the first nonsteroidal, insecticidal BAH (RH-5489)[1] revealed a broad spectrum of activity against lepidopteran, coleopteran and dipteran insects. Extensive work with this new class of chemistry, which involved synthesis guided by structure–activity relationship (SAR) analysis of over 4000 BAH analogs, led to the discovery of the three insecticides selected for commercialization at R&H: tebufenozide, methoxyfenozide and halofenozide. The synthetic routes of these three insecticides and chromafenozide, discovered by Nippon Kayaku and Sankyo, are described by Dhadialla and Ross.[5] The structures of 20E, the first BAH insecticidal prototype (RH-5489) and the first four commercial BAH insecticides are shown in Figure 6.1.

The activity bioassays used to characterize the performance of the BAH insecticides ranged from whole-insect toxicity to insect-cell-based and

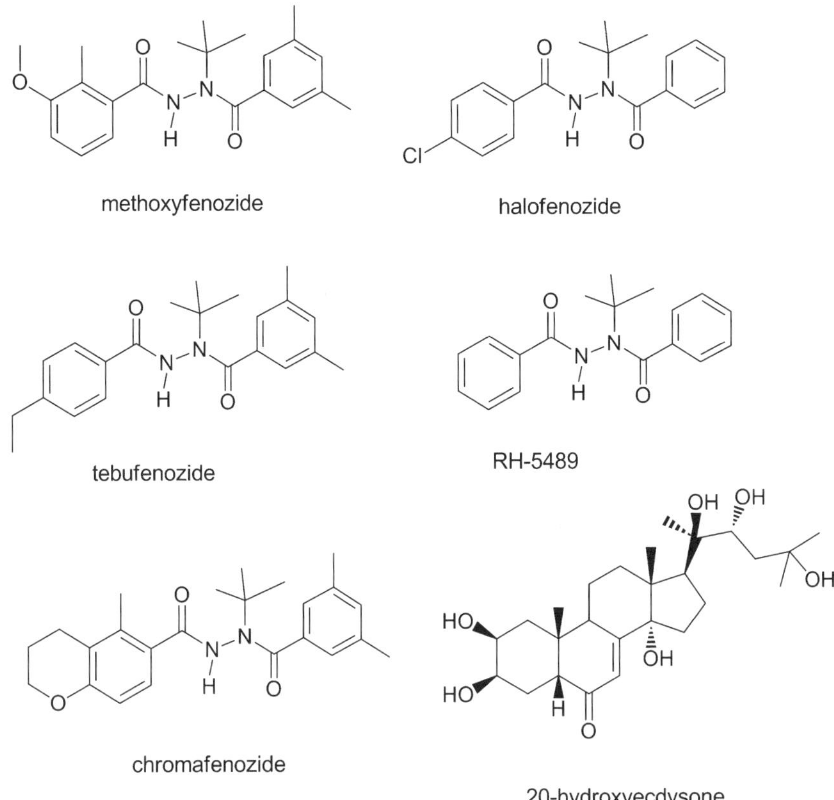

Figure 6.1 Chemical structures of the insect molting hormone (20E), the first BAH prototype (RH-5489) and the first four commercial BAH insecticides (tebufenozide, halofenozide, methoxyfenozide and chromafenozide).

competitive ecdysone receptor (EcR) ligand-binding radiometric assays.[3,5] The SAR results indicated the importance of the core hydrazine structure with two phenyl-substituted rings outside the two negative centers and a bulky, conformation-influencing lipophilic group located asymmetrically between the two negative centers.

SAR studies conducted by Dinan and Hormann[4] also indicated that six-membered aryl groups on either side of the two negative centers favored activity against lepidopteran and coleopteran insects. Activity on lepidopteran larvae could be further enhanced by modifying the substitutions on the A-ring at the 4 position with 1-2-carbon-containing lipophilic groups, or with 2,3- or 2,(3,4)-ring patterns. Substitutions on the B-ring were found to be less specific, though 2-, 2,5-, 3,5- or 3,4,5-positions were favorable.[4] The main difference between the three BAH insecticides discovered at R&H and chromafenozide is that in chromafenozide, the A-ring is fused at the 3-4-positions with a six-membered oxygen and a carbon-containing heterocycle (see Figure 6.1).

6.3   Mode of Action

The mode of action of the BAH insecticides has been described at the whole insect, tissue, cell and molecular levels.[3,5] The BAH insecticides manifest their toxicity by binding to the EcR present in all arthropods. Despite the broad susceptibility of arthropods to this mode of action, methoxyfenozide, tebufenozide and chromafenozide are especially toxic to lepidopteran larvae and have some sublethal effects on lepidopteran adults (see section 6.4). The main reason for this selective toxicity is the extremely high binding affinity of methoxyfenozide and tebufenozide to lepidopteran EcRs.[2,3,5] The insect steroid molting hormone and the potent phytoecdysteroid ponasterone A, bind to EcRs from different insect orders with similar affinities. Methoxyfenozide and tebufenozide bind to the lepidopteran EcRs with affinities that are equivalent to that of ponasterone A (K_d of 1.4 nM; ≈100-fold more active than 20E), but not to EcRs from Coleoptera such as the yellow mealworm, *Tenebrio molitor* L.; from Orthoptera such as the migratory locust, *Locusta migratoria* (L.); and Hemiptera such as the silverleaf whitefly, *Bemisia argentifolii* Bellows & Perring; even at 10 mM concentrations. While the two insecticides bind to EcR in Diptera such as *Drosophila melanogaster* (Meigen), the binding affinities are weak and toxicologically inconsequential. Radiolabeled BAH bound specifically to the EcR complex from the spruce budworm, *Choristoneura fumiferana* (Clemens), but not to the EcR complex from *D. melanogaster*.[3,5]

The interaction of biologically active BAH analogs to EcR was observed by examining crystal structures of the heterodimeric EcR complex in the tobacco budworm, *Heliothis virescens* (F.), in the presence and absence of ponasterone A and a BAH analog.[6]

The novel mode of action of nonsteroidal BAH insecticides, their insect-toxic selectivity and other characteristics (discussed in sections 6.4–6.10) make these insecticides important tools for insecticide resistance management (IRM) and integrated pest management (IPM) programs.

6.4   Effects on Insect Population Dynamics and Individuals

The BAH insecticides can affect insect population dynamics in several ways other than by direct action on lepidopteran larvae, including sublethal effects on larvae and adults that can alter insect reproduction and reduce the overall population. These characteristics of the BAH insecticides are reviewed in this section.

6.4.1   Sublethal and Ovicidal Effects on Adults and Larvae

6.4.1.1   Sublethal Effects

Sublethal effects of BAH compounds have been reported for many key pest species. These studies investigated the sublethal effects on larvae through topical treatment or ingestion. Species investigated for sublethal effects include corn earworm, *Helicoverpa zea* (Boddie); *C. fumiferana*; codling moth, *Cydia pomonella* (L.); European grapevine moth, *Lobesia botrana* (Dennis & Schiffermüller); the European corn borer, *Ostrinia nubilalis* (Hübner); several armyworm species, *Spodoptera* spp.; and others.[7–19] Depending on the BAH compound and insect species, the sublethal effects can vary and may include delayed developmental rates, reduced larval and pupal weights, pupal and adult deformities and reproductive effects on adults that developed from larvae surviving exposure to BAH insecticides.

Studies also have been reported on sublethal effects of tebufenozide or methoxyfenozide that can occur from exposure of adults *via* treated surfaces, topical or oral applications.[10,20–30] In such studies, reductions in fecundity and/or fertility were reported on Egyptian cotton leafworm, *Spodoptera littoralis* (Boisduval); *C. pomonella*; obliquebanded leafroller, *Choristoneura rosaceana* (Harris); bean shoot borer, *Epinotia aporema* (Walsingham); *L. botrana*; vine moth, *Eupoecilia ambiguella* (Hübner); tufted apple bud moth, *Platynotia idaeusalis* (Walker); oriental fruit moth, *Grapholita molesta* (Busck); and others.

Sublethal effects can vary depending on the specific BAH insecticide used, insect species, gender exposed and duration of exposure. In these studies, methoxyfenozide consistently caused stronger effects than tebufenozide. For example, female and/or male adult *C. pomonella*; redbanded leafroller, *Argyrotaenia velutinana* (Walker); and *C. rosaceana* exposed to surfaces treated with either tebufenozide or methoxyfenozide at field rates demonstrated a significant reduction in fecundity and fertility compared to untreated adults.[22,23] Methoxyfenozide significantly reduced adult moth fertility compared with tebufenozide. Depending on species, the negative influence of tebufenozide on fecundity and fertility was female-based, while the negative effects of methoxyfenozide on adult fecundity affected males and females. Prolonged exposure to methoxyfenozide also significantly reduced the longevity of adult male moths.[22,23] Other studies have reported significant reduction in fecundity and fertility when either male or female *C. pomonella* adults were exposed to treated surfaces for as little as one hour.[25]

In addition to the effects on fecundity and fertility, exposure of adults to either methoxyfenozide- or tebufenozide-treated surfaces appears to negatively affect the sexual behavior of adults in some species. For example, male *C. pomonella* exposed to methoxyfenozide-treated surfaces were not as responsive to treated or untreated calling females. Conversely, females exposed to methoxyfenozide-treated surfaces were just as attractive to untreated males as were untreated females. It appears that a male's ability to respond to a calling female is more negatively affected by the ecdysone agonist than a female's ability to call and attract males.[31] The disruption of adult insect chemical communication resulting from exposure of adults to methoxyfenozide-treated surfaces also has been reported for *A. velutinana, C. rosaceana* and *G. molesta*.[32,33]

6.4.1.2 Ovicidal Effects

The ovicidal effects of tebufenozide and methoxyfenozide have been reported in several studies for key insect pests including *O. nubilalis*; southwestern corn

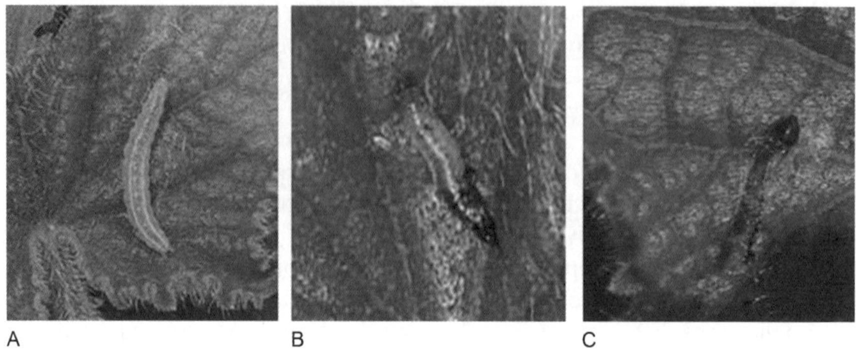

A   B   C

Figure 6.2   Symptomatology of methoxyfenozide-treated melonworm, *Diaphania hyalinata* (L.): (A) at application time, (B) 24 h after application, and (C) 36 h after application.

borer, *Diatraea grandiosella* Dyar; *C. pomonella*; *E. aporema*; *L. botrana*; *E. ambiguella;* and others.[14,15,25–28,34,35] Methoxyfenozide is a more potent ovicide than tebufenozide; however, the extent of ovicidal activity may vary depending on species and whether eggs were laid on a treated surface *vs.* treatment occurring after oviposition. In most cases, eggs laid on treated surfaces appear to be more susceptible than eggs exposed to treatment after oviposition.

6.4.2   Speed of Kill on Larvae

Speed of kill varies with the BAH compounds and depends on several factors. The main factor determining speed of kill is the amount of larval feeding, which determines the amount of toxicant ingested. Feeding is influenced by several factors with ambient temperature being one of the most important. Depending on temperature, intoxicated larvae may remain alive for several days after exposure. Longer kill times have been observed under cooler ambient temperatures because the larvae are not actively feeding and metabolizing the toxicant. When temperatures increase, feeding increases and time to kill decreases. Regardless of the speed of kill, minimal plant damage is consistently recorded after ingestion of BAHs, which is indicative of rapid feeding cessation (three to 14 hours) after ingestion.[5] Figure 6.2 shows the intoxication symptoms observed on melonworm, *Diaphania hyalinata* (L.), treated with methoxyfenozide.

6.5   Environmental Effects

6.5.1   Introduction

Methoxyfenozide is highly selective to lepidopteran species, which, when combined with low application rates and a novel mode of action, makes it a valuable product to manage pests in this order. It is compatible for use with most beneficial insects including pollinators, has low toxicity to mammals, is only moderately toxic to fish and is practically nontoxic to birds. In 1998, R&H was awarded the US EPA's (United States Environmental Protection Agency) Presidential Green Chemistry Award for the invention and commercialization of the BAHs as a new chemical family of insecticides exemplified by methoxyfenozide (Intrepid®), tebufenozide (Confirm®) and halofenozide (Mach 2®). These compounds became the first insecticides to be awarded this prestigious award, which has been given to only four insecticidal technologies since its establishment in 1995 through 2010.[36] The Presidential Green Chemistry Challenge Awards Program provides recognition for innovations in cleaner, more affordable and smarter chemistry.[37] Sections 6.5, 6.6 and 6.7 discuss in more detail the environmental effects, human toxicity and ecotoxicology of methoxyfenozide as a representative compound of the BAH insecticides.

As described in detail for the spinosyns in Chapter 5, DAS also has invested in proprietary studies on methoxyfenozide, which in turn enable understanding of fate and risk for nontarget exposures. Areas of research include mammalian toxicity, ecotoxicology, environmental fate and crop residue field studies. Several regulatory decision documents for methoxyfenozide are in the public domain including a global evaluation of toxicology, metabolism and residues by the Food and Agriculture Organization/World Health Organization (FAO/ WHO) Joint Meeting on Pesticide Residues (JMPR).[38,39] These documents provide reviews of the large database of information produced for methoxyfenozide and reflect its registration history.[38–43]

6.5.2 Metabolism and Bioaccumulation

6.5.2.1 Plant Metabolism

Methoxyfenozide is metabolized in plants by the formation of a free phenol and products that result from the oxidation of the alkyl substituents. In addition, two polar carbohydrate conjugates have been identified in rotational crops. The parent compound is the residue of primary concern for dietary exposure assessments and tolerance setting. Studies carried out to elucidate the metabolic pathway of methoxyfenozide in plants are described in further detail in this section.

The metabolism, distribution and elimination of ^{14}C-methoxyfenozide (labeled in each ring or in the *t*-butyl position in Figure 6.3) were studied in apples, grapes, cotton and rice.[38] In apples and grapes, the half-life of methoxyfenozide was estimated to be <30 days with the decline in residues primarily as a result of fruit growth. Methoxyfenozide was not significantly metabolized in apples, with the majority of the residue remaining as methoxyfenozide. Two metabolites were present at <2% of the residue: RH-131364 and RH-131157 (see Figure 6.4). In grapes, methoxyfenozide remained the major component, with four minor components (each <4%) identified as the glucose conjugate of RH-117236 and RH-131364 and two tentatively identified components (see Figure 6.4). Analysis of a leaf sample also showed the majority of the residue

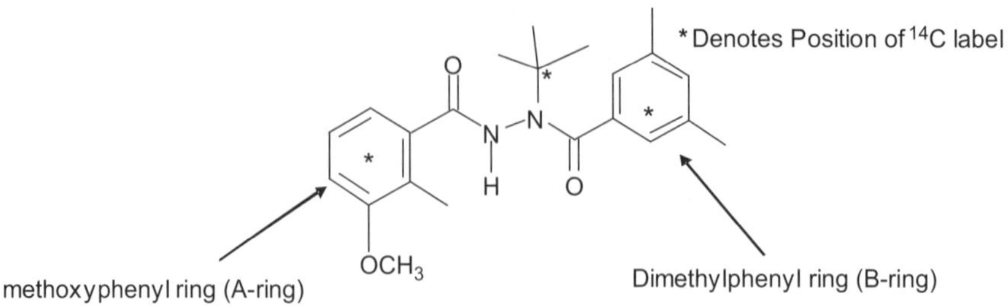

Figure 6.3 Structure of methoxyfenozide indicating position of radiolabel.

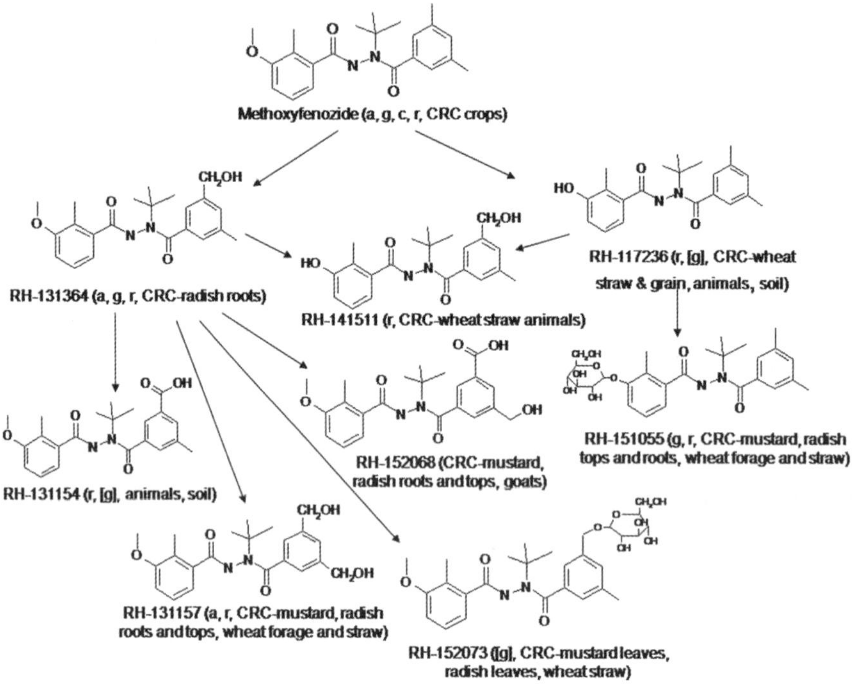

Figure 6.4 Proposed metabolic pathway for methoxyfenozide in plants (a = apples, g = grapes, c = cotton, r = rice, CRC = confined rotational crop study and [ ] indicates tentative identification).

remained as unchanged methoxyfenozide. Similar results were noted in cotton plants where methoxyfenozide was not significantly metabolized, with the majority of the residue in mature bolls and plant samples remaining as methoxyfenozide. In rice treated with foliar applications of methoxyfenozide at the pre-flag leaf and post-flowering stages, the major component of the residue was methoxyfenozide, with 52–59% and 65–69% of the residue in grain and straw, respectively. RH-117236 (see Figure 6.4) was detected at 3.2–7.5% of the residue in grain and ≈3% in the straw. Other metabolites comprised <5% of the radioactivity in grain or straw.[38]

6.5.2.2 Confined Rotational Crops

Mustard, radish and wheat crops were planted at 30, 90 and 365 days (nominal) into a sandy loam soil previously treated with either ring-labeled or *t*-butyl-labeled methoxyfenozide at a rate of 2.2 kg ai (active ingredient) ha^{-1}. Immature and mature crops were sampled. The total radioactive residue levels were <0.05–0.3 mg kg^{-1} in all crops, except wheat forage and straw, which were 1–3 mg kg^{-1}. In general, total radioactive residue levels decreased significantly with increasing plant-back intervals.[38]

Extensive metabolites were observed in all crops indicating translocation of methoxyfenozide and/or soil degradates into the plants. The metabolites identified in the rotated crops contained the dibenzoylhydrazide structure. Radish roots contained the highest levels of nonmetabolized methoxyfenozide. Major metabolites were the A-ring phenol and B-ring alcohol and glucose-, malonylglucosyl- and disaccharide- conjugates of these, some of which are shown in Figure 6.4.

6.5.2.3 Animal Metabolism

The metabolism of methoxyfenozide is well understood for regulatory purposes, with the parent compound and a glucuronide metabolite being the primary residues in animals. Animal feeding studies with diets manufactured with methoxyfenozide-treated commodities have demonstrated that residues are low in tissues, milk and eggs.

The metabolism, distribution and elimination of ^{14}C-methoxyfenozide (labeled in each ring or in the *t*-butyl position) were studied in lactating goats ($\approx$50 ppm dietary equivalents for seven consecutive days) and laying hens ($\approx$60 ppm in the diet for seven consecutive days).[38]

In goats, the majority of the residue was excreted in the feces (74–84% of the dose) with an additional 5–7% of the dose recovered in the urine. In the edible tissues, total radioactive residues were highest in the liver (0.26–1.2 mg kg^{-1}), followed by kidney (0.045–0.2 mg kg^{-1}) and fat (0.018–0.053 mg kg^{-1}). Muscle contained <0.025 mg kg^{-1}. Residues in milk were low ($\leq$0.018 ppm) and reached a plateau within 24–36 hours of dosing. Methoxyfenozide was the major residue in fat, muscle and milk. Liver and kidney showed a large percentage of RH-141518 and trace amounts of other glucuronides.

In hens, the majority of the dose was recovered in the excreta. Less than 0.03% of the dose was recovered in the eggs and tissues where metabolite RH-141518 represented a major component of the residue in addition to methoxyfenozide.

In rats,[39,40] methoxyfenozide (labeled in each ring or in the *t*-butyl position) was rapidly and moderately well absorbed following a single oral administration, with peak blood or plasma levels achieved within 15–30 min and 62–70% of the dose systemically absorbed. Methoxyfenozide is extensively metabolized by rats and eliminated in the bile, feces and urine. Very little radioactivity was released as volatile components such as carbon dioxide (CO_2). Metabolism involves demethylation, hydroxylation and glucuronidation. A total of

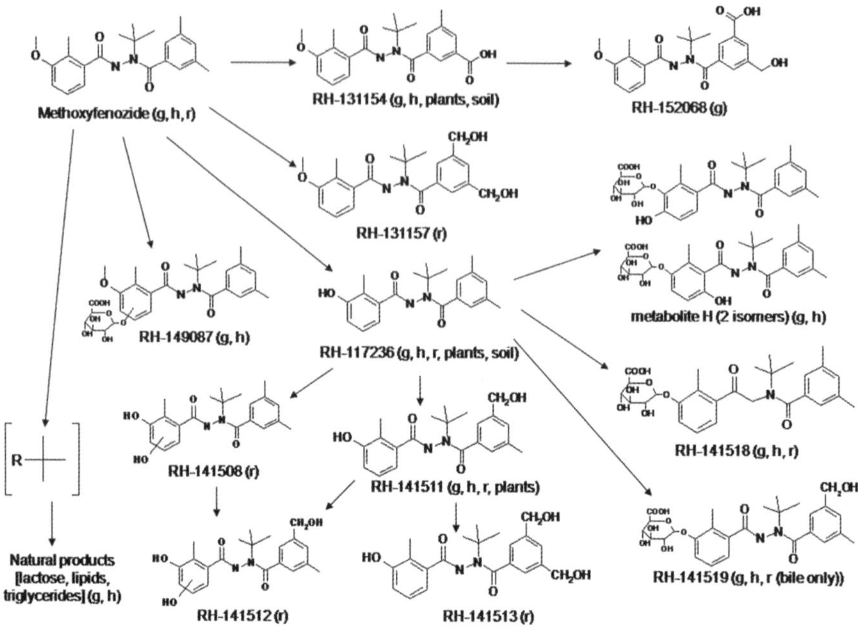

Figure 6.5 Proposed metabolic pathway for methoxyfenozide showing metabolites identified in animals, plants and soil (g = goat, h = hen and r = rat).

31 metabolites were isolated from rat urine and feces (25 of which were identified), and 24 metabolites were characterized in bile (12 of which were identified and 4 of which were only found in bile). There were seven major metabolites present at >5% of the administered dose in feces plus urine, including those shown in the metabolic pathway in Figure 6.5. Parent methoxyfenozide was found in feces, but was not present in urine or bile.

The metabolism of methoxyfenozide is similar in poultry and ruminants, and qualitatively the same in rats. The residues were concentrated in the fat relative to the muscle by factors of $\approx 2\text{--}5$, which is consistent with the log P_{ow} for methoxyfenozide of 3.7. The proposed metabolic pathway in animals is presented in Figure 6.5.

6.5.2.4 Metabolism Summary

Studies in a variety of plants and animals demonstrated that methoxyfenozide was not extensively metabolized, the exception to this was observed in rats as discussed in section 6.5.2.3.[38,39] The major component of the plant residue was parent methoxyfenozide. The majority of methoxyfenozide was excreted by goats and hens, and in the remaining residue methoxyfenozide was the major component of goat milk, fat and muscle, and hen fat and skin. The A-ring phenol glucuronide was the major component in goat and poultry liver and kidney, and poultry eggs. The parent methoxyfenozide constitutes the residue definition in most countries, with the exception of the US where residues in animal tissues include metabolites.

6.5.2.5 Bioaccumulation

The log P_{ow} for methoxyfenozide (3.72)[38] suggests the potential for bioaccumulation of the molecule, however results from bioaccumulation studies have demonstrated the rapid elimination of methoxyfenozide from fish with a 50% depuration estimate of ≈ 0.3 days.[40] In the European Union (EU), the bioconcentration factor (BCF) endpoint in fish is 11.0 (ref. 41), which is remarkably low when compared to <500, the value which is indicative of a low level of bioacumulation.[44]

6.5.3 Fate and Behavior in the Environment

Laboratory studies suggest that methoxyfenozide may be moderately persistent in soil, but degradation is more rapid under field conditions. Regarding soil sorption (mean K_{OC} of 402 ml g^{-1}), methoxyfenozide may be classified as having slight to moderate mobility, but environmental modeling does not predict surface runoff or leaching concerns.

6.5.3.1 Biotic Degradation

6.5.3.1.1 Soil Degradation. Under laboratory aerobic conditions, the degradation of methoxyfenozide is relatively slow.[40] In laboratory studies based on US EPA guidelines, ^{14}C-labelled methoxyfenozide was incubated in four US soils kept in the dark at 25 °C for 365 days. The DT_{50} values (dissipation time for 50% loss) obtained were greater than the duration of the study indicating that methoxyfenozide degraded slowly under lab conditions (336 to >365 days). The decline pattern for the slowest degrading soil suggested that viable soil conditions for aerobic metabolism were not maintained. Therefore, these laboratory studies are unlikely to represent typical aerobic field degradation. No metabolites were found at levels >10% of applied radioactivity (AR) in these studies; however, three minor metabolites were detected. Two of these metabolites are closely related to the structure of methoxyfenozide: the respective carboxylic acid (RH-131154) and a hydroxy analog (RH-117236) of the parent structure (see Figure 6.5). The unextractable residue from these laboratory studies was between 12–27% of the AR after 120 days.

Under field conditions, degradation was significantly faster than under laboratory conditions.[40] Studies were conducted using US EPA guidelines at four sites in the US, with methoxyfenozide applied to bare soil. The DT_{50} values for methoxyfenozide ranged from 92–327 days (arithmetic mean 177 days). Studies also were conducted under EU guidelines at six sites in France, Germany, United Kingdom (UK) and Italy.[41] Using non-first-order kinetics, the DT_{50} values ranged from 39–133 days with the longest value calculated using a second-order kinetic model. Using first-order kinetics, methoxyfenozide still showed faster degradation under field soil conditions compared with the laboratory results, with field DT_{50} values ranging from 121–231 days (geometric mean 181 days). For modeling purposes in the EU, a normalized value of 95 days is used based on a geomean field normalized value corrected for temperature and field moisture (20 °C and pF 2).

Under anaerobic conditions in the laboratory, the degradation of methoxyfenozide is slow with mineralization <0.1% of the AR over 120 days. Unextractable residues reached a maximum of 13% of the AR after 120 days. The hydroxy metabolite (RH-117236; see Figure 6.5) appeared after 40 days under anaerobic conditions, but it degraded rapidly when the soils were returned to aerobic conditions.

Laboratory soil photolysis studies show that methoxyfenozide has a minor route of degradation *via* exposure to sunlight with no major photodegradates formed.[41]

6.5.3.1.2 Aquatic Degradation. For evaluating aquatic degradation, ^{14}C-labelled methoxyfenozide was incubated in the dark at 25 °C for 365 days in two soil–water systems using methods based on US EPA test guidelines.[40] A California clay/paddy sediment–water system and a Texas sandy loam sediment–irrigation water system were used. Methoxyfenozide degraded with a DT_{50} value of 86 days in the clay/paddy sediment–water system using data from the first 60 days. The DT_{50} value in the sandy loam sediment–irrigation water system was 145 days calculated over the first 60 days. In both systems, the DT_{50} values over the duration of the study were longer (>365 days) and the decline in microbial viability of the system over the latter part of the study may have accounted for the difference. Methoxyfenozide also was incubated in the dark at 20 °C for 120 days in two German sediment–water systems, one using loamy sand sediment and the second using loam sediment, using methods based on the Society of Environmental Toxicology and Chemistry test guidelines.[41] DT_{50} values for the whole system ranged from 159–274 days using first order kinetics. DT_{50} values in the water phase ranged from six to eight days using a first-order multicompartment model.

The major metabolite in both US and EU aquatic degradation studies was the hydroxy metabolite (RH-117236; see Figure 6.5), which was formed up to a maximum of 16% of the AR in the whole system with 12–13% in the sediment phase. Ultimately, residues were mineralized to CO_2 (3–6% of the AR) and non-extractable residues (10–67% of the AR).

6.5.3.2 Abiotic Degradation

6.5.3.2.1 Hydrolysis. Studies following the US EPA and Organization for Economic Cooperation and Development (OECD) guidelines for hydrolysis at 25 °C in the dark showed that methoxyfenozide is not degraded by hydrolysis. It was stable in sterile buffers at pH 5, 7 and 9 (ref. 40).

6.5.3.2.2 Aqueous Photolysis. Methoxyfenozide was stable to aqueous photolysis in sterile deionized water buffered at pH 7 in studies based on US

EPA guidelines. In irradiated natural pond water (25 °C, with 12 hours light/ 12 hours dark cycles from a lamp simulating natural sunlight) some degradation was seen (DT_{50} 866 days in darkness, DT_{50} 77 days with irradiation).[40] The slow degradation by photolysis did not result in major metabolites, but seven minor degradates were detected.

6.5.3.3 Sorption

Sorption of methoxyfenozide was evaluated in five US soils and four EU soils in two separate studies.[41] The soils ranged in pH from 5.8–8.1 with organic carbon from 0.12–2.85 % and covered a range of soil types. Methoxyfenozide adsorption coefficients based on organic carbon content (K_{fOC}) ranged from 200–922 (arithmetic mean value 402 ml g^{-1}), with Freundlich adsorption coefficient values ranging from 0.93–1.06 (arithmetic mean value 0.98).

6.5.3.4 Groundwater

In EU field studies there were no residues of methoxyfenozide >LOQ (Limit of Quantification; LOQ = 3 µg kg^{-1}) found below the 30 cm soil horizon, suggesting minimal leaching from the surface applied material with most of the mass retained in the top 0–10 cm horizon. In US field studies,[40] some movement (<5% at 76–91cm after 30 days) of methoxyfenozide down the soil horizons was observed in the four sites, with the most rapid movement in a deep sand (89–95% sand down to 91 cm), where the residues at 76–91 cm occurred 30 days after application.

6.5.3.5 Surface Water

In water–sediment systems under aerobic conditions in the dark, only minimal mineralization of methoxyfenozide was observed,[41] representing <6% of AR. The main route of methoxyfenozide dissipation was through transfer to the sediment phase. There was no major metabolite in the water column.

6.6 Human and Mammalian Toxicology

The endpoints and study results for mammalian toxicology are summarized in Tables 6.1–6.3.

Table 6.1 Acute toxicity of methoxyfenozide.

Type of study	Acute study results	Acute toxicity values	References
oral	no mortalities, clinical signs of toxicity, or body	LD_{50} >5000 mg kg^{-1}	39–42
dermal	weight effects	LD_{50} >5000 mg kg^{-1}	
inhalation		LC_{50} >4.3 mg l^{-1}	
eye	slightly irritating	slightly irritating	
skin	non-irritating, non-sensitizing	non-irritating, non-sensitizing	

Table 6.2 Short- and long-term mammalian toxicological studies.

Study type	Results	$NOEL^a$ (M/F) ($mg\,kg^{-1}\,day^{-1}$)	References
short-term	<u>28 days</u> **rat dermal:** no treatment related systemic effects; no skin irritation. **dog (with 4 week recovery):** changes in hematological parameters were present after 4 weeks of treatment; complete recovery from effects was demonstrated after 4 weeks of recovery.	<u>28 days</u> **rat dermal:** 1000 **dog:** not applicable	39,40,42
	<u>90 days</u> **rat:** increase liver weight and liver hypertrophy at mid- and high-doses; minimal decrease in hematology parameters at high-dose. **mouse:** decreases in body weight gain in high-dose males and females. **dog:** no adverse effects.	<u>90 days</u> **rat:** 69/72 **mouse:** 428/589 **dog:** 198/209	
	<u>1 year</u> **dog:** slight to moderate hematological effects at mid- and high-doses; increased liver and thyroid weights at high-dose (but no histopathological correlation to these changes), histopathological changes in liver, spleen and bone marrow considered secondary to the mild met-hemoglobinema present at this dose. EU and US EPA did not consider the mid-dose effects to be adverse.	<u>1 year</u> **dog:** 10/13 EU/EPA: 106/111	
long-term	<u>24 months</u> **rat:** hematologic effects, liver weight increase, hepatocyte hypertrophy, follicular cell hypertrophy of thyroid at mid- and high-doses; males: decreased survival; females: decreased body weight, increased adrenal weights, chronic progressive glomerulonephropathy of kidney, hyperplasia of the renal pelvic epithelium, and uremic changes of multiple organs at high dose.	<u>24 months</u> **rat:** 10/12	39–42
	<u>18 months</u> **mouse:** no treatment related effects.	<u>18 months</u> **mouse:** 1020/1354	

^{a}No observable effects level.

Table 6.3 Other mammalian toxicological studies and regulatory endpoints: Developmental and Reproductive Toxicology (DART), genotoxicity, neurotoxicity, Acceptable Daily Intake (ADI) and Acute Reference Dose (ARfD).

Study type	Results	NOEL[a] (M/F) ($mg\,kg^{-1}\,day^{-1}$)	References
DART	<u>2-gen reproduction (rat)</u> **parents:** increased liver weight and hepatocyte hypertrophy at mid and high dose. In EU, the effects at the mid dose were not considered adverse.	<u>2-gen reproduction (rat)</u> **parents:** 16/19 EU: 153/181	39,40,42
	reproductive performance: no adverse effects. **offspring:** no adverse developmental effects.	**reproductive performance:** 1552/1956 **offspring:** 1552/1956	
	<u>developmental toxicity</u> **rat:** no maternal or developmental effects at high dose.	<u>developmental toxicity</u> **rat:** 1000	
	rabbit: no maternal or developmental effects at high dose.	**rabbit:** 1000	
genotoxicity	3 *in vitro* and 1 *in vivo* toxicity studies: negative.	all studies negative	39–42
neurotoxicity	**acute:** no evidence of neurotoxic effects at high dose.	**acute:** 2000	39–42
	subchronic: no evidence of neurotoxic effects at high dose.	**subchronic:** 1318/1577	
ADI	**NOELs from the rat and/or dog chronic studies:** 10 mg kg^{-1} day^{-1}		
	uncertainty factor: 100	**ADI:** 0.1	39–42
ARfD	**Australia and US EPA:** no acute effects observed.	**EU:** 0.2 mg kg^{-1} day^{-1}	39–42
	EU: 2-week dog methoxyfenozide study NOEL of 18 mg kg^{-1} day^{-1}	**JMPR:** 0.9 mg kg^{-1} day^{-1}	
	JMPR:[b] single dose tebufenozide study NOEL 89 mg kg^{-1} day^{-1}		

[a]No observable effects level.
[b]Joint FAO/WHO Meeting on Pesticide Residues.

6.7 Ecotoxicology

6.7.1 Aquatic Toxicity

The endpoints for several aquatic invertebrate species are summarized in Table 6.4.

Chronic studies indicated that methoxyfenozide is slightly toxic to the water flea, *Daphnia magna*; and moderately toxic to the mysid shrimp, *Mysidopsis bahia*. The greatest toxicity was observed in midge larvae, *Chironomus riparius* Meigen, with a NOEC (No Observed Effect Concentration) for adult emergence of 18 μg ai l^{-1}, based on the initial overlying water concentration. Results of a microcosm study demonstrated similar toxicity of methoxyfenozide to *C. riparius*[40] as observed in laboratory studies. However, the study also provided evidence that *C. riparius* populations could recover once methoxyfenozide concentrations in water fell sufficiently.

Given the mode of action of methoxyfenozide as a 20E agonist, toxicity to aquatic plants and algae (or terrestrial plants) would not be expected and data from US EPA guideline studies on the green alga, *Pseudokirchneriella subcapitata* (formerly referred to as *Selenastrum capricornutum*), confirms that assumption. Methoxyfenozide was tested as technical grade active (120 hours exposure) and as the 240SC formulation (240 g methoxyfenozide per liter of Suspension Concentrate) (96 hours exposure). Neither form of methoxyfenozide was toxic to *P. subcapitata* up to the solubility limits of the active (3.4 mg kg^{-1}), and formulated methoxyfenozide (107 mg product l^{-1}).[41]

Table 6.4 Summary of endpoints for aquatic invertebrate species.

Test	Species	Test substance	Time-scale	Endpoint	Toxicity $(mg\,l^{-1})$	References
acute toxicity	*Chironomus riparius*	active substance	not tested	EC_{50}	0.05[a]	41
	Crassostrea virginica	active substance	96 h	EC_{50}	1.3	40, 43
	Daphnia magna	240SC formulation	48 h	EC_{50}	420[b]	41
		active substance	48 h	EC_{50}	3.7	41
	Mysidopsis bahia	active substance	96 h	LC_{50}	1.3	40, 43
chronic toxicity	*Chironomus riparius*	active substance	28 d	NOEC	0.018[c]	41
	Daphnia magna	active substance	21 d	NOEC	0.39	41
	Mysidopsis bahia	active substance	37 d	NOAEC	0.051	40

[a]Based on acute to chronic ratio using data from daphnids.
[b]Product.
[c]EU endpoint; derived value.

6.7.2 Terrestrial Invertebrates

Methoxyfenozide is highly selective to the target lepidopteran species as discussed in section 6.3, but has no impact on most beneficial insects including bees, predators and parasitoids. Data from tests of the 240SC formulation on several arthropod species, including parasitoid wasps, egg parasitoids, predatory mites, lycosid spiders, green lacewings and predatory bugs, are summarized in the 2004 EU review.[41] The tests were carried out on both adult and juvenile stages at rates of the formulated product that are highly toxic to the target lepidopteran species. These studies monitored several parameters of toxicity for more than one week; only minor effects on survival and reproductive ability of several invertebrate predator and parasite species were observed. There was some effect noted on the predatory mite, *Typhlodromus pyri*, in a laboratory contact test, but further testing in field/semi-field studies showed no adverse effects on mites or eggs.

The results of several studies indicate that methoxyfenozide is virtually nontoxic to adult honey bees, *Apis mellifera* L. ($LD_{50} > 100 \,\mu$g ai bee^{-1} for acute contact and oral toxicity).[41] Additionally, as noted in the Australian evaluation,[40] when *A. mellifera* colonies were fed with a syrup containing methoxyfenozide, no adverse effects on honey bee brood development were found. Similar results from a semi-field study (tunnel test with honeybee broods) showed that the brood was not affected when bees were exposed to a seasonal rate of 480 g methoxyfenozide ha^{-1} using the 240SC formulation.[45] In a recent study by Mommaerts *et al.*,[46] large earth bumblebee, *Bombus terrestris* (L.), workers were exposed to several insect growth regulators through oral and dermal routes. The results indicated that neither of the two BAHs tested, methoxyfenozide or tebufenozide, caused any acute toxicity to the worker bees, nor were any adverse effects on larval development noted. In contrast, higher numbers of dead larvae were scored in the nests where the worker bees were exposed to the juvenile hormone analogs, pyriproxyfen and kinoprene.

The Australian evaluation[40] summarizes studies carried out on the earthworm, *Eisenia foetida* according to OECD guidelines. The 14-day NOEC values of 1213 mg ai kg^{-1} of dry soil for methoxyfenozide and 1250 mg of 240SC formulation kg^{-1} of dry soil indicate that methoxyfenozide is practically nontoxic to earthworms. A 56-day chronic study exposed *E. foetida* to the 240SC formulation of methoxyfenozide and examined reproduction, growth and mortality at concentrations up to ≈ 2.1 mg ai kg^{-1} of dry soil. No adult mortality and no reduction in the number of offspring produced were observed, but a small increase in the body weight of adult *E. foetida* at the highest rate was recorded.

6.7.3 Fish

Methoxyfenozide has low water solubility, which limits the concentrations that can be tested in aquatic toxicity tests. Thus, the LC_{50} endpoints for the various fish tested are greater than the maximum concentration (solubility limit) tested (see Table 6.5). The same is noted for the NOEC values that are at the highest

Table 6.5   Summary of endpoints for fish.

Test	Species	Test substance	Time-scale	Endpoint	Toxicity $(mg\,l^{-1})$	References
acute toxicity	*Cyprinodon variegatus*	active substance	96 h	LC$_{50}$	>2.8	41
		240SC formulation	96 h	LC$_{50}$	>130[a]	41
	Lepomis macrochirus	active substance	96 h	LC$_{50}$	>4.3	41
	Oncorhynchus mykiss	active substance	96 h	LC$_{50}$	>4.2	41
		240SC formulation	96 h	LC$_{50}$	>130[a]	41
	Pimephales promelas	active substance	96 h	LC$_{50}$	>3.8	41
chronic toxicity	*Cyprinodon variegatus*	active substance	32 d[b]	NOEC	2.6	41
	Pimephales promelas	active substance	33 d[b]	NOEC	2.4	41
		active substance	262 d[c]	NOEC	0.53	41

[a]Product.
[b]Early life stage.
[c]Full life cycle.

concentration tested except, in the case of the full fish life cycle study on the fathead minnow, *Pimephales promelas*, where reduced survival was observed. The LC$_{50}$ values obtained in tests using the 240SC formulation were also established at greater than the maximum concentration tested (the solubility limit for the formulated product).

6.7.4   Avian Species

Results from toxicity tests carried out according to US EPA guidelines indicate that methoxyfenozide is practically nontoxic to both bobwhite quail, *Colinus virginianus*; and mallard duck, *Anas platyrhynchos*. The data are presented in Table 6.6.

6.7.5   Other Species

The Australian evaluation[40] presents a summary on the impact of methoxyfenozide on other species (mammals, plants and soil microorganisms). The studies indicate that methoxyfenozide is practically nontoxic to mammals. Phytotoxicity was not expected from the mode of action of this insecticide and this is borne out from more than 10 years of commercial field experience with many crops. Additionally, studies conducted under European and Mediterranean Plant Protection Organization (EPPO) guidelines with methoxyfenozide-active and SC formulations concluded that soil respiration and soil nitrification

Table 6.6 Summary of endpoints for birds.

Species	Test	Test substance	Time-scale	Endpoint	Toxicity	ref.
Colinus virginianus (Northern bobwhite quail)	acute toxicity	active substance	14 d	LD_{50}	>2250 mg a.s. kg^{-1} bw	41
Colinus virginianus and *Anas platyrhynchos* (mallard duck)	dietary toxicity	active substance	8 d	LC_{50}	>5620 ppm	43
Anas platyrhynchos and *Colinus virginianus*	reproductive toxicity	active substance	21 or 22 weeks	NOEC	1000 ppm	41

indices were only slightly affected and methoxyfenozide is of low risk to soil microflora.[40]

6.8 Formulation

6.8.1 Introduction

The formulation of a pesticide is determined by the physical and chemical properties of the active ingredient. Methoxyfenozide is a high-melting (204–205 °C), non-reactive solid (hot concentrated acids degrade the molecule), with low solubility in water and nonpolar solvents.[47] The molecule is particularly suited for mixtures with other insecticides, as its low reactivity and low solubility allow it to be combined with many other compounds.

6.8.2 Formulation Types

According to Crop Life classification,[48] commercial formulations of methoxyfenozide include suspension concentrates (flowable concentrates (SC)), wettable powders (WP) and dustable powders (DP) (see Table 6.7).

6.9 Overview of Global Uses and Labels

6.9.1 Introduction

Methoxyfenozide first was registered in several countries in Central and South America by R&H in 1998 and 1999 for use on pome fruits, stone fruits, fodder and feed commodities. The first registration of methoxyfenozide in the US was in September 2000 for use on pome fruits followed by cotton in the same year. In 2001, methoxyfenozide was acquired by DAS and subsequent

Table 6.7   Methoxyfenozide formulations and trade names.

Registered names	Concentration	Formulation
Intrepid[®], Runner[®], Prodigy[®], Integro[®]	240 g l[−1]	SC
Prodigy[®], Pacer[®]	100 g l[−1]	SC
Falcon[®]	20% w/w	SC
Monceren[®] AD[a]	0.5% w/w	DP (DL[b])
Monceren[®], Runner[®c]	0.5% w/w	DP (DL[b])
Runner[®]	0.5% w/w	DP (DL[b])
Falcon[®]	4% w/w	WP
Intrepid[®], Runner[®]	80% w/w	WP

[a]Additionally it contains 1.5% pencycuron and 0.25% imidacloprid.
[b]DL is a Japanese code for "Driftless" formulation.
[c]Additionally it contains 1.5% pencycuron.

registrations were gained in China and Japan in 2001 and Australia in 2002. Methoxyfenozide was listed for inclusion in Annex I (EU) on 19 January 2005. Currently, methoxyfenozide is registered in more than 50 countries.

The specificity of methoxyfenozide to lepidopteran species has been an important factor in its wide adoption in IPM programs worldwide. Methoxyfenozide is compatible with pollinators and most natural enemies, which affords greater flexibility in application timing in IPM systems. Larvicidal activity is dependent upon ingestion of treated plant material; however, sublethal doses of methoxyfenozide can adversely affect reproduction of some lepidopteran species and its utility for control of lepidopteran eggs also has been documented (see section 6.4).[28] The global use parameters for methoxyfenozide, the widest registered and used inscecticide of the BAH group, are extensively reviewed in this section. For each crop group, examples of crop species for which methoxyfenozide is registered are presented. Pests affected by methoxyfenozide also are listed. Specific information regarding methoxyfenozide use, such as rates and preharvest intervals, can be found on the product label for the country of interest. Pest scientific and common names listed in this section use the nomenclature approved by the Entomological Society of America[49] and the EPPO Plant Protection Thesaurus.[50]

6.9.2   Tree Fruits and Nuts

6.9.2.1   Introduction

Since its introduction, methoxyfenozide has been an effective tool to control key pests in tree fruit and nut crops. Among pome fruit crops, methoxyfenozide is used extensively in apple and pear for control of pests within the Tortricidae, Noctuidae and Gracillariidae families. Also, it has been widely used for the management of navel orangeworm, *Amyelois transitella* (Walker), in US almonds.

6.9.2.2 Pome Fruits

Registration on pome fruits first occurred in 1999 in Argentina, Israel and Mexico. Subsequent registrations have been approved in more than 25 countries. Key target pests include: *C. pomonella*; lesser apple worm, *Grapholita prunivora* (Walsh); *G. molesta*; *C. rosaceana;* and light brown apple moth, *Epiphyas postvittana* (Walker). Pome fruit crops include: apple, crab apple, loquat, pear and quince (see Table 6.8). Due to its adverse impact on fecundity and fertility,[22,23] methoxyfenozide is commonly used at the petal fall stage of growth to manage early infestations of *C. pomonella*; *C. rosaceana;* and apple pandemis, *Pandemis pyrusana* Kearfott, in US apples.

6.9.2.3 Stone Fruits

Since its first registration on stone fruits in Argentina in 1999, additional registrations for methoxyfenozide have been granted in the US and several European countries. The stone fruit crops where methoxyfenozide is used include apricot, cherry, nectarine, peach, plum and prune. Targeted pests include *G. molesta*; *C. pomonella;* and other tortricid species including: fruittree leafroller, *Archips argyrospila* (Walker); omnivorous leafroller, *Platynota stultana* Walsingham; and *A. velutinana* (Walker). Depending on the pest species, application rates of methoxyfenozide are typically between 72–240 g ai ha^{-1} (see Table 6.9). Because of its specificity and low impact on beneficial arthropods, methoxyfenozide has found wide appeal in IPM programs in stone fruits where outbreaks of other pests such as mites and aphids have the potential to cause significant economic impact.

Table 6.8 Methoxyfenozide global use parameters in pome fruits.

Crops[a]	Pests	Rates
apple	Lepidoptera: Noctuidae	3.6–15 g ai hl^{-1} H$_2$O
crab apple	*Lacanobia subjuncta* (Grote and Robinson)	
loquat		72–173 g ai ha^{-1}
pear	Lepidoptera: Tortricidae	
quince	*Archips argyrospila* (Walker)	
nashi[b]	*Archips rosanus* (L.)	
	Choristoneura rosaceana (Harris)	
	Cydia pomonella (L.)	
	Epiphyas postvittana (Walker)	
	Grapholita molesta (Busck)	
	Pandemis pyrusana Kearfott	
	Platynota idaeusalis (Walker)	
	Lepidoptera: Gracillariidae	
	Phyllonorycter blancardella (F.)	
	Phyllonorycter crataegella (Clemens)	

[a]Refer to local methoxyfenozide labels for specific recommendations for crops, pests and rates in the country of interest.
[b]Asian pear.

Table 6.9   Methoxyfenozide global use parameters in stone fruits.

Crops[a]	Pests	Rates

apricot
cherry
nectarine
peach
plum
prune
	Lepidoptera: Gelechiidae
	  Anarsia lineatella Zeller
	
	Lepidoptera: Noctuidae
	  Lithophane antennata (Walker)
	
	Lepidoptera: Notodontidae
	  Schizura concinna (J.E. Smith)
	
	Lepidiotera: Tortricidae
	  Archips argyrospila (Walker)
	  Argyrotaenia velutinana (Walker)
	  Choristoneura rosaceana (Harris)
	  Cydia pomonella (L.)
	  Grapholita molesta (Busck)
	  Grapholita packardi Zeller
	  Grapholita prunivora (Walsh)
	  Pandemis limitata (Robinson)
	  Pandemis pyrusana Kearfott
	  Platynota idaeusalis (Walker)
	  Platynota stultana Walsinghham
	  Spilonota ocellana (Denis &
	    Schiffermüller)
		3.6–12 g ai hl^{-1} H$_2$O
		
		72–240 g ai ha^{-1}

[a]Refer to local methoxyfenozide labels for specific recommendations for crops, pests and rates in the country of interest.

6.9.2.4   Citrus Fruits

The citrus group encompasses several species and hybrids including grapefruits, lemons, limes, oranges and tangerines. Although the first methoxyfenozide registration in citrus fruit was in Bolivia in 2001, use with these fruits was fairly limited until methoxyfenozide was registered in other countries. One of the most important insect pests of citrus fruit controlled by methoxyfenozide is the citrus leafminer, *Phyllocnistis citrella* Stainton. In some areas *P. citrella* can spread *Xanthomonas axonopodis* pv. *citri.*, the bacterium that causes citrus canker. Methoxyfenozide also is effective against citrus peelminer, *Marmara gulosa* Guillén and Davis, which attacks fruit instead of leaves. Other lepidopteran citrus pests controlled by methoxyfenozide include parsley swallowtail (orangedog), *Papilio zelicaon* Lucas and other *Papilio* spp.; citrus cutworm, *Egira curialis* (Grote); variegated cutworm, *Peridroma saucia* (Hübner); *A. argyrospila* and *P. stultana* (see Table 6.10).

6.9.2.5   Tree Nuts

Methoxyfenozide is registered for use against pests such as *A. transitella*, *C. pomonella* and *C. rosaceana* in several tree-nut producing countries. Crops within the tree-nut grouping include almond, beech nut, butter nut, chestnut,

Table 6.10 Methoxyfenozide global use parameters in citrus.

Crops[a]	*Pests*	*Rates*
clementine	Lepidoptera: Gracillariidae	7.2–36 g ai hl^{-1} H$_2$O
grapefruit	*Marmara gulosa* Guillén & Davis	
mandarin	*Phyllocnistis citrella* Stainton	72–280 g ai ha^{-1}
orange		
tangerine	Lepidoptera: Noctuidae	
citrus hybrids	*Egira (Xylomyges) curialis* Grote	
	Peridroma saucia (Hübner)	
	Lepidoptera: Papilionidae	
	Papilio spp.	
	Lepidiotera: Tortricidae	
	Archips argyrospila (Walker)	
	Platynota stultana Walsinghham	

[a]Refer to local methoxyfenozide labels for specific recommendations for crops, pests and rates in the country of interest.

Table 6.11 Methoxyfenozide global use parameters in tree nuts.

Crops[a]	*Pests*	*Rates*
almond	Lepidoptera: Arctiidae	3.6–4.8 g ai hl^{-1} H$_2$O
beech nut	*Hyphantria cunea* (Drury)	
butter nut		67–426 g ai ha^{-1}
chestnut	Lepidoptera: Gelechiidae	
chinquapin	*Anarsia lineatella* Zeller	
hazelnut		
macadamia	Lepidoptera: Notodontidae	
pecan	*Schizura concinna* (J.E. Smith)	
pistachio		
walnut	Lepidoptera: Pyralidae	
	Acrobasis nuxvorella Neunzig	
	Amyelois transitella (Walker)	
	Lepidoptera: Tortricidae	
	Choristoneura rosaceana (Harris)	
	Cnephasia longana (Haworth)	
	Cydia latiferreana (Walsingham)	
	Cydia caryana (Fitch)	
	Cydia pomonella (L.)	

[a]Refer to local methoxyfenozide labels for specific recommendations for crops, pests and rates in the country of interest.

chinquapin, hazelnut, macadamia, pecan, pistachio and walnut. Pests are controlled using application rates ranging from 67–426 g ai ha^{-1} (see Table 6.11). In the US, methoxyfenozide has been an industry standard for the management of *A. transitella* in almonds. Applications are made at the *hull-split* stage of almonds to target eggs and neonate larvae before entering the shell.

Table 6.12 Methoxyfenozide global use parameters in small fruits.

Crops[a]	*Pests*	*Rates*
aronia berry	Lepidoptera: Noctuidae	3.6–9.6 g ai hl^{-1} H$_2$O
blueberry	*Agrotis* spp.	
bushberry	*Helicoverpa zea* (Boddie)	36–275 g ai ha^{-1}
chilean guava	*Spodoptera* spp.	
cranberry		
currant	Lepidoptera: Pyralidae	
elderberry	*Acrobasis vaccinii* Riley	
European	Lepidoptera: Tortricidae	
barberry	*Archips argyrospilus* (Walker)	
grapes	*Archips rosanus* (L.)	
gooseberry	*Choristoneura rosaceana* (Harris)	
honeysuckle	*Clysia ambiguella* (Hubner)	
huckleberry	*Endopiza viteana* Clemens	
jostaberry	*Grapholita packardi* Zeller	
juneberry	*Lobesia botrana* (Dennis & Schiffermüller)	
lingonberry	*Pandemis pyrusana* Kearfott	
mayhaw		
salal		
sea buckhorn		
spanish lime		

[a]Refer to local methoxyfenozide labels for specific recommendations for crops, pests and rates in the country of interest.

6.9.2.6   Small Fruits

Methoxyfenozide is used in management programs for lepidopteran pests of small fruits in numerous countries. Key target pests include *L. botrana* and several species within the Pyralidae, Tortricidae and Noctuidae families. Major crops include blueberries, cranberries, grapes and strawberries (see Table 6.12).

6.9.2.7   Tropical Fruits

The first methoxyfenozide registration for use in tropical fruits was obtained in Israel in 1999 for use in pomegranate. Currently, it is used to control several key pyralid, tortricid and gracillariid pests, such as *M. gulosa*. Methoxyfenozide is used on a variety of tropical fruit crops including acerola, carambola, jaboticaba, kiwi, lychee, longan, Spanish lime, rambutan, pulasan and wax jambu among others (see Table 6.13).

6.9.3   Vegetables

6.9.3.1   Introduction

Methoxyfenozide is widely used to control lepidopteran pests in vegetables because of its fit in IPM programs as well as its differentiated mode of action for

Table 6.13 Methoxyfenozide global use parameters in tropical fruits.

Crops[a]	Pests	Rates
acerola	Lepidoptera: Gracillariidae	3.6–6 g ai hl^{-1} H$_2$O
avocado	*Conopomorpha cramerella* (Snellen)	
black sapote	*Conopomorpha sinensis* (Bradley)	175–280 g ai ha^{-1}
canistel	*Marmara salictella* Clemens	
carambola	*Marmara* spp.	
feijoa		
guava	Lepidoptera: Elachistidae	
jaboticaba	*Stenoma catenifer* Walsingham	
kiwi		
lychee[b]	Lepidoptera: Lycaenidae	
longan	*Deudorix epijarbas* (Moore)	
mamey[c]		
mango	Lepidoptera: Psychidae	
papaya	*Oiketicus kirbyi* (Guilding)	
passion fruit[d]		
pomegranate	Lepidoptera: Pyralidae	
pulasan	*Deanolis albizonalis* Hampson	
Spanish lime	*Noorda albiizonaliis* Hampson	
rambutan		
pulasan	Lepidoptera: Tortricidae	
sapodilla	*Argyrotaenia citrana* (Fernald)	
star apple		
wax jambu		

[a]Refer to local methoxyfenozide labels for specific recommendations for crops, pests and rates in the country of interest.
[b]*Litchi.*
[c]White sapote.
[d]Granadilla.

IRM programs. Methoxyfenozide is used around the world in vegetables to control pests in the lepidopteran families: Crambidae, Gelechiidae, Noctuidae, Pieridae, Plutellidae and Pyralidae.

Due to its specificity for lepidopteran control, methoxyfenozide is widely used in fruiting vegetable crops where pollinators are an important component of production, such as in all cucurbit crops.

6.9.3.2 Cole Crops

The first registration of methoxyfenozide in cole crops (Family Brassicaceae) occurred in Indonesia in 1998; since then, methoxyfenozide has gained registration in 11 countries on four continents. Key cole crops suited to methoxyfenozide use include: broccoli, cauliflower, cabbage and several other Asian vegetables, for the management of pests such as cabbage looper, *Trichoplusia ni* (Hübner); garden webworm, *Achyra rantalis* (Guenée); and imported cabbageworm, *Pieris rapae* (L.) (see Table 6.14).

Table 6.14   Methoxyfenozide global use parameters in cole crops.

Crops[a]	Pests	Rates
bok choy	Lepidoptera: Crambidae	2.3–12 g ai hl^{-1} H$_2$O
broccoli	*Achyra rantalis* (Guenée)	
broccoli raab	*Evergestis rimosalis* (Guenée)	70–280 g ai ha^{-1}
brussels sprouts	*Hellula undalis* (F.)	
cabbage		
cauliflower	Lepidoptera: Pieridae	
cavalo	*Pieris brassicae* (L.)	
Chinese broccoli	*Pieris rapae* (L.)	
Chinese mustard		
cabbage	Lepidoptera: Plutellidae	
collard	*Plutella xylostella* (L.)	
cress		
kohlrabi	Lepidoptera: Noctuidae	
mizuna	*Agrotis* spp.	
mustard greens	*Pseudaletia unipuncta* (Haworth)	
napa	*Spodoptera exigua* (Hübner)	
petsai	*Spodoptera eridania* (Stoll)	
rape greens	*Spodoptera frugiperda* (J.E. Smith)	
	Spodoptera ornithogalli (Guenée)	
	Trichoplusia ni (Hübner)	

[a]Refer to local methoxyfenozide labels for specific recommendations for crops, pests and rates in the country of interest.

Table 6.15   Methoxyfenozide global use parameters in cucurbit crops.

Crops[a]	Pests	Rates
cucumber	Lepidoptera: Crambidae	48–280 g ai ha^{-1}
melon	*Diaphania hyalinata* (L.)	
vegetable sponge	*Diaphania nitidalis* (Stoll)	
watermelon		
	Lepidoptera: Noctuidae	
	Spodoptera exigua (Hübner)	
	Spodoptera eridania (Stoll)	
	Spodoptera frugiperda (J.E. Smith)	

[a]Refer to local methoxyfenozide labels for specific recommendations for crops, pests and rates in the country of interest.

6.9.3.3   Cucurbit Crops

Methoxyfenozide was first registered in South Korea in 1998 to control several lepidopteran pests of economic importance in cucurbit crops. Currently, methoxyfenozide is registered in eight countries for cucurbit crop use. Key cucurbit crops which use methoxyfenozide include: cucumber, melons, vegetable sponge and watermelon, to control pests such as *Diaphania hyalinata* (L.); pickleworm, *D. nitidalis* (Stoll) and several *Spodoptera* spp. (see Table 6.15). Specificity of

Table 6.16 Methoxyfenozide global use parameters in fruiting vegetables.

Crops[a]	Pests	Rates
chili pepper	Lepidoptera: Crambidae	8.4–28.8 g ai hl^{-1} H$_2$O
eggplant	*Neoleucinodes elegantalis* (Guenée)	
paprika		48–600 g ai ha^{-1}
pimento		
sweet pepper	Lepidoptera: Gelichiidae	
tomato	*Rachiplusia nu* (Guenée)	
tomatillo	*Tuta absoluta* (Meyrick)	
	Lepidoptera: Noctuidae	
	Agrotis spp.	
	Helicoverpa armigera (Hübner)	
	Heliothis virescens (Fabricius)	
	Spodoptera exigua (Hübner)	
	Spodoptera eridania (Stoll)	

[a]Refer to local methoxyfenozide labels for specific recommendations for crops, pests and rates in the country of interest.

methoxyfenozide for lepidopteran pests is a key attribute for cucurbit crops where pollinators such as *A. mellifera* are needed to produce good crop yield.

6.9.3.4 Fruiting Vegetables

In 1998, Chile, Colombia and Indonesia became the first countries to register methoxyfenozide in fruiting vegetable crops. Currently, methoxyfenozide is registered in 21 countries on six continents for use in those crops to control lepidopteran pests. Key fruiting vegetable crops that use methoxyfenozide include: tomato, pepper and eggplant, to control pests such as cutworms, *Agrotis* spp.; *H. virescens*; *Spodoptera* spp.; and tomato leaf miner, *Tuta absoluta* (Meyrick) (see Table 6.16).

6.9.3.5 Leafy, Legume and Bulb Vegetables

Methoxyfenozide was first registered in Indonesia in 1998 for the control of lepidopteran pests in leafy vegetables and in Israel in 1999 in legumes and bulb-vegetable crops; it is currently registered in eight countries on four continents for these crops. Key leafy and bulb vegetable crops that use methoxyfenozide include: lettuce, spinach, onions, garlic and leek, to control pests such as *Agrotis* spp.; *Spodoptera* spp.; and *T. ni* (see Table 6.17).

6.9.4 Row Crops

6.9.4.1 Introduction

Methoxyfenozide is widely used for lepidopteran pest control in row crops because of its effectiveness against key pests and its relatively long residual

Table 6.17 Methoxyfenozide global use parameters in leafy, legume and bulb
              vegetables.

Crops[a]	Pests	Rates
Leafy Vegetables	Lepidoptera: Crambidae	2.3–12 g ai hl^{-1} H$_2$O
arugula	*Achyra rantalis* (Guenée)	
chard		70–280 g ai ha^{-1}
chervil	Lepidoptera: Noctuidae	
lettuce	*Agrotis* spp.	
orach	*Pseudaletia unipuncta* (Haworth)	
purslane	*Spodoptera exigua* (Hübner)	
spinach	*Spodoptera eridania* (Stoll)	
	Spodoptera frugiperda (J.E. Smith)	
Legumes	*Spodoptera ornithogalli* (Guenée)	
bean pods	*Trichoplusia ni* (Hübner)	
dry beans		
green bean		
Bulb Vegetables		
florence fennel		
garlic		
green onion		
leek		
onion		

[a]Refer to local methoxyfenozide labels for specific recommendations for crops, pests and rates in the
country of interest.

control. Methoxyfenozide also represents a new mode of action in row crops
globally, making it a tool for IRM or to control pests resistant to other
insecticide modes of action (*e.g.* pyrethroid-resistant Lepidoptera). The main
lepidopteran families controlled with methoxyfenozide in row crops are:
Crambidae, Noctuidae and Pyralididae.

6.9.4.2 Cereals

In 1998, Colombia was the first country to register methoxyfenozide for lepidopteran pest control in rice and corn. Currently, it is registered in 22 countries
on four continents for use in these crops. The main row crops where methoxyfenozide is used include: corn, milo/sorghum, rice, sweet corn and wheat, to
control pests such as *Agrotis* spp.; Asiatic rice borer, *Chilo suppressalis*
(Walker); rice ear-cutting caterpillar, *Mythimna separata* (Walker); South
American white borer, *Rupella albinella* (Cramer); fall armyworm, *Spodoptera
frugiperda* (J. E. Smith); and yellow stem borer, *Scirpophaga incertulas*
(Walker) (see Table 6.18).

6.9.4.3 Oilseeds

Methoxyfenozide was first registered in Colombia and Indonesia in 1998 for
lepidopteran pest control in cotton and soybean, respectively. Currently, it is

Table 6.18 Methoxyfenozide global use parameters in cereal crops.

Crops[a]	Pests	Rates

corn
milo/sorghum
rice
seed corn
wheat
	Lepidoptera: Crambidae
	  Chilo suppressalis (Walker)
	  Scirpophaga incertulas (Walker)
	
	Lepidoptera: Noctuidae
	  Agrotis spp.
	  Mythimna separata (Walker)
	  Spodoptera frugiperda (J.E. Smith)
	
	Lepidoptera: Pyralidae
	  Tryporyza incertulas (Walker)
	
	Lepidoptera: Schoenobiidae
	  Rupella albinella (Cramer)
		48 g ai hl^{-1} H_2O
		
		19.2–240 g ai ha^{-1}

[a]Refer to local methoxyfenozide labels for specific recommendations for crops, pests and rates in the country of interest.

Table 6.19 Methoxyfenozide global use parameters in oilseed crops.

Crops[a]	Pests	Rates

cotton
peanut
soybean
sunflower
	Lepidoptera: Noctuidae
	  Alabama argillacea (Hübner)
	  Anticarsia gemmatalis Hübner
	  Heliothis virescens (F.)
	  Helicoverpa gelotopoeon (Dyar)
	  Helicoverpa zea (Boddie)
	  Rachiplusia nu (Guenée)
	  Spodoptera frugiperda (J.E. Smith)
		14.4–600 g ai ha^{-1}

[a]Refer to local methoxyfenozide labels for specific recommendations for crops, pests and rates in the country of interest.

registered in 16 countries on four continents for use in: cotton, peanuts, soybean and sunflower, to control pests such as cotton leafworm, *Alabama argillacea* (Hübner); velvetbean caterpillar, *Anticarsia gemmatalis* Hübner; *H. virescens* (F.); cotton bollworm, *Helicoverpa gelotopoeon* (Dyar) and *S. frugiperda* (see Table 6.19).

6.9.4.4 Forages

In 1999, Israel was the first country to register methoxyfenozide for lepidopteran pest control in alfalfa. Currently, this active ingredient is also registered in South Arabia and the US for use in forage crops to control pests such as: *Agrotis* spp.; alfalfa looper, *Autographa californica* (Speyer); armyworm, *Pseudaletia unipuncta* (Haworth); and *Spodoptera* spp., among others (see Table 6.20).

Table 6.20 Methoxyfenozide global use parameters in forage crops.

Crops[a]	Pests	Rates

alfalfa
clover
grasses
lupin
mixed forage stand
	Lepidoptera: Crambidae
	   Fissicrambus mutabilis (Clemens)
	   Parapediasia teterrella (Zincken)
	   Pediasia trisecta (Walker)
	
	Lepidoptera: Noctuidae
	   Agrotis spp.
	   Autographa californica (Speyer)
	   Pseudaletia unipuncta (Haworth)
	   Spodoptera exigua (Hübner)
	   Spodoptera eridania (Stoll)
	   Spodoptera frugiperda (J.E. Smith)
	   Spodoptera ornithogalli (Guenée)
	   Spodoptera praefica (Grote)
	
	Lepidoptera: Pieridae
	   Colias eurytheme Boisduval
		14.4–600 g ai ha^{-1}

[a]Refer to local methoxyfenozide labels for specific recommendations for crops, pests and rates in the country of interest.

Table 6.21 Methoxyfenozide global use parameters in tea.

Crop[a]	Pests	Rates

tea	Lepidoptera: Geometridae
	   Ascotis selenaria Denis & Schiffermüller
	
	Lepidoptera: Gracilariidae
	   Gracilaria theivora Wals
	
	Lepidoptera: Noctuidae
	   Spodoptera litura F.
	
	Lepidoptera: Tortricidae
	   Adoxophyes honmai (Yasuda)
	   Homona magnanima Diakonoff
		960–1920 g ai ha^{-1}

[a]Refer to local methoxyfenozide labels for specific recommendations for crops, pests and rates in the country of interest.

6.9.5 Specialty Uses: Tea, Ornamentals and Forestry

Tebufenozide and methoxyfenozide are used across many countries for the control of various Lepidoptera in a number of specialty uses such as tea, forestry and in facilities for production of ornamental plants or plant parts, such as cut flowers or potted plants. Their target-specific activity makes them highly suited for IPM programs in nurseries, glasshouses and silviculture, where pollinators, parasitoids and predators are an important component for maintaining an ecological balance. The uses, pests and rates for methoxyfenozide are summarized in Tables 6.21 and 6.22.

Table 6.22 Methoxyfenozide global use parameters in ornamentals and forests.

Crop or application area[a]	Pests	Rates
Ornamentals	Lepidoptera: Arctiidae	9.6 g ai hl^{-1}
greens	*Hyphantria cunea* (Drury)	
roses		36–120 g ai ha^{-1}
flowers	Lepidoptera: Geometridae	
perennials	*Alsophila pometaria* (Harris)	
outdoor areas	*Ennomos subsignaria* (Hübner)	
residential areas		
Forestry	Lepidoptera: Lasiocampidae	
silviculture	*Malacosoma americanum* (F.)	
nursery trees		
Christmas trees	Lepidoptera: Lymantriidae	
	Euproctis chrysorrhoea (L.)	
	Lymantria dispar (L.)	
	Orgyia leucostigma (J.E. Smith)	
	Lepidoptera: Noctuidae	
	Callopistria floridensis (Guenée)	
	Choristoneura fumiferana (Clemens)	
	Spodoptera spp.	
	Lepidoptera: Psychidae	
	Thyridopteryx ephemeraeformis (Haworth)	
	Lepidoptera: Saturniidae	
	Opodiphthera eucalypti (Scott)	
	Lepidoptera: Thaumatopoeidae	
	Thaumetopoea pityocampa (Schiff)	
	Lepidoptera: Tortricidae	
	Epiphyas postvittana (Walker)	

[a]Refer to local methoxyfenozide labels for specific recommendations for crops, pests and rates in the country of interest.

6.10   Insecticide Resistance Management

The results from synergist studies, metabolism studies and the analysis of field resistance incidents point to oxidative metabolism as the primary resistance mechanism for tebufenozide and methoxyfenozide.[2,5,51] For example, in a methoxyfenozide-resistant beet armyworm strain, *Spodoptera exigua* (Hübner), the rate of oxidative metabolism was 1.4 times greater than a susceptible laboratory strain, resulting in a half-life for radiolabeled methoxyfenozide in resistant larvae that was one-half that in susceptible larvae, and a rate of radiolabel excretion by resistant larvae that was twice that of susceptible larvae.[52]

There is little evidence for a BAH resistance mechanism involving an altered target site in whole insects.[2,5] Subclones of an *S. exigua* cell line resistant to the cell proliferation inhibition effects of methoxyfenozide have been created, and resistance mechanisms other than alteration of the EcR were absent, but it appears that the resistance mechanism selected in these cells is not likely to be selected in insect populations.[53]

There are a number of reports of BAH insecticide cross-resistance with azinphos-methyl in lepidopteran pests of tree fruits, but not with other organophosphate insecticides.[2,5,54,55] Increased acetylcholinesterase activity has been observed in a methoxyfenozide-resistant strain of *S. exigua*, but this was considered the result of prior exposure of this population to organophosphate and carbamate insecticides, and not a result of methoxyfenozide exposure.[52] In a tebufenozide-resistant diamondback moth, *Plutella xylostella* (L.), strain cross-resistance with abamectin was observed, but no cross-resistance was observed with the organophosphate insecticides: trichlorfon, phoxim and acephate.[56] There is little additional evidence of BAH cross-resistance with insecticides possessing other modes of action.

Inheritance of tebufenozide resistance in *S. exigua* is incompletely dominant, autosomal and not sex-linked, and probably multigenic in nature.[51] In a study with a methoxyfenozide-resistant *S. exigua* strain, sex-linkage of resistance was observed in mortality results after 4 days of exposure but not for results after 10 days of exposure.[57]

There is evidence of reduced fitness in BAH-resistant insects. Compared to a susceptible laboratory strain, a tebufenozide-resistant *S. exigua* strain had significantly lower survival between the fourth and fifth instars; prolonged duration of the larval and pupal stages; lower pupal weight; lower pupation rate; and lower fecundity.[51] A tebufenozide-resistant *P. xylostella* strain had significantly lower larval, prepupal and pupal survival rates; lower oviposition rates; lower fecundity; and lower rates of egg hatch produced by treated females.[56]

Given the lack of genetic dominance observed for BAH resistance, the reduced fitness of BAH-resistant insects, and the few cases of cross-resistance with other insecticides, it is very likely that the effectiveness of BAH insecticides can be maintained by following sound IRM practices, including use of multiple IPM tactics and adequate product rotation.

6.11 Conclusions

The unique mode of action of the BAHs, their specificity and high level of efficacy on the target pests and their toxicological and ecotoxicological profiles make this chemistry class a valuable tool for any pest management program. Furthermore, BAH insecticides are ideally suited to programs designed to integrate several modes of action and/or several management techniques such as IRM and IPM programs. Many of the crop systems on which lepidopteran larvae are pests, use hymenopteran species as pollinators to produce harvestable fruit. Since BAH insecticides are nontoxic to such insects, they can easily be incorporated into management of those crops.

By specifically targeting the organism of interest while minimizing impact on non-target organisms and the environment, the BAH insecticides are categorized as *green chemistry insecticides*. One of the most beneficial characteristics of the BAH insecticides is that by having a greater affinity to the EcR than the naturally occurring 20E in the target pests, it makes them highly specific to those pests. Within the target insect species, the BAH insecticides are toxic to the larval stage, which is usually the stage that causes the damage. In adults, the BAH insecticides have some sublethal effects, such as delayed developmental rates and reduced fecundity and fertility, which can result in long-term population reduction.

Insect pests will always threaten crop production and adapt to the methods created to reduce their populations. Green chemistries, such as the BAH insecticides, provide an additional tool that can be rotated with other chemistries within IRM programs or can be incorporated with other methods of control within IPM programs, helping make those programs sustainable for longer periods of time.

The advantageous characteristics of BAH insecticides mean that we can anticipate a continued development in this class of chemicals in the future. An increase in their use by growers is also likely, since current regulations favor products which have a low impact on nontarget organisms and humans, and which are compatible with IPM and IRM programs.

Acknowledgements

The authors would like to thank the following colleagues from Dow AgroSciences for their important input and technical reviews to improve this manuscript: Cheryl Cleveland, Steve Norman, Giovanna Meregalli, Jon Babcock, John Fitt, Don Kelley, William Brewster, Carl Corvin, Nick Simmons and Diane Kindell. A special appreciation is also extended to Rayda Krell for her editorial assistance.

We also acknowledge all other Dow AgroSciences and Dow Chemical colleagues, as well as university and private researchers who have studied BAH chemistry through the years, helping in the understanding and development of these insecticides as important tools for insect management.

Confirm®, Integro®, Intrepid®, Falcon®, Mach 2®, Pacer®, Prodigy® and Runner® are trademarks of Dow AgroSciences L.L.C. Monceren® is a trademark of Bayer.

References

1. A. C.-T. Hsu, in *Synthesis and Chemistry of Agrochemical, II*, ed. D. R. Baker, J. G. Fenyes and W. K Moberg, American Chemical Society, Washington, D.C., USA, 1991, p. 478.
2. T. S. Dhadialla, G. R. Carlson and D. P. Le, *Annu. Rev. Entomol.*, 1998, **43**, 545.
3. T. S. Dhadialla, A. Retnakaran and G. Smagghe, in *Comprehensive Molecular Insect Science,* ed. L. I. Gilbert, K. Iatrou and S. S. Gill, Elsevier, Oxford, UK, 2005, vol. 6, p. 55.
4. L. Dinan and R.E. Hormann, in *Comprehensive Molecular Insect Science*, ed. L. I. Gilbert, K. Iatrou and S. S. Gill, Elsevier, Oxford, UK, 2005, vol. 3, p. 197.

5. T. S. Dhadialla and R. R. Ross, in *Modern Crop Protection Compounds*, ed. W. Kramer and U. Schirmer, Wiley-VCH Verlag GmbH & Co., Weinheim, Germany, 2007, vol. 2, p. 773.
6. I. M. L. Billas, T. Iwema, J.-M. Garnier, A. Mitschler, N. Rochel and D. Moras, *Nature*, 2003, **426**, 91.
7. J. E. Carpenter and L. D. Chandler, *J. Entomol. Sci.*, 1994, **29**, 428.
8. M. M. Adel and F. Sehnal, *J. Insect Physiol.*, 2000, **46**, 267.
9. D. Biddinger, L. Hull, H. Huang, B. McPheron and M. Loyer, *J. Econ. Entomol.*, 2006, **99**, 834.
10. S. Pineda, M. I. Schneider, G. Smagghe, A. M. Martinez, P. Del Estal, E. Viñuela, J. Valle and F. Budia, *J. Econ. Entomol.*, 2007, **100**, 773.
11. S. Pineda, F. Budia, M. I. Schneider, A. Gobbi, E. Viñuela, J. Valle and P. Del Estal, *J. Econ. Entomol.*, 2004, **97**, 1906.
12. B. Carton, G. Smagghe, A. K. Mourad and L. Tirry, *Meded. Fac. Landbouwkd. Toegep. Biol. Wet. Univ. Gent.*, 1998, **63**, 537.
13. M. Sundaram, S. R. Palli, G. Smagghe, I. Oshaaya, Q. L. Feng, M. Primavera, W. L. Tomkins, P. J. Krell and A. Retnakaran, *Insect. Biochem. Mol. Biol.*, 2002, **32**, 225.
14. A. Trisyono and M. Chippendale, *J. Econ. Entomol.*, 1997, **90**, 1486.
15. A. Trisyono and M. Chippendale, *Pestic. Sci.*, 1998, **53**, 177.
16. R. K. Seth, J. J. Kaur, D. K. Rao and S. E. Reynolds, *J. Insect Physiol.*, 2004, **50**, 505.
17. G. Smagghe, T. S. Dhadialla, S. Derycke, L. Tirry and D. Degheele, *Pestic. Sci.*, 1998, **54**, 27.
18. C.-L. Rodríguez Enríquez, S. Pineda, J. I. Figueroa, M.-I. Schneider and A. M. Martínez, *J. Econ. Entomol.*, 2010, **103**, 662.
19. G. Smagghe, D. Bylemans, P. Medina, F. Budia, J. Avilla and E. Viñuela, *Ann. Appl. Biol.*, 2004, **145**, 291.
20. S. Pineda, G. Smagghe, M. I. Schneider, P. Del Estal, E. Viñuela, A. M. Martínez and F. Budia, *Environ. Entomol.*, 2006, **35**, 856.
21. S. Pineda, A. M. Martínez, J. I. Figueroa, M. I. Schneider, P. Del Esta, E. Viñuela, B. Gomez, G. Smagghe, J. Valle and F. Budia, *J. Econ. Entomol.*, 2009, **102**, 1490.
22. X. Sun and B. A. Barrett, *J. Econ. Entomol.*, 1999, **92**, 1039.
23. X. Sun, B. A. Barrett and D. J. Biddinger, *Entomol. Exp. Appl.*, 2000, **94**, 75.
24. R. Alhumada and A. Brieva, Informe Técnico BPF No. 42, Bioforest S.A. Santiago, Chile, 1999.
25. A. L. Knight, *J. Econ. Entomol.*, 2000, **93**, 1760.
26. E. Castiglioni, H. Silva and I. Russi, in *Resumos de XX Cogresso Brasileiro de Entomologia*, 2004.
27. P. J. Charmillot, R. Favre, D. Pasquier, M. Rhyn and A. Scaldom, *Schweiz. Entomol.*, 1994, **67**, 393.
28. F.-J. Sáenz-de-Cabezón Irigaray, V. Marco, F. G. Zalom and I. Peréz-Moreno, *Pest Manag. Sci.*, 2005, **61**, 1133.
29. C. T. Myers and L. A. Hull, *J. Entomol. Sci.*, 2003, **38**, 420.
30. M. D. Reinke and B. A. Barrett, *J. Entomol. Sci.*, 2007, **42**, 457.

31. J. A. Hoelscher and B. A. Barrett, *J. Econ. Entomol.*, 2003, **96**, 623.
32. J. A. Hoelscher and B. A. Barrett, *Entomol. Exp. Appl.*, 2003, **107**, 133.
33. M. D. Reinke and B. A. Barrett, *J. Econ. Entomol.*, 2007, **100**, 72.
34. P. J. Charmillot, A. Gourmelon, A. L. Fabre and D. Pasquier, *J. Appl. Entomol.*, 2001, **125**, 147.
35. S. Pons, H. Reidl and J. Avilla, *J. Econ. Entomol.*, 1999, **92**, 1344.
36. http://www.epa.gov/gcc/pubs/pgcc/winners/dgca98.html (last accessed August 2010).
37. http://www.epa.gov/gcc/pubs/pgcc/presgcc.html (last accessed August 2010).
38. http://www.fao.org/ag/AGP/AGPP/Pesticid/JMPR/Download/2003_eva/methoxyfenozide%202003.pdf (last accessed August 2010).
39. http://whqlibdoc.who.int/publications/2004/924166519X_methoxyfenozide.pdf (last accessed August 2010).
40. http://www.apvma.gov.au/registration/assessment/docs/prs_methoxyfenozide.pdf (last accessed August 2010).
41. http://ec.europa.eu/food/plant/protection/evaluation/newactive/methoxyfenozide_review_report.pdf (last accessed August 2010).
42. http://frwebgate.access.gpo.gov/cgi-bin/getdoc.cgi?dbname = 2002_register &docid = 02-23996-filed.pdf (last accessed October 2010).
43. www.cdpr.ca.gov/docs/registration/ais/publicreports/5698.pdf (last accessed August 2010).
44. http://www.unece.org/trans/danger/publi/ghs/ghs_rev03/English/04e_part4.pdf (last accessed August 2010).
45. http://dsp-psd.pwgsc.gc.ca/collection_2008/pmra-arla/H113-9-2008-15E.pdf (last accessed August 2010).
46. V. Mommaerts, G. Sterk and G. Smagghe, *Ecotoxicology*, 2006, **15**, 513.
47. G. R. Carlson, T. S. Dhadialla, R. Hunter, R. K. Jansson, C. S. Jany, Z. Lidert and R. A. Slawecki, *Pest Manag. Sci.*, 2001, **57**, 115.
48. http://www.croplife.org/files/documentspublished/1/en-us/PUB-TM/4147_PUB-TM_2008_05_01_Technical_Monograph_2_-_Revised_May_2008.pdf (last accessed August 2010).
49. http://www.entsoc.org/pubs/common_names/index.htm (last accessed August 2010).
50. http://eppt.eppo.org/ (last accessed August 2010).
51. B.-T. Jia, Y.-J. Liu, Y.-C. Zhu, X.-G. Liu, C.-F. Gao and J.-L. Shen, *Pest Manag. Sci.*, 2009, **65**, 996.
52. G. Smagghe, S. Pineda, B. Carton, P. Del Estal, F. Budia and E. Viñuela, *Pest Manag. Sci.*, 2003, **59**, 1203.
53. H. Mosallanejad, T. Soin and G. Smagghe, *Arch. Insect Biochem. Physiol.*, 2008, **67**, 36.
54. M. Reyes, P. Franck, P.-J. Charmillot, C. Ioratti, J. Olivares, E. Pasqualini and B. Sauphanor, *Pest Manag. Sci.*, 2007, **63**, 890.
55. J. E. Dunley, J. F. Brunner, M. D. Doerr and E. H. Beers, *J. Insect Sci.*, 2006, **6**, 14.
56. G.-C. Cao and Z.-J. Han, *Pest Manag. Sci.*, 2006, **62**, 746.
57. J. Gore and J. J. Adamczyk, *Florida Entomol.*, 2004, **87**, 450.

Needles in the Haystack: Exploring Chemical Diversity of Botanical Insecticides

MURRAY B. ISMAN[1] AND GRETCHEN PALUCH[2]

[1] University of British Columbia, Faculty of Land and Food Systems, Vancouver V6T 1Z4, Canada; [2] EcoSMART Technologies Inc. c/o Iowa State University, Department of Entomology, Ames, IA 50011, U.S.A.

7.1 Introduction

The scientific literature continues to document the biological diversity of plant secondary metabolites that are known to affect pest insects. Botanicals have a long history of use and are referenced in recordings from ancient civilizations and folklore.[1,2] Likewise, the natural history of insect–plant chemical interactions spans the past 400 million years, or as some authors like to say, terrestrial plants have engaged in chemical warfare against insects and other herbivores over that vast period of time. Therefore, to the non-expert, it would seem intuitively obvious that higher plants should be a source of unique natural chemicals with potential utility for the anthropocentric management of insects and related pests. Indeed, prior to the discovery of the insecticidal properties of DDT in the middle of the 20th century, some of the most important insecticides used in agriculture and related industries were of botanical origin. Almost completely displaced by synthetic insecticides in the DDT era, interest in botanicals was rekindled in the 1970s once the detrimental effects of synthetic insecticides on human and environmental health were fully realized.[3]

RSC Green Chemistry No. 11
Green Trends in Insect Control
Edited by Óscar López and José G. Fernández-Bolaños
© Royal Society of Chemistry 2011
Published by the Royal Society of Chemistry, www.rsc.org

At the same time, interest in the chemistry of plant secondary metabolites and their ecological roles blossomed, spawning a voluminous scientific literature on the effects of plant allelochemicals on insects that is growing exponentially at present. For example, according to the Web of Science database, a search for "plant extract and insecticidal activity" in mid-2010 generated a list of 62 journal papers over the five year interval from 2005–2009; searching for "essential oils and insecticidal activity" yielded 115 papers over the same interval (= 23 papers per year), and the key words "botanical insecticide" yielded 116 papers.[4] Much of the prior research in this field has been reviewed extensively, in volumes such as *Insecticides of Plant Origin*,[5] *Phytochemicals for Pest Control*,[6] *Botanical Pesticides in Agriculture*,[7] *Phytochemical Biopesticides*[8] and *Biopesticides of Plant Origin*.[9] Given the effort and resources expended in this area of scientific investigation over the past 35 years, it would be reasonable to ask – has much of this knowledge found its way to the actual practice of pest management? And if not, why not?

The simple answer to the first question is that this vast body of scientific inquiry has generated only a handful of commercially viable pest management products (*viz.* azadirachtin, essential oils), at least in North America, Western Europe and Japan. At the same time, some areas of pest management have been transformed through the genetic modification of crop plants, the discovery of new and powerful insecticides of microbial origin, and the discovery of new and potent synthetic insecticides with novel modes of action and dramatically reduced risks to human health and the environment. Botanical products rarely match conventional synthetic or microbial insecticides in terms of absolute efficacy, speed of action, or cost – considerations of the utmost importance to farmers whose crop production practices have long included conventional pesticides.[3] This is not to say that botanical insecticides cannot be useful in highly mechanized or intensively managed agricultural systems. In certain crop–pest contexts, botanicals have been shown to have an efficacy comparable to conventional pesticides as stand-alone products, but the former products may have greater utility in tank mixes with conventional products, in rotation with conventional products (to mitigate the development of insecticide resistance in pest populations), or for early season application in conjunction with augmentative biological control when pest pressures are low.[10]

In regions and countries where product standardization and consistency, and regulatory scrutiny are not as rigorous (compared to the G7 countries), many more botanical insecticides are being produced and used in agriculture. For example, several companies in China produce insecticides based on the quinolizidine matrine, obtained from the roots of *Sophora flavescens* (Fabaceae), and these products are exported to several other Asian countries (see Table 7.1). In India it is often mixed with other botanical products, synthetic insecticides or microbial insecticides, although it can be used as a stand-alone product.[11] Another botanical insecticide produced in China contains nicotine and toosendanin, the latter a limonoid obtained from the bark of the Chinaberry, *Melia azedarach* (Meliaceae).[12] In Thailand, more than a dozen locally produced botanical insecticides are used, including those based on

Table 7.1 Botanical insecticides currently in use on a commercial basis.

Pesticide	Source plant(s)	Major constituent(s)*	Mode(s) of action against insects	Region where cultivated
		Major products		
Pyrethrum **LD$_{50}$ = 1400 mg kg^{-1}	Tanacetum cinerariaefolium	Pyrethrins, jasmolins, cinerins (chrysanthemic acid esters)	Axonic poisons (sodium channel agonists)	Kenya, Tanzania, Ecuador, Australia
Neem LD$_{50}$ = >5000 mg kg^{-1}	Azadirachta indica	Azadirachtins, other limonoid triterpenes	Moulting inhibitors (ecdysone antagonists), antifeedants	India, West and East Africa, Central America, Brazil
Essential oils LD$_{50}$ = 4400–5100 mg kg^{-1} (limonene)	many, esp. families Lamiaceae, Myrtaceae, Lauraceae, Poaceae	Monoterpenes and sesquiterpenes	Neurotoxins (octopamine agonists), repellents	worldwide
Nicotine LD$_{50}$ = 424–475 mg kg^{-1}	Nicotiana species	Nicotine, related alkaloids	Neurotoxin (acetylcholine agonist)	worldwide
		Minor products		
Rotenone LD$_{50}$ = 39.5–102 mg kg^{-1}	Derris, Lonchocarpus	Rotenone, related isoflavones	Mitochondrial cytotoxin	Southeast Asia, South America
Ryania LD$_{50}$ = 1200 mg kg^{-1} (powdered stems)	Ryania speciosa	Ryanodine (alkaloid)	Neuromuscular poison (calcium channel agonist)	Caribbean
Sabadilla LD$_{50}$ = >5000 mg k^{-1}	Schoenocaulon officinale	Cevadine, veratridine (alkaloids)	Axonic poisons (sodium channel agonist)	Latin America
Quassia LD$_{50}$ = 800 mg kg^{-1} (quassin)	Quassia, Aeschrion, Picrasma	Quassin (triterpene lactone)	unknown	Venezuela, Brazil
Matrine LD$_{50}$ = 1200 mg kg^{-1} (Sophora alkaloids)	Sophora flavescens	Matrine (quinolizidine alkaloid)	unknown	China
Toosendanin Mouse LD$_{50}$ = 250–500 mg kg^{-1}	Melia azedarach	Toosendanin (limonoid)	Antifeedant, stomach poison	China

*see Figures 7.2 and 7.3 for structures of these major constituents.
**LD$_{50}$ values are representative of rat acute oral toxicity, unless otherwise noted.[71–75]

lemongrass oil (*Cymbopogon* species; Poaceae), saponins from tea (*Camellia sinensis*; Theaceae), and *Stemona* alkaloids (Stemonaceae).[13] In many (perhaps most) tropical and subtropical countries in Asia, Africa and Latin America, a wide variety of indigenous crude plant preparations are often used to thwart insect pests in fields, in stored food and in human habitations. Indeed, it can be argued that use of these indigenous botanicals by subsistence farmers is of far greater value than the use of more refined botanical insecticides in more advanced agricultural systems in the Northern Hemisphere.[14]

7.2   Bioactivity of Plant Natural Products to Insects

The discovery and development of conventional insecticides has predominantly focused on substances that are acutely toxic to insects and generally fast-acting. As a consequence, the vast majority of conventional insecticides target the insect nervous system, with a few acting as mitochondrial poisons or through interference with other essential metabolic processes. Plant natural products that act on insects in this fashion are relatively rare – perhaps a hundred or so among the more than 100 000 plant natural products thus far isolated and elucidated. These are truly then "needles in the haystack". In contrast, hundreds if not thousands of plant substances have been shown to be biologically active against insects, at least based on observations in the laboratory. The reason for this broad discrepancy lies in the fact that most plant secondary compounds thought to function as defensive agents for the plants that produce them have far more subtle effects on insects – as inhibitors and deterrents rather than acute toxins. Examples of these different mechanisms of action are provided in the following sections.

7.2.1   Acute Insecticides

Not surprisingly, some of the best known botanical insecticides and those with the longest use history are those that are truly insecticidal, *i.e.*, that kill insects shortly after contact (see Table 7.1).[3] The most extensively used botanical to date remains pyrethrum, extracted from the dried flower heads of *Tanacetum cinerariaefolium* (Asteraceae). As classical axonic neurotoxins, the pyrethrins and naturally occurring analogues (see Figure 7.1) are famous for their fast knockdown effect on flying insects. Indeed it was only the rapid photolability of these compounds that limited their broader use in agriculture, and thus stability of the molecules became the key goal for development of the synthetic pyrethroid insecticides (see Chapter 3). Another long used botanical is nicotine (from foliage of *Nicotiana* spp. and related *Anabasis* spp. [Solanaceae]) (see Figure 7.1). Nicotine, an alkaloid, is a synaptic neurotoxin that served as the progenitor of the now widely used neonicotinoid synthetic insecticides (see Chapter 4). As a botanical it has fallen out of favour in recent years owing to the extreme risk it poses to humans.

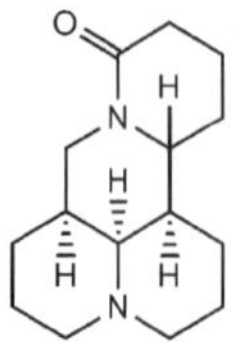

Matrine

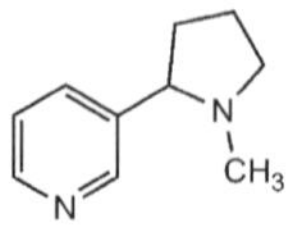

Nicotine

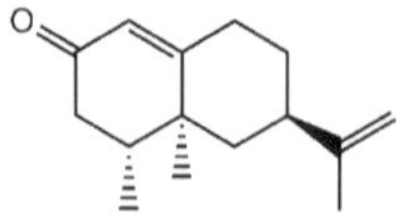

Nootkatone

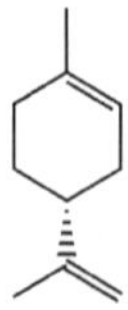

Limonene

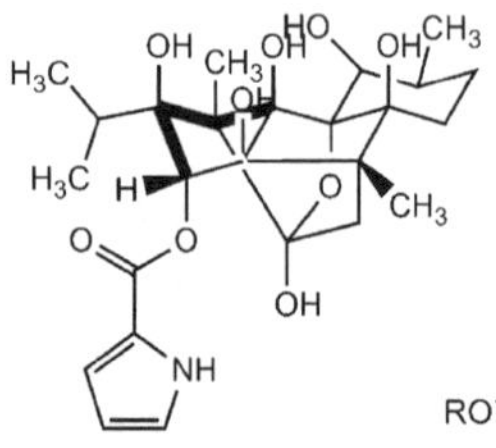

Ryanodine

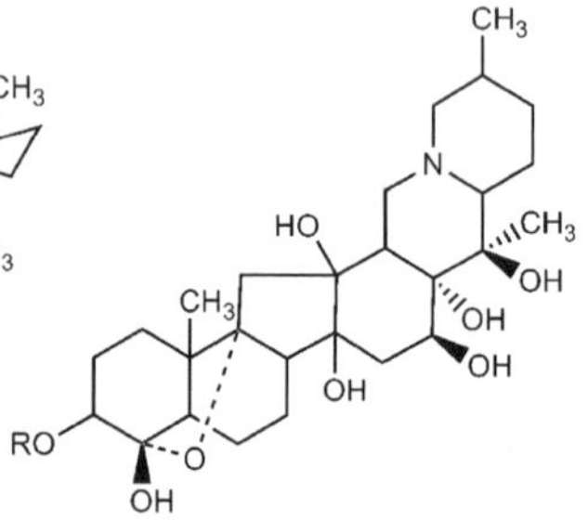

Veratridine: R = veratric acid
Cevadine: R = angelic acid

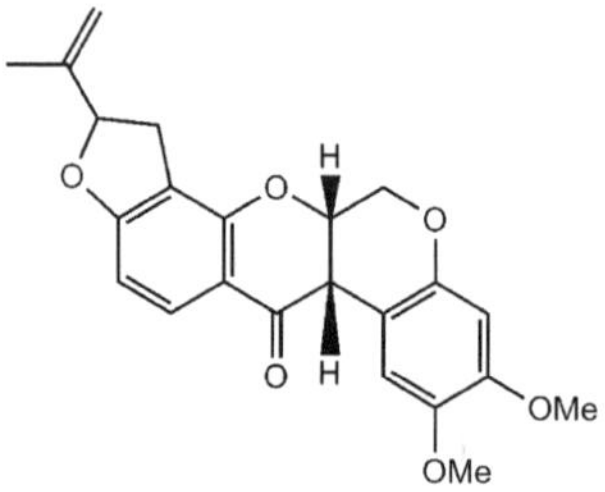

Rotenone

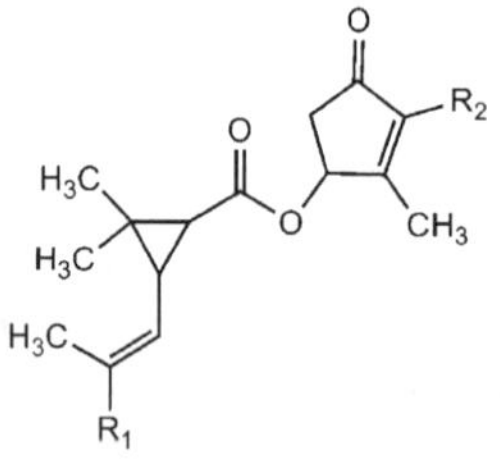

Pyrethrin I: R1 = CH₃; R2 = CH₂CH=CHCH=CH₂
Pyrethrin II: R1= CO₂CH₃ ; R2 = CH₂CH=CHCH=CH₂
Cinerin I: R1 = CH₃; R2 = CH₂CH=CH₃
Cinerin II: R1 = CO₂CH₃; R2 = CH₂CH=CH₃
Jasmolin I: R1 = CH₃; R2 = CH₂CH=CHCHCH₃
Jasomlin II: R1 = CO₂CH₃; R2 = CH₂CH=CHCHCH₃

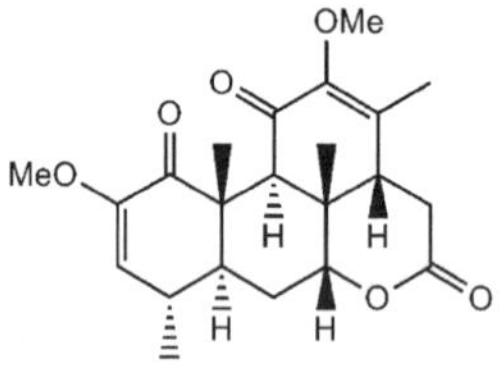

Quassin

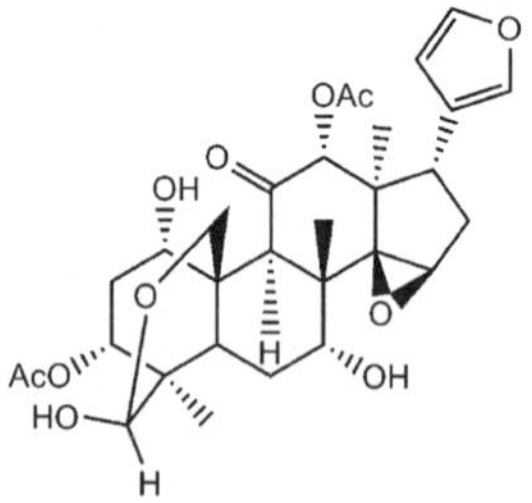

Toosendanin

Figure 7.1 Structures of the major constituents of botanical insecticides in current use. (See Table 7.1 for details).

Several other botanicals that have seen some commercial use are insecticidal – in that they kill insects after a single exposure – but with less rapid action. Among these are rotenone (see Figure 7.1), an isoflavonoid from the rhizomes of *Derris* species (Fabaceae) that is a cytotoxin targeting energy production in mito-chondria. Rotenone enjoys some minor use as a consumer insecticide, but the major use of this material is a commercial fish poison. Sabadilla refers to the seed powder of the South American lily *Schoenocaulon officinale* (Liliaceae) and consists of a mixture of ceveratrum alkaloids with a neurotoxic mode of action not unlike that of the pyrethrins. Ryanodine (see Figure 7.1), an alkaloid found in the wood of the South American shrub *Ryania speciosa* (Flacourtiaceae), has been used to a limited extent, primarily in organic production. It has a mode of action based on poisoning of neuromuscular junctions. Lastly, quassin (see Figure 7.1) also merits mention as a botanical insecticide. It was originally obtained from the wood and bark of the Central American tree *Quassia amara* (Simaroubaceae), and now is also obtained from closely related shrubs in the genera *Aeschrion* and *Picrasma*. Like many other botanicals, quassin fell out of favour in the 1950s with the advent of synthetic insecticides, but one company in Germany is reintroducing it to the market as an alternative product for use in organic food production.[15]

Leading a resurgence in interest in botanical insecticides were products based on seed extracts and oils from the Indian neem tree, *Azadirachta indica* (Meliaceae), and the outstanding insect antifeedant/insect growth regulator, azadirachtin (see Figure 7.2).[16] Though known for centuries in India, the biological activities of this substance were not investigated and documented in the international scientific literature until the 1970s, and the first neem-based insecticide was introduced into the US market around 1990. At present, refined neem-based products are used in the US, Europe and India; crude extracts are used in Africa, Latin America and throughout Asia. Azadirachtin works primarily through interference with molting by blocking the synthesis and release of ecdysteroids in insects, but numerous other neuroendocrine effects have been reported. Owing to these modes of action, azadirachtin/neem products take considerable time to kill exposed insects, although many phytophagous pests suspend feeding shortly after exposure, providing the desired crop protection effect. The relatively high cost of refined neem products appears to have been a major impediment to their success in North America and Europe.[17]

The other group of botanicals that have rekindled interest in this category are insecticides based on plant essential oils. The exempt status of certain oils used as flavouring agents in foods and beverages (*viz.* those obtained from cinnamon, cloves, mints, rosemary and thyme) from registration as pesticide active ingredients in the USA facilitated the development and commercialization of insecticides based on these oils a decade ago.[18] Monoterpenoids and sesquiterpenoids, major constituents of the oils, are neurotoxic to a wide range of

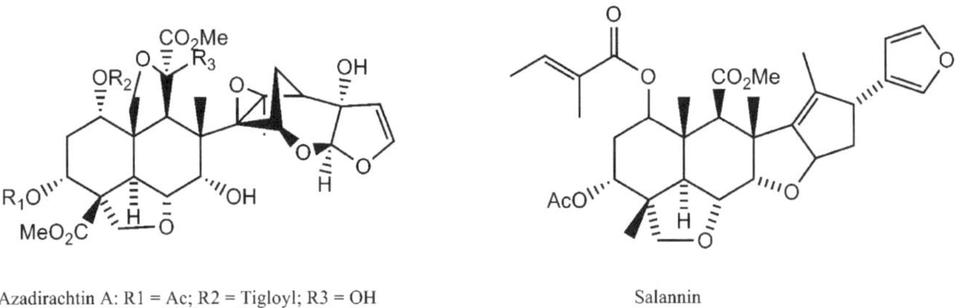

Azadirachtin A: R1 = Ac; R2 = Tigloyl; R3 = OH
Azadirachtin B: R1 = Tigloyl ; R2 = H; R3 = H

Salannin

Figure 7.2 Structures of azadirachtin A and B, and salannin, major constituents of neem seeds (*Azadirachta indica*) and botanical insecticides based on neem seed extracts.

insects and mites. While their effects on the neuromodulator octopamine have been demonstrated in some insects, there is emerging evidence that certain of the terpenoids may target other sites in the insect nervous system.[19] They can be very fast acting, even faster than some conventional neurotoxic insecticides, albeit at doses orders of magnitude greater than conventional products. Nonetheless their relative safety to humans and other nontarget species, short environmental persistence, and the public's general familiarity with them (as flavourings, fragrances and in aromatherapy) makes them ideally suited for consumer uses in and around the home and on companion animals.

7.2.2   Inhibitors of Growth, Development and Reproduction

That many plant extracts suppress the growth of insects in a laboratory setting is little more than an explanation and confirmation of one of the most obvious "non-events" in biology – the observation that insects do not defoliate terrestrial plants on a widespread scale, except in highly disturbed and anthropogenic ecosystems, *e.g.*, crop monocultures. This non-event is attributed to plant secondary chemistry evolved to deter or otherwise thwart herbivores, of which insects are the largest class. This is also the common explanation for the fact that most plants are unsuitable as hosts for most herbivorous insects – host-plant specialists are more common taxonomically speaking than generalists.

According to one authority, demonstrating that plant secondary compounds are toxic, repellent or growth-inhibiting to herbivores has become a minor industry supporting professional ecologists. One could easily extrapolate that comment to include natural product chemists, in part because demonstrations of bioactivity better justify their search for novel molecules, and in part because insects are convenient (and arguably economically significant) organisms upon which laboratory bioassays can be conducted. And among bioassays with insects, those that purport to demonstrate inhibition of insect growth are the easiest – simply add a plant extract to the insect's normal diet (artificial or otherwise) and compare growth to controls maintained on untreated diet – although the interpretation of results is not as simple. Why? Because a reduction in insect growth can be a consequence of behaviour (feeding deterrence or anorexia), of physiological malaise (sublethal poisoning or impaired nutrient utilization), or a combination of the two. So while it is relatively easy to find plant extracts or allelochemicals that interfere with insect growth in the laboratory, what proportion of these actually show bioactivity at a level that suggests potential uses in crop protection?

We have screened numerous plant extracts over the past 30 years in search of new botanical insecticides; in our experience, the proportion of extracts that are potent inhibitors of larval growth (based on noctuid caterpillars feeding on artificial media to which plant extracts are added) is quite small. Our standard screening concentration has been 1000 ppm (0.01%) fresh weight, and we consider extracts that reduce larval growth by 90% or more to be relatively

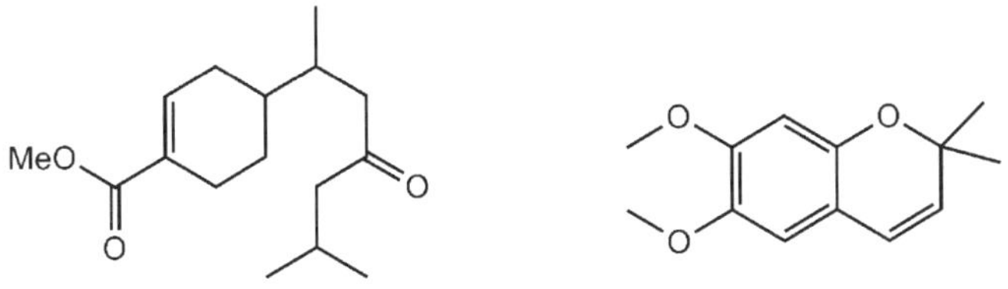

Figure 7.3   Structures of the natural insect growth regulators juvabione and precocene II.

potent.[20] Using these criteria, we found that only four of 50 extracts of *Aglaia* species from Southeast Asia were potent growth inhibitors,[21] only one of eight extracts of *Annona squamosa* from Indonesia were potent,[22] and only two of eight extracts of *Trichilia* species from Costa Rica were potent.[23] Among 18 wood extracts from tropical timber species in Indonesia and Malaysia, only one was potent.[24] It should also be noted that the observed bioactivity can in some cases be attributable to a single chemical or closely-related suite of chemical substances (*e.g.* azadirachtin in neem seed extracts),[25,26] but in other cases may be a consequence of several chemicals acting in concert (*e.g.* rosemary, *Litsea* essential oils).[27,28]

Genuine inhibitors of insect development and/or reproduction are probably as rare as those plant substances that are acute toxins. The best known examples are those plant natural products that interfere with the endocrine control of development. These include the juvenile hormone (JH) mimicking substance juvabione from Balsam fir (see Figure 7.3), the anti-JH substance precocene II from the bedding plant *Ageratum houstonianum* (see Figure 7.3), and the aforementioned azadirachtin from neem (*Azadirachta indica*) that blocks molting and reproduction by preventing the synthesis of prothoracico-tropic hormone that in turn stimulates synthesis and release of ecdysteroids (=moulting hormone). Many studies purport to show developmental aberrations in pupal or adult holometabolous insects (especially moths) after feeding as larvae on diets containing specific plant extracts. However dietary imbalances (especially limitation of dietary fatty acids) can have similar effects, so the "developmental" effect on the insect could be very indirect.

7.2.3   Inhibitors of Feeding and Oviposition

It is safe to say that more than a thousand natural products isolated from plants have been demonstrated to deter feeding by one or more insects based on laboratory bioassays. Countless crude plant extracts have similarly been shown to deter plant feeding. The subject of plant-based insect antifeedants has been reviewed by several authors. Indeed, the frequency with which plant extracts or substances thereof have been shown to deter insect feeding suggests that this may be the most common mechanism through which terrestrial plants avoid depredation by herbivorous insects in natural ecosystems. In spite of the diversity of chemicals and plant sources with feeding deterrent bioactivity, and the general enthusiasm amongst the scientific community for their potential use as nontoxic crop protectants, hardly any such products have seen commercial-scale use.[29]

Interest in insect feeding deterrents and their use in crop protection was in large part stimulated by the discovery of azadirachtin, the active constituent of neem seed extracts. First isolated in the late 1960s, azadirachtin remains the most potent antifeedant for the desert locust (*Schistocerca gregaria*) discovered to date.[16] The profound activity of this compound in the locust and other specific pests, combined with the systemic action of this compound in some plants augured well for its use as a nontoxic crop protectant. Unfortunately, subsequent studies demonstrated wide interspecific variation in the antifeedant action; for example, the migratory grasshopper (*Melanoplus sanguinipes*) of North America can eat azadirachtin-treated foliage with impunity.[30] Furthermore, insects initially strongly deterred by azadirachtin are capable of rapidly habituating to this compound,[31] and the systemic action may be limited to certain crop plants.[32]

In fact, the efficacy of azadirachtin as a crop protectant based on its antifeedant action has yet to be rigorously and unambiguously demonstrated under field conditions, even though many pest management practitioners assume this to be the primary mechanism of action. In our opinion, it is the physiological actions of azadirachtin on the insect endocrine system that more likely account for the efficacy of commercial neem products currently used in North America, Western Europe and elsewhere.[17] Several other plant extracts and isolated compounds have potent antifeedant action against certain pest species,[29] but none of these have seen commercial use.

A number of plant substances that deter insect feeding additionally deter oviposition by adult insects[33] and this mechanism could also potentially be exploited for crop protection. However, development of products based on oviposition deterrence is in its infancy and demonstration of this action at the field scale is needed.

7.2.4   Repellents

Plant preparations representing a large number of species have seen traditional use as repellents against blood-feeding insects and other arthropods, for protection of livestock and other domestic animals, and for the protection of stored grain and legumes. For example, a recent study in Laos indicated 15 species of plants used to repel mosquitoes, lice, bedbugs, mites and ticks.[34] Scientific literature in this field, in particular related to repellent effects of plant essential oils to culicine and anopheline mosquitoes,[35] is growing rapidly and has recently been reviewed.[36] In large measure this is driven by the pantropical resurgence of the mosquito-borne diseases malaria and dengue, and the introduction and spread of the mosquito-borne West Nile virus in North America.

It should be noted that the term "repellent" is often used indiscriminately. By definition, a repellent causes an *oriented* movement *away from* the source. In practice, any substance that reduces mosquito bites in human trials is considered a repellent; when in fact, reduced biting may be the outcome from any of several behaviours, including: feeding deterrence and close-range agitation that prevents the insect from alighting.

Commercially successful botanical insect "repellents" in North America include those based on citronella or lemongrass oil (from *Cymbopogon* spp., Poaceae), lemon eucalyptus (from *Corymbia citriodora*, Myrtaceae), catnip (*Nepeta cataria*, Laminaceae), wild tomato (from *Lycopersicon hirsutum*, Solanaceae), and mixtures of oils of rosemary (*Rosmarinus officinalis*, Labiateae), cinnamon (*Cinnamomum zeylandicum*, Lauraceae) and the monoterpene, geraniol. In general terms products based on these oils can be as efficacious as the conventional products DEET (*N,N*-diethyl-*m*-toluamide) and icaridin, although the latter often provide longer lasting protection when applied to human skin. Health concerns related to the heavy use of DEET, especially on children, has created market opportunities for insect repellent products based on plant oils, at least in North America and Europe.

The exact mode of action of repellents, including DEET, remained controversial for over 50 years. Hypotheses included the masking of lactic acid (a known attractant) in human skin, and "jamming" of the mosquito's olfactory system. However a recent study demonstrated that DEET stimulates a specific olfactory receptor neuron in the antennae of the mosquito *Culex quinquefasciatus* that actually repels flying mosquitoes.[37] Analogous olfactory receptor neurons have yet to be discovered that respond to the monoterpenes and sesquiterpenes in plant essential oils.

In many countries plant materials have long been used to protect stored grain,[9] presumably owing to volatile compounds providing fumigant (*i.e.*, toxic) and/or repellent actions. Evidence that plant essential oils repel or at least deter other agricultural pests (spider mites and caterpillars) has also recently been presented.[38]

7.3 Practical Considerations: Challenges to the Commercialization of Botanical Insecticides

Natural sources continue to supply new lead compounds and identify novel modes of insecticidal action. Select botanical extracts have benefited from modern advances in formulation chemistry and application technologies, which can enhance the activity and expand the use of insecticidal compounds.[36] However, utilization of the plant source materials continues to be limited to a select few botanicals, primarily pyrethrum, neem, and essential oils/limonene/citronella; lesser used botanicals include ryania, sabadilla, rotenone, and nicotine. Many of these materials are highly valued in niche markets, such as organic food production, but there is also a growing demand for products that are described as biorational, green, natural, and/or eco-chemical.[1] There are inconsistencies in the definition of these terms, but they are often used to communicate the natural origin of the source material, and convey a reduced risk to human health and the environment, as well as an environmental benefit. However, the popular perception that "natural" or "green" is synonymous with "healthy" or "safe" merits close scrutiny,[39] and even the assumed environmental benefits of "eco-friendly" pesticides may have been overstated.[40] Irrespective of this, the continued demand for plant-based technologies raises important issues surrounding availability, quality and performance of natural source materials that must be addressed for successful commercialization.[41]

7.3.1 Botanical Extracts

The majority of botanical extracts will contain a mixture of compounds produced by a source plant that is continually interacting with its environment. Thus, the plant extract is a sample from a dynamic environment where biotic and abiotic factors fluctuate, affecting the functional and delivery aspects of biologically active compounds. In this regard, plant extracts used for the production of insecticides are no different from coffee beans, cocoa beans or wine grapes. Appropriate interpretation of the extract/blend properties must take into account insecticidal activity, as well as the influence of the environment on the host plant. These components are important to apply to product development for a target pest species and specific end-user application. This approach to insecticide active ingredient formulation and delivery presents challenges to the traditional product development model. A modernized approach to botanicals requires knowledge of the physical, chemical and biological properties of the desired extracts and source plants, which can serve as a basis for ensuring supply and quality performance standards in production.

7.3.2 Quality of Source Materials

Essential to the process of selecting source materials that meet quality standards are studies of composition analysis, physical property, stability, biological activity of the extract, interaction among the compounds in the extract, and knowledge of the range of toxicological and environmental risks associated with the active ingredient extract. A large amount of data is required to fully address all of these issues, and represents a major area of focus for advancements in botanical product development.

Quality control standards for neem-based products continue to develop as these products are in demand. Neem extracts are currently sourced from seeds of the Indian neem tree, *Azadirachta indica* A. Juss. Product quality controls standards are in some cases limited to detection of contaminant levels, and a primary active or marker compound. Azadirachtin is one of such compounds in neem extracts, which is known to act as an antifeedant and insect growth regulator. Salannin (see Figure 7.2) and other limonoid components of neem seed extracts are also reported in the literature to have some antifeedant/ insecticidal activities,[42,43] albeit with much less potency.[24] In addition, research has shown that use of neem seed extracts, in contrast to pure azadirachtin, can provide delay of insecticide resistance.[44] Given the status of quality control standards for neem seed raw materials that presently focus on detection of azadirachtin A and azadirachtin B (3-tigloylazadirachtol, see Figure 7.2),[45] there is potential for improvement to address functionality of added compounds, and set concentration standards for optimal performance of the plant extract.[46]

Raw materials for essential oil-based insecticides represent another category of botanical extracts that can vary in quality and performance. Insecticide active ingredients can include oils extracted from multiple source plants, including: clove (*Eugenia caryophyllata* L.), peppermint (*Mentha piperita* L.), lemongrass (*Cymbopogon citrates* Stapf.) and common lambsquarter (*Chenopodium album* L.).[47] Interestingly, essential oil active ingredients commonly show evidence of synergy among the blends' individual components.[27,28] This presents a challenge for selection of essential oil source materials and quality control in production. Importance of essential oil production standards, including specification of blend composition and methods of analysis, are well developed for the food and flavour industries.[48] Other industries are in the process of improving quality standards for essential oil source materials that utilize these compounds in herbal medicine and dietary supplements.[49,50] Some approaches establish an activity profile or chromatographic fingerprint,[51,52] employing common analytical techniques to trace biological activity of the essential oils, and could be applied to set standards for essential oil-based insecticides.

7.3.3 Supply and Production

Quality of botanical raw materials depends on the desired chemical composition and intended use. Extracts can range from simple to complex blends of compounds that are typically obtained directly from raw plant materials (blossoms, leaves, stems, roots, rhizomes, seeds, buds, fruits, nuts, resins, *etc.*). The more complex blend requirements can have limited commercial potential as a result of the detailed range specifications for selected compounds. Other important factors that affect supply and drive up costs include verification of oil authenticity and production practices, both of which are important to meet certain regulatory and market standards.

Components of supply chain management are also an important consideration that can limit insecticide production. Stable supplies of botanical extracts are dependent on environmental and biological factors that affect plant growth; cultivation practices; production and post-production practices (drying processes, timing of harvest season, extraction procedures, *etc.*); as well as methods of handling and shipping. Politics and economic factors can also impact the supply raw materials and extracts, which have occurred in the past with select botanicals.[53,54] The cumulative weight of all these factors on cost and supply of raw materials can ultimately affect the quality and consistency of the botanical extract active ingredient(s) put into insecticide product development.

Given the commercial potential of neem and pyrethrum, alternative means of supply have been developed for these materials, including notable improvements such as those in tissue cultures[55–57] and microbial fermentation pathways.[57,58] The costs of large-scale production using these techniques continue to remain high – and economically unfeasible – given the current state of botanical insecticide markets.

Commercial potential of essential oil extracts in the food and beverage industry has given way to automated production of essential oils,[48] which is achieved through blending of synthetic compounds and/or addition of pooled samples from previous distillations and fractionations. While automated production of essential oils helps standardize the process and provide adequate supply of source materials, it does not deliver authentic essential oils that are often highly prized for their blend complexity and sensory perception. Authentication of essential oils also carries regulatory and commercial value, driving development of industry standard methods of analysis, such as stable isotope analysis.[58,59]

7.3.4 Regulatory Status

In the United States, pesticides are regulated by the Environmental Protection Agency (EPA) under the Federal Insecticide, Fungicide and Rodenticide Act (FIFRA) and the Federal Food, Drug and Cosmetic Act (FFCDA). Pesticides fall into chemical or conventional pesticides, antimicrobials or biopesticide categories and must undergo review to ensure that the pesticide will not result in unreasonable adverse effects on humans, the environment and non-target species. The passage of the Food Quality Protection Act of 1996 (FQPA) gave way to new safety and environmental standards, largely motivated by a National Academy of Sciences report covering diets of infants and children.[60] Maintaining the trend toward development of safer pest management tactics, the EPA put forth a Reduced-Risk Initiative and Policy in 1993 for conventional pesticides and established a separate Biopesticides and Pollution Prevention Division in 1994 to facilitate accelerated registration of biological pesticides.[61,62] Biological pesticides include biochemicals from natural plant sources, as well as those produced by microorganisms and genetically modified plants (plant-incorporated protectants). Furthermore, some plant materials and essential oil extracts qualify as "minimum risk" ingredients that are exempted from EPA registration under section 25(b) of FIFRA, based on several factors, including: their widespread use, nontoxic mode of action and limited persistence in the environment.[63] Since botanical pesticides fit into the biopesticide and minimum risk pesticide categories, companies developing these products are incentivized through reduced registration timeframes and lower data requirements. However, because these botanicals are produced naturally, intellectual property protection for these companies' efforts can be challenging.

Despite this risk, the U. S. system in place for registering biopesticides (microbials, biochemicals, and plant protectants) and the exemption of minimum risk pesticides have provided means and incentives for product development and commercialization of botanicals. International biopesticide registrations are reportedly fewer based upon differences in regulatory policies in place,[64] challenges/risks with respect to foreign patent protection, and performance standards.[14] Some of these markets are starting to see a growth in demand for biopesticide or reduced risk control options, which is being driven by regulatory bodies,[65–67] as well as expanded end-user knowledge of implementation programs to manage resistance, desire to minimize chemical residues, and improvements in crop yield and quality.

7.3.5 Performance Standards

Many of the current regulatory requirements and test procedures in place are standardized for insecticides containing one active ingredient compound. Since a single botanical extract serves as one active ingredient, there can be many moving parts for active ingredients that are inherently complex mixtures. This presents a significant challenge when approaching insecticide product registration for a botanical extract, as applies to product chemistry, toxicology, environmental stability and non-target organism data requirements.[68]

Two important aspects of product development are the identification of biologically active compound(s) contained in the extract and verification of their insecticidal mode(s) of action. Given the large body of work that exists on botanicals, this can involve studies of contact insecticides, insect growth regulators, feeding and oviposition deterrents, repellents and fumigants. Also useful are data on the defensive roles of select compounds/blends in nature,[69] which can provide valuable insight into how certain broad spectrum extracts are utilized effectively by plants without impacting non-targets.[70] These data can help establish chemical composition standards that are tied to activity and thus, through analytical analysis during production, can provide some guarantee of product performance.

7.4 Conclusions

Given the historical significance of botanicals in pest management and recent commercial development on select extracts, there is potential for new growth of plant-based technologies. Over the next 20 years, much of the botanical insecticide market growth will largely depend upon the stable supply of source plant material, establishment of extract quality and production standards, and delivery of dependable product performance. In addition to these factors, regulatory bodies can substantially impact market changes and growth in the number of botanical insecticides registered or approved for market introduction.

To be honest, it is unrealistic to expect botanical insecticides to assume a major share of the insect control market, at least in highly industrialized agriculture. In large part this is because the multinational chemical companies have responded to the regulatory challenge to produce synthetic insecticides with reduced risks to human health and the environment, as the current generation of conventional products attest. In addition, microorganisms and products thereof have proven to be a useful source of new insecticides with minimal non-target effects (see Chapter 8). Given that botanical insecticides currently hold little more than 1% of the global insecticide market, however, the scope for growth among these – both toxins and behaviour-modifying products – is substantial. Finally, as mentioned previously, the opportunities for deploying botanical insecticides is vastly greater in developing countries, especially in sub-Saharan Africa, Latin America and Southeast Asia, where plant diversity is greatest, the cost of conventional pest control products is severely limiting, and regulatory barriers are fewest.[14]

References

1. N. K. Dubey, R. Shukla, A. Kumar, P. Singh and B. Prakash, *Curr. Sci.*, 2010, **98**, 479.
2. D. M. Secoy and A. E. Smith, *Econ. Bot.*, 1983, **37**, 28.
3. M. B. Isman, *Annu. Rev. Entomol,.* 2006, **51**, 45.
4. ISI Web of Knowledge (Thompson Reuters), http://wokinfo.com, http://apps.isiknowledge.com/summary.do?qid=1&product=WOS&SID=2BNP5hfdia5nh99ok41&search_mode=GeneralSearch (last accessed July 24th, 2010).
5. J. T. Arnason, B. J. R. Philogene and P. Morand, in *Insecticides of Plant Origin*, American Chemical Society, Washington, DC, USA, 1989.
6. P. A. Hedin, R. M. Hollingworth, E. P. Masler, J. Miyamoto and D. G. Thompson, in *Phytochemicals for Pest Control*, American Chemical Society, Washington, DC, USA, 1997.
7. A. Prakash and J. Rao, in *Botanical Pesticides in Agriculture*, Lewis Publishers, Boca Raton, FL, USA, 1997.
8. O. Koul and G. S. Dhaliwal, in *Phytochemical Biopesticides*, Harwood Academic, Amsterdam, The Netherlands, 2000.
9. C. Regnault-Roger, B. J. R. Philogene and C. Vincent, in *Biopesticides of Plant Origin*, Lavoisier Tech & Doc, Paris, France, 2005.
10. M. B. Isman, S. Miresmailli and C. Machial, *Phytochem. Rev.*, in press.
11. N. Rajmohan, Kanchipuram, India, personal communication.
12. Z. Jiang, Yangling, China, personal communication.
13. Thailand Department of Agriculture, 2008, personal communication.
14. M. B. Isman, *Pest Manag. Sci.,* 2008, **54**, 8.
15. H. Kleeberg, Lahnau, Germany, personal communication.

16. H. Schmutterer, in *The Neem Tree,* Neem Foundation, Mumbai, India, 2nd edn, 2002.

17. M. B. Isman, in *Neem: Today and in the New Millenium,* ed. O. Koul and S. Wahab, Kluwer Academic, Dordrecht, The Netherlands, 2004, p. 33.

18. M. B. Isman, *Crop Protection*, 2000, **19**, 603.

19. E. Enan, *Insect Biochem. Mol. Biol.*, 2005, **35**, 309.

20. M. B. Isman in *Biopesticides: Use and Delivery*, ed. F. R. Hall and J. J. Menn, Humana Press, Totowa, NJ, USA, 1999, p. 139.

21. C. Satasook, M. B. Isman, F. Ishibashi, S. Medbury, P. Wiriyachitra and G. H. N. Towers, *Biochem. Syst. Ecol.* 1994, **22**, 121.

22. J. A. Leatemia and M. B. Isman, *Phytoparasitica*, 2004, **32**, 30.

23. D.A. Wheeler, M. B. Isman, P. E. Sanchez-Vindas and J.T. Arnason, *Biochem. Syst. Ecol.*, 2001, **29**, 347.

24. M. B. Isman, P. J. Gunning and K. M. Spollen, in *Phytochemicals for Pest Control*, ed. P. A. Hedin, R. M. Hollingworth, E. P. Masler, J. Miyamoto and D. G.Thompson, American Chemical Society, Washington, DC, USA, 1997, p. 27.

25. M. B. Isman, O. Koul, A. Luczynski and J. Kaminski, *J. Agric. Food Chem.*, 1990, **38**, 1406.

26. M. B. Isman, H. Matsuura, S. MacKinnon, T. Durst, G. H. N. Towers and J. T. Arnason, in *Phytochemical Redundancy in Ecological Interactions*, ed. J. Saunders, P. Barbosa and J. Romeo, Plenum Press, New York, NY, USA, 1996, p. 155.

27. S. Miresmailli, R. Bradbury and M. B. Isman, *Pest Manag. Sci.*, 2006, **62**, 366.

28. Z. Jiang, Y. Akhtar, R. Bradbury, X. Zhang and M. B. Isman, *J. Agric. Food Chem.*, 2009, **57**, 4833.

29. M. B. Isman, *Pestic. Outlook*, 2002, **13**, 152.

30. D. E. Champagne, M. B. Isman and G. H. N. Towers, in *Insecticides of Plant Origin*, ed. J. T. Arnason, B. J. R. Philogene and P. Morand, American Chemical Society, Washington DC, USA, 1989, p. 95.

31. M. K. Bomford and M. B. Isman, *Entomol. Exp. Appl.*, 1996, **81**, 307.

32. D. T. Lowery, M. B. Isman and N. L. Brard, *J. Econ. Entomol.*, 1993, **86**, 864.

33. Y. Akhtar and M. B. Isman, *J. Chem. Ecol.*, 2003, **29**, 1853.

34. H. de Boer, C. Vongsombath, K. Palsson, L. Bjork and T. G. T. Jaenson, *J. Med. Entomol.*, 2010, **47**, 400.

35. Y. Trongtokit, Y. Rongsriyam, N. Komalamisra and C. Apiwathnasorn, *Phytotherapy Res.*, 2005, **19**, 303.

36. L. S. Nerio, J. Olivero-Verbel, E. Stashenko, *Bioresource Technol.*, 2010, 101, 372.

37. Z. Syed and W. J. Leal, *Proc. Natl. Acad. Sci. U. S. A.*, 2008, **105**, 13599.

38. M. B. Isman and S. Miresmailli, in *Recent Developments in Invertebrate and Vertebrate Repellents*, ed. G. Paluch and J. R. Coats, American Chemical Society, Washington, DC, USA, 2010, in press.

39. J. T. Trumble, *Am. Entomol.*, 2002, **48**, 7.

40. C. A. Bahlai, Y. Xue, C. M. McCreary, A. W. Schaafsma and R. H. Hallett, *PLoS One*, 2010, **5**, e11250.
41. S. Guleria and A. K. Tiku, in *Integrated Pest Management: Innovation-Development Process*, ed. R. Peshin and A. K. Dhawan, Springer Science + Business Media, B. V., The Netherlands, 2009, p. 317.
42. W. Kraus, S. Baumann, M. Bokel, U. Keller, A. Klenk, M. Klingle, H. Pohnl, and M. Schwinger, in *Natural Pesticides from the Neem tree and Other Tropical Plants,* ed. H. Schmutterer and K. R. S. Ascher, GTZ: Eschborn, Germany, 1987, p. 111.
43. L. Lin-Er, J. J van Loon and L. M. Schoonhoven, *Physiol. Entomol.*, 1995, **20**, 134.
44. R. Feng and M. B. Isman, *Experientia*, 1995, **51**, 831.
45. M. R. Forim, M. F. G. F. Silva, Q. B. Cass, J. B. Fernandes and P. C. Vieira, *Anal. Methods*, 2010, **2**, 860.
46. N. K. Maniania, and A. M. Varela, in *International Symposium on Biopesticides for Developing Countries*, ed. U. Roettger and R. Muschler, Centro Agronómico Tropical de Investigación u Emseñanza, CATIE, 2004, section 2, pp. 117–126.
47. F. E. Dayan, C. L. Cantrell and S. O. Duke, *Bioorg. Med. Chem.*, 2009, **17**, 4022.
48. C. Frey, in *Natural Flavors and Fragrances*, ed. C. Frey and R. Rouseff. ACS Symposium Series, American Chemical Society, Washington, D.C., USA, 2005, p. 3.
49. J. H. Cardellina II, *J. Nat. Prod.*, 2002, **65**, 1073.
50. V. S. Srinivasan, *Life Sci.*, 2006, **78**, 2039.
51. Y. Liang, P. Xie and K. Chan, *J. Chromatogr., B: Biomed. Appl.*, 2004, **812**, 53.
52. J. H. Zeng, G. B. Xu and X. Chen, *Med. Chem. Res.*, 2009, **18**, 158.
53. J. Njoki, Kenya: Pyrethrum players work on reviving ailing industry, *Business Daily (Nairobi)*, 2009, www.Allafrica.com.
54. M. Kimani, East Africa feels blows of Kenyan crisis, *Africa Renewal*, 2008, **22**, 3.
55. G. Hrazdina, *J. Agric. Food Chem.* 2006, **54**, 1116.
56. E. J. Allan, J. P. Eeswara, S. Johnson, A. J. Mordue (Luntz), E. D. Morgan and T. Stuchbury, *Pest Manag. Sci.*, 1994, **42**, 147.
57. J. A. Teixeira da Silva, *Biotech. Adv.*, 2003, **21**, 715.
58. S. Takahashi, Y. Yeo, B. T. Greenhagen, T. McMullin, L. Song, J. Maurina-Brunker, R. Rosson, J. P. Noel and J. Chappell, *Biotechnol. Bioeng.*, 2006, **206**, 170.
59. F. Hammerschmidt, G. E. Krammer, L. Meier, D. Stöckigt, S. Brennecke, K. Herbrad, A. Lückhoff, U. Schäfer, C. O. Schmidt and H. Bertram, in *Authentication of Food and Wine,* ed. S. Ebeler, G. Takeoka and P. Winterhalter, American Chemical Society, Washington, D.C., USA, 2006, p. 87.
60. National Research Council, *Pesticides in the Diets of Infants and Children*, National Academy Press, Washington, D.C., USA, 1993, p. 13.

61. W. B. Wheeler, *J. Agric. Food Chem.*, 2002, **50**, 4151.
62. J. Anderson, A. Leslie, S. Matten, and R. Kumar, Weed Technol., 1996, **10**, 966.
63. Environmental Protection Agency, *Code of the Federal Registrar*, Pesticides; Exemption of Certain Substances from Federal Insecticide, Fungicide, and Rodenticide Act Requirements, Sept. 15th 1994.
64. D. Chandler, W. Grant, J. Greaves, G. Prince and M. Tatchell, *Outlooks Pest Manag.*, 2008, **19**, 77.
65. S. Milmo, Europe set to crack down on pesticides, *Nature News*, April 2009.
66. E. Kähkönen, T.Hirvonen and K. Nordström, *J. Bus. Chem.*, 2010, **7**, 69.
67. Government of Ontario, *Ontario regulation 63/09 – Toronto (Ontario)*, available from: http://www.e-laws.gov.on.ca/html/source/regs/english/2009/elaws_src_regs_r09063_e.htm (accessed January 3rd, 2009).
68. N. K. Maniania and B. Löhr, in *International Symposium on Biopesticides for Developing Countries*, ed. U. Roettger and R. Muschler, Centro Agronómico Tropical de Investigación y Enseòanza, CATIE, 2004, sec. 3, p. 194.
69. J. Gershenzon and N. Dudareva, *Nat. Chem. Biol.*, 2007, **3**, 408.
70. M. Euler and I. T. Baldwin, *Oecologia,* 1996, **107**, 102.
71. J. R. Bloomquist, in *Radcliffe's IPM World Textbook*, ed. E. B. Radcliffe, W. D. Hutchison and R. E. Cancelado, University of Minnesota, St. Paul, MN, http://ipmworld.umn.edu (accessed October 31st, 2010).
72. Environmental Protection Agency, *Biopesticide Active Ingredient Fact Sheets,* http://www.epa.gov/oppbppd1/biopesticides/ingredients/index.htm (accessed October 31st, 2010).
73. D. H. Phua, W. Tsai, J. Ger, J. Deng and C. Yang, *Clin. Toxiocol.*, 2008, **46**, 1067.
74. U.S. Consumer Product Safety Commission, Final Report Study of Aversive Agents, 1992, http://www.cpsc.gov/library/foia/foialook.html (accessed October 31st, 2010).
75. Y. P. Zhu, *Chinese Materia Medica: Chemistry, Pharmacology, and Applications*, Harwood Academic Publishers, Amsterdam, The Netherlands, 1998, p. 107.

Towards a Healthy Control of Insect Pests: Potential Use of Microbial Insecticides

ALEJANDRA BRAVO,[1] M. CRISTINA DEL RINCON-CASTRO,[2] JORGE E. IBARRA[3] AND MARIO SOBERÓN[1*]

[1] Instituto de Biotecnología Universidad Nacional Autónoma de Mexico, Ap. Postal 510-3 Cuernavaca, 62250 Mor, Mexico; [2] División de Ciencias de la Vida, Universidad de Guanajuato, Km. 9 Carr. Irapuato-Silao, 36500 Irapuato, Gto, Mexico; [3] CINVESTAV-IPN, Unidad Irapuato, Ap. Postal 629 Irapuato 36500, Gto, Mexico

8.1 Introduction

Biological control of insect pests by microbial pathogens has been recognized as an important alternative to the use of chemical insecticides. All biological control strategies including bacteria, virus and fungi are highly specific to their insect targets and biodegradable, making them environmental compatible. However, their commercial use in crop protection is currently limited, representing a maximum of only 2% of the total insecticidal market. This marginal use of microbial pathogens is mainly due to their narrow spectrum of action that enables them to kill only certain insect species. Their limitations also include a low persistence in the environment and the requirement of precise application practices, since some of these pathogens are specific to young insect larval stages or are sensitive to sun radiation. It is interesting to note that some

RSC Green Chemistry No. 11
Green Trends in Insect Control
Edited by Óscar López and José G. Fernández-Bolaños
© Royal Society of Chemistry 2011
Published by the Royal Society of Chemistry, www.rsc.org

characteristics of microbial-based insecticides that make them environmentally friendly had hampered a wider use in agriculture. In contrast, chemical insecticides have a broad spectrum of action against many insect species, persist in the environment and kill insects despite their developmental stage. However, the use of chemical pesticides for pest control has led to several problems such as the pollution of the environment, an increase in human diseases such as cancer and several immune system disorders, and the selection of insect resistant populations and important outbreaks of secondary pests.[1] For these reasons, there is an increasing concern in society for lowering the use of chemical insecticides and moving to safer practices in crop protection, such as the development of new environmentally friendly bioinsecticides.

Microbial pathogens of insects have been identified and studied for many years and some of them have been used for the development of commercial insecticides. In this chapter, we will review the potential use of some of these microbial pathogens, such as bacteria, viruses and fungi, with emphasis on their mode of action, insect specificity, risk assessments for the environment and also key application practices to assure the development of safer insect control products in agriculture.

8.2　Entomopathogenic Bacteria

Biological insecticides based on entomopathogenic bacteria represent 95% of the microbial insecticides market and are based mainly on one bacterium species: *Bacillus thuringiensis* (Bt). There are several Bt-based products for the control of mosquitoes that transmit human diseases, and also for the control of the most important lepidopteran and coleopteran crop and forest pests. Insecticide products are also commercially available based on two other entomopathogenic bacteria: *Serratia entomophila* and *Bacillus sphaericus* for the control of grass grub and mosquitoes, respectively.

Most of the entomopathogenic bacteria invade their host after ingestion by the larvae causing the disruption of the midgut tissue, followed by septicemia. In most cases these bacterial pathogens rely on insecticidal toxins during the pathogenic process but also produce an array of virulence factors that contribute to the killing of the insect.[2–4]

8.2.1　*Bacillus thuringiensis*

Bacillus thuringiensis (Bt) is a member of the *Bacillus cereus* group of bacteria that also include *B. cereus* and *B. anthracis* among others.[2] Bt is distinguished from the other members of the *B. cereus* group for its entomopathogenic properties due to the production of insecticidal δ-endotoxin proteins during the sporulation phase of growth-forming crystal inclusions (see Figure 8.1). These δ-endotoxins, named Cry and Cyt, are highly specific to their target insect, innocuous to humans, other vertebrates and plants, and completely biodegradable. Therefore, they represent a viable alternative for the control of

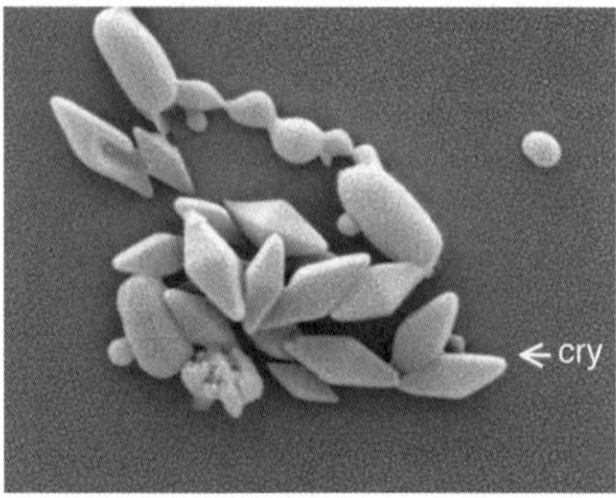

Figure 8.1 Scanning electron microscopy observation of the spore and crystal mixtures produced by a *B. thuringiensis* subsp kurstaki strain that produces Cry1A crystal proteins. Arrows point to crystal inclusions (cry).

insect pests in agriculture and disease vectors of importance in human public health.[2,5]

8.2.1.1 Diversity of Bt Toxins

Cry toxins produced by *Bacillus thuringiensis* are classified by their primary amino acid sequence and have been divided in at least 67 groups (Cry1 to Cry67) comprised by more than 500 different gene sequences (see http://www.biols.susx.ac.uk/ Home/Neil_Crickmore/Bt/index.html).[2,5] The δ-endotoxins are organized in four main groups that are not related phylogenetically: the three domain Cry toxins (3d-Cry), the mosquitocidal-like Cry toxins (Mtx-like), the binary-like Cry toxins (Bin-like) and the Cyt toxins. It is proposed that each of these groups of δ-endotoxins may have a different mechanism of action.[5] Besides these Cry and Cyt toxins that are produced upon sporulation, some Bt strains produced additional insecticidal toxins during their vegetative stage of life named VIP toxins (for vegetative insecticide proteins). Three different VIP proteins have been characterized VIP1/VIP2 that is a binary toxin and VIP3. VIP3 has been successfully produced in transgenic plants.[6]

In relation to the Cry toxins produced by Bt, it is interesting to note that the Mtx-like and the Bin-like Cry toxins have some similarity with the Mtx or Bin toxins produced by *B. sphaericus,* although in the case of *B. sphaericus* these toxins are toxic against mosquitoes and in Bt they are toxic to coleopteran larvae.[5] The 3d-Cry toxins represent the biggest group of Cry proteins classified in more than 40 groups and many subgroups (Cry1Aa, Cry1Ab, Cry2Aa...). The 3d-Cry toxins are composed of three distinct domains.[2,5] Interestingly, members of this family of proteins have been found in two other bacterial species: *Paenibacillus popilliae* (Cry18) and *Clostridium bifermentans* (Cry16 and Cry17).[7,8] *P. popilliae* is the etiological agent of the milky disease in scarabaeid larvae and *C. bifermentans* has mosquitocidal activity, although it is not clear the role of their Cry toxins in the pathogenesis of this bacterium.

8.2.1.2 Bt Commercial Products

Bt products are presented in different types of formulations depending on their target pests, such as liquid concentrates, wettable powders, ready-to-use dusts and granules. Most of these products are composed of spore-crystal preparations derived from wild-type strains as *B. thuringiensis* var. *kurstaki* (Btk) HD1 and HD73 strains that express some Cry1A and Cry2Aa proteins (*i.e.* Biobit, Condor, Cutlass, Dipel, Full-Bac, Javelin, M-Peril and MVP products) that are highly effective in the control of many common leaf-feeding caterpillars, including pests on vegetables, cereals and cotton.[9] Some Btk-based products have been successfully used in Canada and USA forests for the control of Lepidopteran defoliator pests of conifers, such as web worms, tent caterpillars and other forest caterpillars.[10] Other products based on other Bt strains such as *B. thuringiensis* var. *aizawai* HD137 (*i.e.* Certan, Agree and Xentari), which produces slightly different Cry toxins such as Cry1A, Cry1B, Cry1C and Cry1D, are used for the control of other caterpillars such as the Indian meal moth larvae in stored grains. For the control of beetles, products based on *B. thuringiensis* var. *san diego* and *B. thuringiensis* var. *tenebrionis*, which produced Cry3 toxins (*i.e.* M-Trak, Foil and Novodor) are available.[11] Finally, products for the control of mosquitoes have been developed based on *B. thuringiensis* var. *israelensis* (Bti) containing Cry4, Cry10, Cry11 and Cyt1Aa toxins (*i.e.* Vectobac, Teknar, Bactimos, Skeetal, and Mosquito Attack).[12,13] Nevertheless, Bti is mostly toxic to *Aedes* and *Culex* species but shows moderate toxicity against *Anopheles spp.*, a vector of malaria. Additional screening of Bt strains active against *Anopheles spp.* larvae could provide means to control this important vector. However, the most extensive use of Cry proteins on crop protection against insect pests came with the development of transgenic crops that produce Cry toxins (Bt-Crops). More than 200 million hectares have been planted with Bt-crops since their release in 1996.[14] In transgenic plants the Cry protein is produced continuously, protecting the insecticidal toxin from degradation and making it reachable to chewing and boring insects. The extensive use of these insect-resistant Bt-crops has decreased considerably the use of chemical pesticides in the fields.[15] Interestingly, the use of Bt-cotton in countries like China, Mexico and India has shown a significant positive effect on the final yield, with a significant reduction in the use of chemical pesticides.[15,16]

Despite the large number of *cry* genes characterized so far, only few Cry toxins have been developed as commercial products or Bt-crops. However, the potential use of these toxins for insect control is enormous.

8.2.1.3 Mode of Action of Bt Toxins

To date, the three-dimensional structures of eight different members of the 3d-Cry toxins with different insect specificity have been solved. Figure 8.2 shows a representative structure of one member of 3d-Cry group, the Cry8Ea, that was the most recent 3d-Cry structure to be identified.[17] Domain I, a seven α-helix

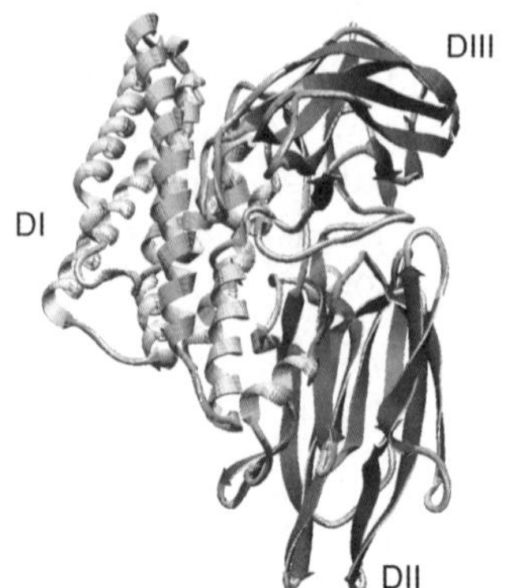

Figure 8.2 Three-dimensional structure of Cry8Ea (PDB 3ebA) insecticidal toxin produced by *Bacillus thuringiensis*. Domain I (DI) is an α-helix bundle, domain II (DII) consists of three anti parallel β-sheets with exposed loop regions, and domain III (DIII) is a β-sandwich.

bundle, is implicated in membrane insertion, toxin oligomerization and channel formation.[5] Domain II consists of a beta-prism of three anti-parallel β-sheets packed around a hydrophobic core, and domain III is a β-sandwich of two anti-parallel β-sheets. Domains II and III are implicated in insect specificity by mediating specific interactions with several receptor proteins.[5] The structure of other 3d-Cry members (Cry1Aa, Cry2Aa, Cry3Aa, Cry3Ba, Cry4Aa and Cry4Ba) is highly similar to Cry8Ea, despite the low amino acid sequence similarity, and share conserved residues located in the internal regions of domain I and III, as well as in interdomain contacts. This suggests that all members of the 3d-Cry family may share a similar mechanism of action.[5] It is widely recognized that 3d-Cry toxins exert their toxic effect by forming pores in the insect midgut epithelium cells, resulting in cell lysis and disruption of the midgut epithelium.[2,5] The mode of action of these group of proteins has been mostly studied in lepidopteran insects but more recently also in coleopteran and dipteran insects.[18] The 3d-Cry toxins are produced as protoxins that need to be dissolved and processed proteolytically by insect proteases to release the active toxic fragment. Two groups of protoxins are produced by different Bt strains: some of them are large protoxins (such as Cry1Aa of 130 kDa) and some are short protoxins of 70 kDa (such as Cry2Aa). However, in both cases the activated toxins have a similar size of 60 kDa. Large protoxins lose half of the *C*-terminal end and 20 to 50 amino acids of the *N*-terminal end, while short protoxins are processed only at the *N*-terminal end. The activated toxin is composed of the three structural domains mentioned above. The activated toxin goes through complex sequential binding events with different insect gut proteins leading to membrane insertion and pore formation. In the case of the lepidopteran insect, *Manduca sexta,* it has been proposed that the first binding interaction of the activated Cry1Ab toxin occurs through exposed amino acid regions of domain II (specifically through loop 3) and domain III (through strand β-16) of the toxin, with at least two glycosylphosphatidylinositol (GPI)-anchored proteins identified as alkaline phosphatase (ALP) and aminopeptidase N (APN), which are highly abundant low affinity binding sites for the toxin (Kd 100 nM).[5,19,20] These binding interactions help to concentrate the activated toxin in the microvilli membrane of the midgut cells where the toxin binds with high affinity through exposed domain II loops 2, 3 and alpha-8 to a low abundant transmembrane protein identified as a cadherin receptor (Kd 1 nM).[5,19,20] This high affinity binding interaction with cadherin facilitates further proteolytic cleavage of the toxin of the *N*-terminal end including helix α-1 of domain I. This proteolytic cleavage induces the formation of a toxin pre-pore oligomer.[5,19,20] The oligomeric structure of the toxin shows an important increase of 200 fold in its affinity to GPI-anchored receptors ALP and APN involving domain II loop 2 region (Kd = 0.5 nM).[19] The binding of the pre-pore to the GPI-anchored proteins leads finally to insertion into the membrane causing pore-formation and cell lysis.[5] Figure 8.3 depicts the molecular events that lead to Cry toxin membrane insertion and pore formation.

As mentioned previously, Cry toxins are highly specific and only kill a narrow spectrum of insect species. Insect specificity of these proteins is determined mainly by the specific recognition of insect midgut proteins, although in some cases proteolytic activation and solubilization of Cry protoxins could affect susceptibility in certain insect species.[21] As mentioned previously, domain II and domain III of Cry toxins are the structural determinants of specificity of Cry toxins. Example hybrid toxins that involve domain III or domain II loops exchange between different Cry toxins with a correlative change on insect

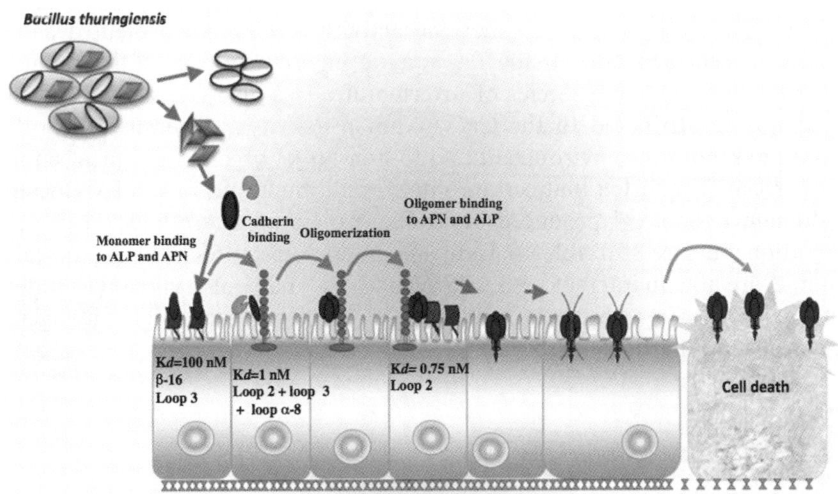

Figure 8.3 Mode of action of *Bacillus thuringiensis* Cry toxins. Below the toxin receptor interaction are the apparent binding affinities (Kd) and the toxin regions involved in each toxin interaction with the midgut receptors.

specificity have been reported.[22–25] In addition, the analysis of the phylogenetic relationships of the isolated domains of members of the 3-Domain Cry family revealed that domain III swapping occurred during evolution of these proteins, suggesting that *in vivo* recombination has been a strategy for increasing novel specificities.[26]

8.2.1.4 Public Concerns on the Use of Bt Products

Cry toxins have proven to be safe to the environment through their use in crop and forest protection, as well as in the control of insect born diseases for more than 50 years. However, if new insecticidal products based on different Cry toxins are to be developed, risk assessments have to be undertaken since each toxin is expected to have a particular narrow spectrum of action that could potentially affect other non-target organisms.

Currently, one of the major concerns of Cry toxins is their potential effects on non-target organisms. Public attention concerning the safety of Cry toxins first came to the fore with the massive spray of Bt for control of forest defoliators in Canada.[27] After worldwide use of Bt-crops, it was reported that Bt-maize pollen was toxic to Monarch butterfly larvae, and more recently there have been reports on the possible effects of Bt-maize debris on aquatic inverte-brates.[28,29] Nevertheless, Cry1A toxins used in transgenic plants have been extensively shown to be specific against target insects and safe to non-target organisms.[30] In the case of the Monarch butterfly, follow-up studies showed that low expression of Bt toxin genes in the pollen of most commercial trans-genic hybrids, no overlap of pollen shed and larval activity, and limited overlap in distribution of Bt-maize and milkweed made field risk to the Monarch populations negligible.[31] Finally, studies on the effect of Bt-maize debris on aquatic invertebrates revealed that the activity of Cry1Ab protein declined rapidly in senesced corn tissue on aquatic environments and the putative adverse effects on some species of invertebrates was also not significant, and could not be attributed to the Cry1Ab toxin activity.[32] Overall, the Cry1A toxins are safe to the environment and to non-target organisms. Furthermore, it was recently reported that certain single point mutations of Cry1Ab located in domain I α-helix 4 produced non-toxic proteins that are affected in pore formation but are still able to form oligomeric structures. These particular mutants are able to interact with wild type toxins and could function as potent anti-toxins of Cry1Ab that could help to protect potentially endangered organisms in a particular ecosystem.[33]

8.2.2 Bacillus sphaericus

Some *Bacillus sphaericus* strains are toxic to mosquito species such as *Culex spp.* and *Anopheles spp.* but lack activity against *Aedes aegypti*.[34] *B. sphaericus* also produces crystal inclusions upon sporulation, which are formed by two equimolar components of a binary toxin, BinA and BinB. In addition to the Bin toxin, *B. sphaericus* produce at least three different non-related Mtx toxins during its vegetative phase growth.[34] In contrast to Cry toxins, Bin toxin does not completely disrupt the midgut epithelium but causes mainly swelling of mitochondria and the endoplasmic reticulum, with enlargement of the vacuoles of midgut cells.[34,35] As in the case of Cry toxins, Bin toxins are solubilized and proteolytically activated by the insect gut proteases, the BinA of 42 kDa and BinB of 51 kDa are activated to proteins of 39 and 43 kDa, respectively.[34,36] It has been shown that BinA has its own insecticidal activity by contrast with BinB, but the BinA toxicity is greatly enhanced in the presence of BinB. This synergistic effect of BinB on BinA activity is mostly due to the fact that BinB binds to specific receptor molecules, enhancing BinA binding to the brush border membrane epithelium.[37] The receptor of Bin toxin was identified as a GPI-anchored α-glucosidase in *C. pipiens*, *C. quinquefasciatus* and *An. gambiae*.[38–40] Although Bin toxin has been shown to form pores on brush border membrane vesicles isolated from *Culex spp.*,[41] it is believed that one or the two Bin components may be internalized into the cell to exert its toxic effect.[34] It is still unknown what could be the intracellular targets and the effect of Bin toxin leading to cell death. The Bin toxin insect specificity relies mainly on specific receptor recognition, since there is a correlation in the lack of Bin binding to *Aedes* midgut epithelium, and in the lack of toxicity and resistance to Bin toxin with mutations that affect the GPI-anchored α-glucosidase in *Culex pipiens*.[34,42]

Much less is known about the mode of action of Mtx toxins, where Mtx1 is a 100 kDa protoxin that upon proteolytic activation by midgut proteases releases two protein fragments of 27 and 70 kDa that remain associated. Mtx2 and Mtx3 are proteins of 32 and 36 kDa, respectively.[34] In the case of Mtx1, the 27 kDa component shows sequence similarity with ADP-ribosylating enzymes and is capable of ADP ribosylation of different *E. coli* proteins.[43] It is believed that the Mtx1 70 kDa component binds to a specific unknown midgut receptor and somehow facilitates the translocation of the 27 kDa component into the cell, where it exerts its toxic activity by ADP-ribosylation of some key cellular components resulting in cell death.[44]

Additional insecticidal toxins produced by *B. sphaericus* were recently identified.[45] The genome sequencing of *B. sphaericus* strain C3-41 revealed the presence of an additional MTX2/MTX3-like protein (Mtx4).[46] In addition, there is a 53 kDa protein, sphaericolysin, which has insecticidal activity when injected into *Blattella germanica* with a low toxicity to *Spodoptera litura*.[47] Finally, a binary toxin (Cry48/Cry49) was identified on a *B. sphaericus* strain that shows toxicity to a *Culex* population highly resistant to the BinA/BinB toxin.[48] Interestingly, Cry49 is related to BinA/BinB proteins, while Cry48 belongs to the 3-D family of Cry toxins.[48]

8.2.3 *Serratia entomophila* and *Photorhabdus luminescens*

Serratia entomophila is a gram negative enterobacterium which is the causal agent of the amber disease of the grass grub, *Costelytra zealandica*.[49] This disease is characterized by the clearing of the gut after bacteria colonization, resulting in a characteristic amber color. Larval death occurs after a prolonged period of time as a result of the bacterial invasion of the haemo-coel.[49] The toxicity determinants of *Serratia* to the grass grub are plasmid born and were identified as a three component toxin, SepA, SepB and SepC.[50] Besides these toxins, *S. entomophila* also synthesizes other anti-feeding components that are responsible for cessation of feeding in the infected larvae.[51] The SepABC toxins show extensive similarity to TcABC toxins previously identified in *Photorhabdus* and *Xenorhabdus* bacteria.[52] *Photorhabdus* and *Xenorhabdus* are gram negative bacteria that live in symbiosis with nematodes that attack insects by invasion of the haemocoel.[3] Once inside the insect haemocoel, the bacteria are released from the nematode-liberating bacterial toxins causing septicemia, which results in the killing of the insect. Both the nematodes and the bacteria replicate on the insect cadaver finishing with the re-uptake of the bacteria by the nematodes, resulting in considerable offspring that seek out new insect hosts.[3] *Photorhabdus luminescens* contains several Tc loci that are believed to code for different TcABC toxin complexes with different insect specificity.[3,53] However, insect specificity relies on the TcA component since the *Xenorhabdus* TcA components have different insecticidal activity against specific lepidopteran insects.[54] It has been shown that the TcA component has its own insecticidal activity and can confer insect resistance when expressed in transgenic plants or to recombinant *Escherichia coli* cells.[54] However, it has been shown that the toxicity of TcA is greatly enhanced when produced along with the TcBC component in *E. coli*.[55,56] The molecular mechanism by which the TcABC (or SepABC) toxin kills the insect remains unknown.

In addition to Tc toxins, *Photorhabdus* produces a range of different insecticidal proteins that have been identified by genome sequencing or by complementing insect toxicity of *E. coli* cells by *Photorhabdus* genomic DNA clones.[56] The novel toxins identified include: PirAB, a binary toxin that is toxic to lepidopteran and dipteran insects;[3] PVCs (Photorhabdus Virulence Cassettes), a multiprotein complex shown to act as molecular syringes to deliver toxic effectors; and Mcf toxins ("makes caterpillar floppy") that act by inducing apoptotic cell death on target cells.[57,58] It has been suggested that the different toxins produced by *Photorhabdus* could provide means for showing activity against different insect hosts.[56]

8.2.4   Conclusions

As mentioned previously, a major break through in the reduction of chemical insecticides in agriculture came with the release of transgenic crops producing Bt Cry toxins.[5] However, a major threat for this technology in the near future is the development of resistance due to increased selection pressure. In fact two different insect species have developed resistance to Cry toxins and Bt-spray products in the field, and several other insect species have developed resistance to transgenic crops, specifically to Bt-cotton.[59] However, insect resistance management techniques are currently being implemented, such as the use of refuges, rotation of different cultivars expressing different Cry toxins or gene stacking of different insecticidal toxins with different modes of action.[60,61] The identification and characterization of the mode of action of new insecticidal toxins from different entomopathogenic bacteria will therefore provide new alternatives for insect control either in transgenic plants or as sprayable products. It has been suggested that transgenic Bt plants could be a key component in integrated pest management strategies, since the reduction in the use of chemical insecticides with a broad insect spectrum could result in a higher abundance of natural enemies, such as parasitoids or predators.[62] However, there is still only a limited number of crops that have been developed for insect resistance using transgenic methodology with a reduced number of Cry toxins. Furthermore, transgenic crops are not a choice for certain agricultural practices, such as organic agriculture, and not all countries have embraced the release of transgenic plants for crop production. Therefore we still need to develop new sprayable products based on different entomopathogenic bacteria as an alternative to the use of chemical insecticides.

8.3   Entomopathogenic Viruses

Entomopathogenic viruses represent important biological control agents. There is a wide diversity of insect viruses in nature, and each particular family shows some specific features which may become important for the development of biological control agents. One particular group of insect viruses, the baculoviruses, is a very diverse group and has been isolated almost exclusively from insects (mainly from species within the orders Lepidoptera, Coleoptera, Diptera, and Hymenoptera). Only a few have been isolated from crustacean and spider species. In recent years, baculoviruses have also gained in importance as eukaryotic expression vectors. This is due to the large number of studies published on their molecular biology, which has led to significant advances in the use of baculoviruses, not only as biological control agents, but also in genetic therapy, genetic engineering, protein expression, and molecular and evolutionary virology.

Early taxonomists were unaware of microbes, so viruses were not included in the first attempts to classify the living microorganism. Even with the discovery of microbes, viruses still remained hidden to the human eye until the beginning of the 20th century. Once the great diversity of viruses became evident, attempts to include them within the living organism classification were only artificial. It became apparent that viruses lack some important characteristics of other organisms and that they may be as simple as a pack of nucleic acid surrounded by a protein capsule. Therefore, viruses are not included within the three-domain classification, but they follow some of the hierarchies used in the classification taxonomy, such as order, family and genus. Thus, examples of orders are Caudovirales, Mononegavirales and Nidovirales, examples of families are Poxviridae, Baculoviridae, and Iridoviridae, and examples of genera are Nucleopolyhedrovirus and Granulovirus. Additionally, some viruses are widely known by their short names, such as NPV for Nucleopolyhedrovirus, EPV for Entomopoxvirus, or Cypovirus for cytoplasmic polyhedrosis virus. Nowadays, virus classification relies on the type of nucleic acid, morphology of the viral particle, presence of an envelope or an occlusion body, host species, among others. New sequencing tools now allow a comparison of complete genomes of viruses, which makes the phylogenetic analysis between species more reliable.

8.3.1 Baculoviruses

Viruses within the family Baculoviridae contain large chains (80 a 130 kbp) of circular dsDNA, within a rod-shaped viral particle. Virions are occluded within occlusion bodies (OBs) as shown in Figure 8.4, known either as polyhedra or granules, constituted by a protein (polyhedrin or granulin).[63] There are two genera within the family: *Nucleopolyhedrovirus*, typically known as nuclear polyhedrosis viruses or NPVs; and *Granulovirus*, typically known as granulosis viruses or G*Vs*. NPVs are further divided into single nuclear polyhedrosis viruses (or SNPVs) and multiple nuclear polyhedrosis viruses (or MNPVs), depending on the number of nucleocapsids within each envelope. NPVs replicate only in the nuclei of infected cells, their OBs may be as large as 15 μm in diameter and contain many enveloped virions. GVs, on the other hand, contain only a singly enveloped nucleocapsid per granular OB, and therefore they are smaller than the NPV OBs (0.2 to 0.5 μm).

Some years ago, the so-called non-occluded viruses (NOV) were also included within the Baculoviridae family, but the International Committe on

Figure 8.4 Baculoviruses occlusion bodies (OB) isolated from *Autographa californica*. The virus was originally isolated from the alfalfa looper. The OBs are composed of a protein matrix (polyhedrin for the nucleopolyhedrovirus, or granulin for the granulovirus) and are responsible for the primary infection of the host.

Taxonomy of Viruses (ICTV) recently decided to take them out and create the new genus *Nudivirus*.[64] This is a group of rod-shaped virions which are not occluded within an OB. The type-species of *Nudivirus* attacks the palm scarab, *Oryctes rhinocerus*, and constitutes an excellent example of the use of a virus with the classical biological control technique.

Some baculovirus genomes have been modified in order to obtain viral strains with foremost insecticidal properties, especially of AcNPV, BmNPV and HearSNPV.[65] For this purpose, hormones, enzymes and insecticidal genes have been used to modify baculovirus genomes.[66] With regard to the use of insecticidal toxins, a toxin gene Belt, encoding insectotoxin-1, obtained from the scorpion *Buthus eupeus*, was tested but the recombinant AcNPV obtained had no effect on either larval paralysis or on larval death.[67] In contrast, genes coding for three arthropod venoms have been introduced into the AcNPV genome. One is the neurotoxin AaHIT from the scorpion *Androctonus australis*, which causes paralysis and death to a great variety of insects. Recombinant AcNPV virions expressing this toxin caused a 25% reduction in the median lethal time (LT_{50}).[68] Another recombinant organism with this AaHIT gene was obtained, but introduced into the genome of *Helicoverpa armigera* SNPV, with a reduction of 17–34% in the median survival time (ST_{50}).[69] When recombinant viruses carrying *Pyemotes tritici* neurotoxin TxP-1 were used to infect greater wax moth larvae, *Galleria mellonella* (Lepidoptera: Pyralidae), paralysis was observed two days after infection. However, mortality was caused by the viral infection rather than by the toxin, with no significant difference with the parental viral strain.[70]

On the other hand, some toxins of animal origin were integrated into the AcNPV genome.[71] These were: the µ-Aga-IV venom from the spider, *Agelenopsis aperta*, and the As II and Sh1 toxins from the sea anemones, *Anemonia sulcata* and *Stichodactyla helianthus*. All three toxins show neurotoxic effects on insects. The efficiency of each recombinant virus varied both from each other and according to the insect species tested. The first toxin showed the highest activity on *S. frugiperda* larvae, with a 37% reduction in its lethal time to kill 50% of the tested population (LT_{50}).[71] While the As II and Sh1 toxins were more efficient against *T. ni* larvae, with reductions in LT_{50} of 38.4% and 36%, respectively.[71] In another example, the gene coding for the juvenile hormone esterase was integrated into the AcNPV genome. The hydrolysis of this hormone by esterase enzyme caused the triggering of a molt in the insect, as well as a stop in feeding. When recombinant viruses expressing this gene infected *T. ni* larvae, a reduction in feeding was observed, but no effect was detected on the LT_{50}.[72] Genes expressing toxins from other sources, such as *B. thuringiensis* δ-endotoxins, have also been integrated into the AcNPV genome.[73] Genes expressing Cry protoxins from *Bt aizawai* and *Bt. kurstaki* were successfully expressed in recombinant viruses. Even, the typical *Bt* bipyramidal crystals were observed in the infected cell cultures. However, when recombinant viruses were tested against lepidopteran larvae, no difference was observed either in the LT_{50} or the LC_{50} (mean lethal concentration).[74]

It is important to mention that a gene from one baculovirus has been introduced into another baculovirus. This is the *vef* gene from the TnGV that was introduced into the AcNPV genome. The *vef* gene codes for the enhancin enzyme, which degrades the peritrophic membrane present in the insect midgut lumen thereby facilitating the contact of the virions with the midgut epithelial cells. When the recombinant virus was tested against *T. ni* larvae, a reduction of 22% in the LT_{50} was observed.[75] The recombinant AcMNPV-enMP2 virus, expressing the MacoNPV enhancin gene under control of its native promoter was also developed and characterized.[76] In *T. ni* larvae, the median lethal dose (LD_{50}) of this recombinant virus was 4.4 times lower than that of AcMNPV wild-type virus. Conversely, a *H. armigera* single-nucleocapsid nucleopolyhedrovirus (HaSNPV) was used to drive the expression of an insect-selective scorpion toxin (AaIT) at the egt gene locus of HaSNPV. Laboratory bioassays indicate that the median survival time (ST_{50}) of *H. armigera* larvae was reduced 17–34% after infection with HaSNPV-AaIT in comparison to that of the wild-type virus.[69] Finally, an AcMNPV recombinant was engineered to express the insect-selective toxin IT2 from the scorpion, *Leiurus quinquestriatus*. This recombinant virus elicited the response significantly faster than the common progenitor wild-type virus. The EC_{50} (effective concentration, EC) values of wild type AcMNPV, vAP10IT$_2$ and vAPcmIT$_2$ to 3rd-instar larvae of *Trichoplusia ni* were 1.00, 0.19 and 0.17 polyhedral inclusion bodies (PIBs) mm^{-3}, respectively.[77]

Some tests using recombinant baculoviruses have been focused mainly on their prevalence in the field, rather than on their performance as bioinsecticides. This is due to some concerns related to the spread of the recombinant viruses in the environment and on the possibility of horizontal transmission of the transgene to other organisms. Still, very strict regulatory and safety requirements must be followed in many countries before the release of any genetically modified organisms, including viruses.

Recently, the potentially adverse effects of a recombinant AcMNPV (AcAaIT) were analyzed in rabbits and fishes. In the former, no marked changes in production of some enzymes were recorded. In addition, immunohistochemical observation of tissues such as stomach, intestine, liver, kidney, brain, spleen and lung showed only slight changes. In the latter, no mortality was found in treated (AcAaIT) or untreated fish during the experimental period.[78]

8.3.2 Entomopoxviruses

There are two subfamilies within the family Poxviridae: Entomopoxvirinae and Chordopoxvirinae. The first subfamily includes viruses found only in insects, and contains very large (130 to 320 kb) dsDNA genomes within ellipsoidal virions of 250 to 300 nm in size, occluded within egg-shaped occlusion bodies (OBs) called spheroids. Entomopoxviruses (EPVs) replicate exclusively in the cytoplasm of infected cells. There are three groups of EPVs: group A infects coleopteran species; group B infects species within the orders Lepidoptera and Coleoptera; and group C infects dipteran species. Lately, a fourth group (group D) which infects insect species within the order Hymenoptera has been proposed, but it still requires official recognition by the ICTV.

8.3.3 Cypovirus

Insect-infecting viruses within the family Reoviridae are commonly known as cytoplasmic polyhedrosis viruses, or cypoviruses (CPV). The great majority of these viruses attacks insects within the order Lepidoptera, although sporadically viruses have been found infecting species of Diptera, Hymenoptera and Coleoptera. Virions are occluded within anisometric OBs which may be as large as 10 μm, similar in size to those found in NPVs, but they are formed only in the cytoplasm of the host cell. Besides their potential as biological control agents, cypoviruses have been recently studied for their potential as protein nano-carriers, due to the crystalline arrangement of their polyhedrin.[79] The occluded virions are icosahedral in shape showing 12 lateral projections. Their genomes are made out of dsRNA molecules, segmented in 10 to 12 fragments of 19 to 32 kbp each. This feature is used in the classification at the species level.[80]

8.3.4 Iridovirus

Iridescent viruses or iridoviruses belong to the family Iridoviridae, which includes five genera, two of them (*Iridovirus* and *Chloriridovirus*) specific to invertebrates, mainly insects. Virions are icosahedral in shape, of 120 to 300 nm in size, and are not occluded in OBs. Their genomes are made out of circularly permuted and terminally redundant dsDNA,[81] of 105 to 212 kbp in size.[82] They are frequently found infecting important crop pests as well as some insects, that are vectors of human diseases, such as mosquitoes. Iridoviruses may cause both lethal and sublethal infections in their insect host, and their distinctive characteristic during the infection is the presence of iridescent colors shown in the infected tissues, mainly in the antenna segments, legs, prolegs and mouth parts.

8.3.5 Mechanism of Viral Infection

Most insect viruses gain access to the insect host by ingestion of contaminated food, as none of them are able to penetrate the insect tissues through the insect cuticle (exoskeleton). Still, there are some other alternative ways of infection, such as the internal or external contamination of the host eggs, and by the contaminated ovipositor of parasites.[83] Particularly for baculoviruses, insect hosts ingest the OBs present on the food (leaves, for instance), and once in the insect midgut the OBs are dissolved due to the highly alkaline environment of the gut (pH 9 or higher) releasing the occluded virions.[84] Once the virions are released, their envelopes are fused to the membrane of the midgut epithelial cells and the bare nucleocapsids are able to penetrate the cell and reach the nucleus where the virus replicates forming a generation of non-occluded virions. The bare virions bud out of the host cell, surrounded in part by the host cell membrane and disseminate throughout the whole insect body cavity (haemocoel) by the insect blood (hemolymph), infecting susceptible tissues and causing a systemic infection. The budded virions enter the cells and replicate in the nuclei, producing a second generation of virions, most of them occluded in polyhedra which are formed *de novo*, but others bud out of the cell and infect other cells. Once the infection spreads out in the whole body, the insect is killed and remains attached to the plant by their prolegs, head-down, and millions of OBs are released to the environment as the insect cuticle (exoskeleton) is degraded, constituting new inocula for other susceptible insects (see Figure 8.5).[84]

Infections caused by entomopoxviruses are very similar, as they start by the ingestion of sphaeroids by the susceptible insects. Sphaeroids are also degraded in the insect midgut, and virions are released and infect the susceptible fat body cells, where the virus replicates.[85] Infected insects become weak and eventually die in 12 to 15 days post-infection, although some may survive up to 25 to 30 days. Cypovirus infections, on the other hand, are restricted to the midgut epithelial cells, causing mostly chronic infections on the gut of lepidopteran and dipteran (mosquitoes) insects.[86] Cypovirus OBs are also degraded in the insect midgut, where the icosahedral virions are released and fused to the microvilli membrane of the midgut cells before they penetrate and replicate in the cytoplasm. Infected larvae show a whitish coloration in the midgut and later tend to

Figure 8.5   *Trichoplusia ni* infected with baculovirus TnSNPV. In baculoviruses, infected insects become whitish in color because of the massive infection of the fat body. The infected larva crawls up and hangs head down from its crochets in an inverted "V" position. A greyish to creamy liquid is released, where billions of OBs are suspended, which facilitates the spread of inocula in the field.

dry up. As for the iridoviruses, they replicate both in the cytoplasm and in the nuclei of the cells infected by pinocytosis;[87] however, the assembling (paracrystalline arrangement) of the virions particles occurs only in the cytoplasm. Virions replicate mostly in the mosquito fat body, although other tissues are eventually susceptible. Infected larvae show iridescence in several parts of the body, with orange, blue-green or turquoise tonalities. Frequently, iridoviruses are not lethal to the infected larvae, causing chronic infections.

8.3.6 Commercial Use of Entomopathogenic Viruses

Several strategies have been followed for the commercial use of insect viruses as biological control agents of pests. The selection of the most suitable strategy depends on the target species, in terms of feeding habits, behavior and the type of damage caused. The goal expected after the application of the viral agent should also be taken into account, such as reducing the plant damage (in terms of yield or cosmetic damage) or the elimination of the insect pest population.

The inoculation strategy is widely used to induce artificial epizootics on the natural pest population by the introduction of viral inocula. If successful, the pest population densities will decrease before the pest reaches its economic threshold. This strategy is based on the presumption that the virus is already present in the environment, either because it is part of the natural biotic factors or because it was introduced earlier. In this case, its natural levels are too low to start an epizootic, and an application of high levels of inocula may trigger the epizootic. This strategy has proven to be useful in forest environments, where a basal inoculum remains in the soil until the next season which complements the effect of an early application. Some examples where this strategy has been used include: the control of the gypsy moth (*Lymantria dispar*) in Northeast United States, Central and East Europe, the Mediterranean region and Japan using the product Gypchek; the control of the Douglas-fir tussock moth (*Orgya pseudotsugata*) in the USA using TM-Biocontrol-1; and the control of the European pine sawfly (*Neodiprion sertifer*) using Neochek-S on more than 20 000 ha over the last 30 years in the USA.[88] Considering all the forest pests, inundative applications of these products are out of the question and therefore planned applications on specific areas are used to trigger early epizootics in large areas of forest.

Viruses have also been used under the classical biological control strategy. After the introduction of a virus into the susceptible pest population, the virus may colonize and become established as a regulating biotic factor of the pest population, without requirement of further virus releases. One example where this strategy has been successful is in the control of the rhinoceros beetle (*Oryctes rhinoceros*) in several South Pacific islands with *Nudivirus*.[89] Indeed, the strategy was so successful that the pest has been eradicated in many of these islands, eliminating the damage caused to coconut and oil palms.[75] As with classical strategies, the inoculative release of the virus inocula can be carried out: either by simple spraying of the product (similar to the inundative strategies described below); by releasing infected hosts to contaminate healthy individuals; or by releasing contaminated parasitoids or predators (including birds) whose contaminated droppings will spread the inocula onto the pest's food. The use of contaminated baits is another technique for spreading the inocula, as adults may become contaminated and later transfer the infection to their offspring and mates. Contamination of seedlings may be another approach to spreading the viral infection, by stressing the insect population (*e.g.* by using a low-dose insecticide application), a sublethal or a covert infection may emerge as a lethal factor.[89]

The inundative strategy is based on the use of the virus as a bioinsecticide. That is, the virus is produced at industrial levels and is applied every time the pest is causing (or is expected to cause) economic damage. This strategy has been successfully incorporated into IPM programs, as long as basic information of the pest's biology, population density, economic threshold, *etc.* is known, in order to optimize its application and to program other control alternatives. This strategy has the advantage of using the same application technology used for the spraying of chemical insecticides, such as additives, spraying techniques, machinery (*i.e.* ground or aerial), *etc.*, so no additional or special equipment is required.

Although the inundative strategy is used worldwide, the development of viral bioinsecticides has occurred mostly in Europe and North America, as the first viral insecticide, Elcar, was used in the United States against the cotton bollworm, *Helicoverpa zea*. Today there are further CpGV (granulovirus obtained from *Cydia pomonella*) products on the European market: Granupom (Probis GmbH), Carpovirusine (Arysta LifeScience) and Cyd-X (Certis). In Europe, CpGV is successfully applied on more than 100 000 ha per year in organic and integrated pome fruit production. In spite of some constraints regarding the use of viral insecticides, a lack of technical information about the potential of these control agents and some reticence to use unknown agents in the field, a few Latin American countries have also developed some viral bioinsecticides. Countries such as Brazil, Peru and Bolivia produce and use their own viral bioinsecticides. In fact, one of the most successful cases for the use of baculoviruses is the control of the velvet caterpillar, *Anticarsia gemmatalis* (Lepidoptera: Noctuidae), in Brazil using its own NPV. For more than 20 years, this pest has been kept under the economic threshold almost exclusively by the spraying of the AgNPV (Baculovirus anticarsia EM), on an average of two million ha of soybean, annually.[90]

Another example is the control of potato tuberworm, *Phthorimaea operculella*, with granulovirus, PoGV, in Peru.[91] This pest is highly harmful to potato crops as it feeds on the aerial part of the plant, as well as on the tubers (see Figure 8.6). The International Potato Center (CIP) has coordinated the production of the PoGV, especially for the use of low income producers in the Andean highland territories. Its use is focused on the protection of the tuber "seeds" by covering them with a dust formulation ("Baculovirus de la polilla de la papa") before storage.[92]

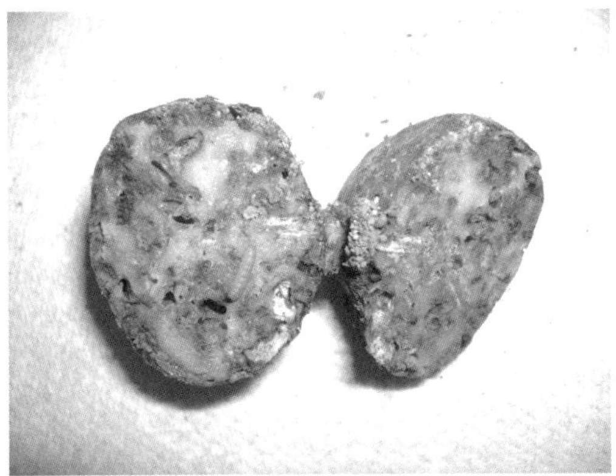

Figure 8.6   *Phthorimaea operculella* feeds on the potato tubers. In Peru the potato tuberworm, *Phthorimaea operculella* (Lepidoptera: Gelechiidae), has been efficiently controlled with its own GV (PoGV). Producers dust potatoes just after harvest and before storage with a dry formulate of PoGV, provided mostly by government agencies.

One of the best known examples for the commercial use of a baculovirus-based bioinsecticide is the successful control of the *Heliothis-Helicoverpa* complex by the commercial product Elcar, based on the *Heliothis virescens* NPV, HvNPV. This is an important complex of pests because of its wide host range, which includes cotton, soybean, maize, sorghum, tomato, tobacco, among others.[89] Elcar was the first viral insecticide registered in the USA. Another baculovirus widely used as a bioinsecticide is the *Autographa californica* NPV, AcNPV, mostly used against the cabbage looper, *T. ni* (Lepidoptera: Noctuidae) in crops such as broccoli, cabbage, cauliflower, and lettuce, among others.[92] This is the only species of baculovirus known to have a wide host range, which includes 23 species of Lepidoptera, most of them within the family Noctuidae. This is the main reason for its potential as a bioinsecticide. In fruit-producing systems, the *Cydia pomonella* granulovirus, CpGV, has been developed to control this important pest of apples, which has proven to be effective on codling moth larvae from USA, Canada, Europe and Australia. Several insect viruses have been developed and registered for use as insecticides. Most are specific to a single species or a small group of related forest pests, for example the gypsy moth, Douglas-fir tussock moth, spruce budworm and pine sawfly. They are not commercially available but are produced and used by the United States Forest Service. Other insect viruses investigated for use as insecticides include those that infect the alfalfa looper, soybean looper, armyworms, cabbage looper and imported cabbageworm. Although some of these viruses have been formulated and applied in field tests, none has been registered or sold commercially. Both the codling moth GV (Decyde®) and the *Heliothis* NPV (Elcar®) were at one time registered by the US EPA and produced commercially, but these products are no longer registered or available.

8.3.7 Factors Affecting the Efficiency of Viral Insecticides

Virulence of viruses varies not only between virus species but also between strains within the same species. Hence, baculoviruses such as the TnNPV or the HvNPV tested against *T. ni* or *H. virescence* larvae, respectively, show very high virulence levels (LC_{50} estimated for the wild-type TnNPV was 0.402 OB mm^{-2}),[75] contrary to species such as the SfNPV against *Spodoptera frugiperda*, whose LC_{50} is comparatively very high (LC_{50} value calculated for SfMNPV was 114 OB mm^{-2}).[93] Obviously, those species with high virulence levels show the highest potential to be developed as control agents. Likewise, virulence variability between strains within a species occurs in nature, and selection of geographically separated strains with higher virulence is feasible.

Environmental factors are also very important during the use of insect viruses under field conditions. High temperature and, most of all, solar radiation may affect viruses negatively in the field. Baculoviruses are easily inactivated by the UV radiation from sun light, in periods as short as a few hours; that is the reason why most viral formulations include sun protection additives such as uric acid or Tinopal, which increase the mean life and hence the activity of field application of viral bioinsecticides. This is also the reason why viral insecticide applications should be carried out early in the morning or late in the afternoon.

Another important factor that should be taken into account to increase the efficiency of a viral bioinsecticide is the proper knowledge of the target pest population. Dense populations of last instar larvae are very difficult to control with viral insecticides, as older larvae are much less susceptible to the virus. Therefore, viruses are very good candidates to be incorporated into IPM programs, where the pest populations are tightly monitored throughout the season. So, early applications of viral insecticides targeting young larvae should optimize its efficiency, as these larvae are very susceptible. Furthermore, feeding habits of the target pest should be considered. Cryptic feeding habits such as those of borer or miner insects will hamper the ingestion of a killing dose. Another factor that may interfere with the efficiency of viral insecticides is the chemical content of target species' host plant. Although more evidence is still required, tannins, acids and other chemicals may inactivate the virus applied on the plant surface.

Finally, it is important to mention that entomopathogenic viruses show a great potential as biological control agents against many insect pests, as they represent a real, economically viable and technologically feasible alternative, whose efficiency has been proven, in many cases for many years, by agricultural producers and companies, as well as government agencies, in many developed and developing countries. Still, there is much to do to find and select highly potent strains, to develop new and more efficient production technologies and to expand the market for viral insecticides.

8.4 Entomopathogenic Fungi

Within the great diversity of niches and habitats occupied by fungi in nature, some can be found as parasites and pathogens of a great number of living organisms, including insects. Entomopathogenic fungi are those causing infective diseases in insects, and may constitute important biotic factors that control insect population densities within certain levels in nature.[94] There are more than 700 species of entomopathogenic fungi reported to date, from obligate to opportunistic pathogens; some showing great potential to be developed as bioinsecticides, others with features more suited to use in classical biological control approaches. There are many examples of the successful use of entomopathogenic fungi, but still much work has to be done to achieve their real potential.

Reports of insects affected by pathogens have been recorded as far back as 4700 years ago. It was in the 18th century, however, that the first fungus, a species of *Cordyceps*, was actually identified and described, due to its large size and ease of visualization. Muscardines were reported earlier, but their association with a fungus was unknown, until the development of the microscope. A fungus was also the first microorganism to be associated with the development of a disease. During the early 19th century, the Italian Agostino Bassi studied the white muscardine on the silkworm larvae. His detailed studies clearly proved that this fungus, later known as *Beauveria bassiana*, caused an infective disease on the silkworms, and he even suggested the use of the fungus as an insect control agent. All this happened many years earlier than the development of the germ theory of disease by Louis Pasteur. It was also a fungus that was the first entomopathogen to be used in the field to control an insect pest.[95] In 1879, the eminent Russian scientist Ilya Mechnikov (1908 Nobel laureate), discovered, studied, produced and applied the green muscardine (later known as *Metarhizium anisopliae*, see Figure 8.7) to control the white grub, *Anisoplia austriaca,* and the weevil, *Cleonus punctiventris*.[96] Some other cases of the use of entomopathogenic fungi, mostly muscardines, were recorded during the early 20th century, both in Europe and United States. To date, many fungus-based bioinsecticides or mycoinsecticides are commercially available, and many others have been used for inoculative or classical biological control.

8.4.1 Diversity of Entomopathogenic Fungi

More than 90 genera and 700 species of fungi have been reported to be etiological agents in insect diseases. The term "fungus" was redefined when molecular tools indicated that some "fungi" were no longer considered within this group. This was the case for Oomycets, now associated to brown algae (Chromista or Stramenopila) rather than to fungi,[97] making the chitin (and not the cellulose) content of their cell walls a definitive feature of true fungi, which belong to the kingdom Mycota. This new classification somewhat hampers the definition of entomopathogenic fungi, as a few of them belong to the Class Oomycetes.[98] So, for practical purposes only, we will consider insect-infecting oomycetes within the entomopathogenic fungi, and for the same reason, the recent integration of Microsporidia within the fungi will not be included in this review,[99] as their diversity and mode of action is totally different to what "used to be" fungi.

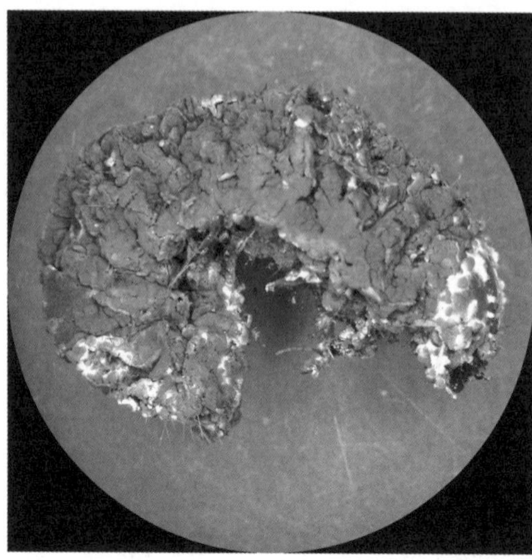

Figure 8.7   White grub (Coleoptera: Scarabaeidae) covered by a mat of mycelia, sporangia, and conidia of the green muscardine, *Metarhizium anisopliae*. Under appropriate conditions, the fungus inside the mummified cadaver grows out through the insect cuticle and sporulates to form a mat of infecting conidia.

Entomopathogenic fungi have been organized in four major classes: Chytridiomycetes, Oomycetes, Zygomycetes and Ascomycetes. There are two genera of Basidiomycetes infecting insects; however, rather than a pathogenic relationship, these show a commensalistic relationship with the insect, causing no serious deleterious effect. One additional consideration is that the class Deuteromycota is now considered within the class Ascomycota. Although most Deuteromycota (Hyphomycetes) only show the asexual (anamorphic) reproductive phase and lack the typical ascospores, asci, and asocarps of Ascomycota, their molecular characterization has shown an unambiguous relationship between both groups.

8.4.1.1   Chytridiomycota

These are the most primitive fungi, as their habitat is still aquatic and their spores are motile by a single flagellum. For obvious reasons, the few entomopathogenic chytrids infect aquatic insects such as mosquito and blackfly larvae. Among the best known example of entomopathogenic chytrid is *Coelomomyces* spp., attacking a great variety of mosquito larvae. Its life cycle is complex, as they need an intermediate host, mostly a microcrustacean, to complete the alternate sexual and asexual cycle.[100] It is an obligate pathogen, meaning that its *in vitro* production has not been developed, and it is this feature, as well as the complexity of its cycle, which has hampered its use as a biological control agent for mosquitoes. Other genera of chytrids, such as *Coelomycidum* and *Myriophagus* have been reported as pathogens of mosquito and blackfly larvae.

8.4.1.2 Oomycota

As mentioned above, this class is no longer within the kingdom Mycota but, still, it contains a few species that infect insects. As their phylogenetic relatives (brown algae), oomycetes are aquatic and therefore the entomopathogens attack aquatic insects, as well as some crustaceans. Similar to the chytrids, oomycetes form motile spores, but with two flagella. They also form resting spores that can survive to extreme environmental conditions. The best known species of entomopathogenic oomycete is *Lagenidium giganteum*. This is a species that infects mosquito larvae, and due to its ability to grow in a semi-artificial medium, as well as form resting spores, this pathogen has been used as a commercial biological control agent of mosquitoes.[101]

8.4.1.3 Zygomycota

This group is a well defined member of the Mycota kingdom, whose main characteristic is the formation of a resistant spore (zygospore) by fusion of two complementary hyphae (gametangia). One of the most diverse groups of entomopathogenic fungi belongs to the zygomycetes, the order Entomophthorales, with more than 200 species, although other orders include some few entomopathogenic species, too. Entomophthoralean fungi are well known because they forcibly discharge their conidia (asexual spores), as a means to improve their dispersion, increasing the possibility to reach a susceptible insect. Many cases of epizootics caused by entomophthoralean fungi have been recorded, and their main use as biological control agents rely on this attribute. Also important is the formation of resting spores and zygospores, which allow them to remain viable in the environment. Unfortunately, most entomophthoralean fungi are unable to grow well on artificial media and hence the possibility to be developed as bioinsecticides is limited, and their conidia are also highly susceptible to desiccation. Many genera are included in this order, some with a high level of specificity, such as *Entomophthora, Entomophaga, Erynia, Zoophthora, Strongwellsea,* and *Neozygites*, to mention some.[102]

8.4.1.4 Ascomycota

This group is known because of the ascocarps (fruiting bodies) which contain asci, and each ascus (sporangium) containing eight ascospores (sexual spores). Only a few genera are capable of infecting insects; however, the genus, *Cordyceps*, contains more than 300 entomopathogenic species. As mentioned above, historically this was the first described entomopathogen, due to its size, which may be as large as 30 cm in length or more. Another well known ascomycete is *Ascosphaera*, a pathogen of honeybees. As mentioned above, based on molecular information (mostly on the sequencing of rDNA), the old phylum Deuteromycota, containing only anamorphs (asexual phase), is now part of the ascomycetes, even if no teleomophic (sexual) phases are known. That makes ascomycetes the most diverse and most developed group of fungi used as

Figure 8.8 The white muscardine, *Beauveria bassiana*, growing on a mummified cadaver of a *Leptinotarsa texana* adult beetle. Highly sclerotized cuticle structures such as elytra and pronotum are important barriers for the fungus to grow through.

biological control agents against a wide variety of pests. This group, still known as Hyphomycetes, contains more than 40 entomopathogenic genera, including the widely used *Beauveria, Metarhizium, Paecilomyces, Nomuraea, Hirsutella, Lecanicillium* (formerly *Verticillium*), and *Culicinomyces*, among many others. Their conidia (asexual spores) are the main component of a great variety of bioinsecticides, most of them cultured on organic wastes or byproducts, such as bran or low quality grains. Since the classification of Hyphomycetes was artificial, it will slowly fade, as the molecular tools will identify their teleomorphic counterparts, if they exist. In fact, the white muscardine, *Beauveria bassiana* (see Figure 8.8) is now more frequently found in the literature as *Cordyceps bassiana*.[103]

8.4.2 Mode of Action of Entompathogenic Fungi

Contrary to most entomopathogens, which infect insects by oral uptake of the inoculum, fungi infect their hosts by penetrating the insect cuticle (exoskeleton). The following is a generalized description of a fungal infection, which may vary according to the fungus species. Once a spore (conidium) settles on the surface (exoskeleton) of a susceptible insect, it is thought that some chemical signals such as lipids secreted by the cuticle, trigger the germination of a germ hypha. This germ hypha grows shortly until a larger structure is formed called an "appressorium", which is firmly attached to the cuticle and starts the invasion *via* a penetration hypha. This hypha is known to secrete a cocktail of enzymes, mostly proteases and chitinases, which degrades the main components of the cuticle, allowing the hypha to penetrate through the multilayer exoskeleton. Once the penetration hypha reaches the haemocoele, filled with haemolymph, it grows and breaks into small fragments called blastospores. The blastospores keep growing and dividing until all the tissues of the insect host are depleted and the insect becomes a compact pack of mycelium, looking dry like a mummy. If the environmental humidity is high enough, the inner mycelium in the cadaver starts growing out through the cuticle, forming a mat of mycelium that covers the entire surface of the cadaver. Eventually, the apical hyphae start their differentiation into conidiophores (asexual sporangia) and conidia are formed, leaving the insect cadaver covered by a fine dust made of millions (up to 10^9) of mature conidia. These become the inoculum for further infection of other individuals, either in the short or long term.[104]

Variations to this general description occur, such as: the formation of conidia in the inner cavity of the insect cadaver by some strains adapted to dry conditions; the formation of wall-less syncytia in the haemocoel in some entomophthoralean fungi, instead of blastospores; the production of a great variety of toxins during the growth in the insect, which either assist in the pathogenesis or restrict saprophytic competition with other microorganisms;[105] the production of sticky conidia in some entomophthoralean fungi, which favors their attachment to a potential host; and many other variations associated with the specific biology of each species.

8.4.3 Commercial Use of Fungi as Bioinsecticide

Entomopathogenic fungi have been successfully used as biological control agents against a great number of insect pests. Similar to other biological control agents, such as parasitoids and predators, fungi have been used following inundative and inoculative strategies, as well as classical biological control. The inundative strategy is based on the use of fungi for the repetitive and massive release of the inocula to control existing pest populations, similar to the use of chemical insecticides. For inundative strategies, massive (industrial) amounts of fungal spores must be produced and formulated as bioinsecticides (mycoinsecticides).[106] Once a fungal strain has proven high levels of virulence under controlled conditions (through laboratory bioassays against the target pest), this becomes a candidate to be massively produced as a biological control agent. Most fungal bioinsecticides are readily produced in cultures, using some organic byproducts as substrates, either in solid or liquid fermentations, or in a combination of both. Bran, broken rice or corn kernels, and oats are some of the most frequently used substrates to grow entomopathogenic fungi.[107] Substrates may be dispensed in autoclavable plastic bags, glass bottles or shallow pans where they are autoclaved and then inoculated with the fungus. Once incubated under optimum temperature and humidity conditions, the fungus grows and sporulates all over the substrate surface. Spores are then removed from the substrate particles and mixed with a variety of inert compounds such as talc, or the spores may be formulated along with the substrate particles.

There are many commercial products available worldwide, mostly based on Hyphomycetes. The endurance of conidia from Hyphomycetes, their ease of production, long shelf life, persistence in the field, facility of application and relatively wide host range, make them well suited for commercialization.[108] These features are rarely met by entomophthoralean fungi, as they produce much lower amounts of conidia. These conidia are also normally much more fragile and short-living, which makes them unsuitable for the usual formulations and storage procedures. Above all, *in vitro* cultivation of entomophthoralean fungi is very difficult, and limited growth has been achieved on only a few species. Efficient mass production and formulation methods therefore still need to be developed for entomophthoralean fungi. By contrast, many hyphomycetes have been developed as bioinsecticides to control a great number of pests within the orders Coleoptera, Diptera, Homoptera, Lepidoptera, Thysanoptera, Orthoptera and even mites.[109,110] Despite there being more than 40 known entomopathogenic genera of Hyphomycetes, only a few species have been extensively developed as bioinsecticides worldwide. These include the white muscardine, *B. bassiana*; the green muscardine, *M. anisopliae*; the pink muscardine, *Paecilomyces fumosoroseus*; *B. brongniartii*; and *Lecanicillium lecanii*. It is important to note that due to the low technical requirements for producing this type of microbial insecticide (suitable for production on both a cottage and an industrial scale), it is more widely produced in developing countries.[111]

Another advantage of using entomopathogenic fungi as bioinsecticides is the possibility for improvement by genetic manipulation. With the development of reproducible transformation techniques (mostly those based on electroporation, biolistics and PEG-mediated transformation of protoplasts), heterologous genes can now be integrated into the wild-type genomes. Since the first barrier that fungi must overcome to get entry into the insect body is the cuticle or exoskeleton, the first attempts to increase their virulence were focused on the enzymes used by the penetration peg to dissolve the multilayer cuticle. A *pr1*-type protease, found in many entomopathogenic hyphomycetes, was over expressed by the insertion of multiple copies of the gene in a strain of *M. anisopliae*. The speed of kill of the transformed strain, as estimated by its LT_{50}, was reduced, although transformants had very poor sporulation ability. Also, chitinases implicated in the pathogenesis have been the target for genetic manipulation. So far, several chitinases are known from *Beauveria* and *Metarhizium*; however, over expression of these chitinases have failed to increase the fungus virulence.[112] Nevertheless, recent reports have shown that the expression of scorpion neurotoxin AAIT by transgenic *Metarhizium* resulted in increased toxicity to different insect species from different insect orders.[113] Also transgenic *B. bassiana* expressing AAIT and Pr1A showed improved toxicity against different insect species.[114] Finally overexpression of chitinase and Pr1A in *B. bassiana* resulted in higher toxicity.[115]

8.4.4 Use of Fungi under Inoculative and Classical Biological Control Strategies

The inoculative strategy is based on the release of a starting inoculum in the field, so that it can proliferate by itself on the insect population until the establishment of an epizootic. It normally requires the eventual release of the first inoculum at the beginning of a cropping season. This strategy presumes the existence of the fungal pathogen in the field, but its development in the insect population at epizootic levels takes a long time. The development of a natural epizootic is generally associated to certain factors such as transmission ability of the pathogen, insect density and environmental factors (mainly rain and humidity). Natural epizootics caused by *B. bassiana* on the oak katydid, *Pterophylla beltrani*, in northern Mexico devastate the insect population, but only when the rainy season starts and the insect density is so high that the oak community has already been heavily defoliated.[116]

A variant of this strategy is used to expand the region of infestation by spreading inocula into new areas, where there is no fungus. The entomophthoralean fungus, *Entomophaga maimaiga,* has successfully controlled natural populations of gypsy moth, *Lymantria dispar*, in northern USA by spreading resting spores originally collected from areas where the fungus has been successfully established.[117] A similar approach has been followed to successfully control the asparagus cicada, *Diceroprocta ornea*, in northern Mexico by spreading a homogenate of *Cordyceps* sp. spores to new areas (see Figure 8.9).[118] In both examples, production of the fungus *in vitro* is limited or unfeasible.

On the other hand, classical biological control is achieved when there is an introduction of an exotic fungus for permanent establishment and long-term

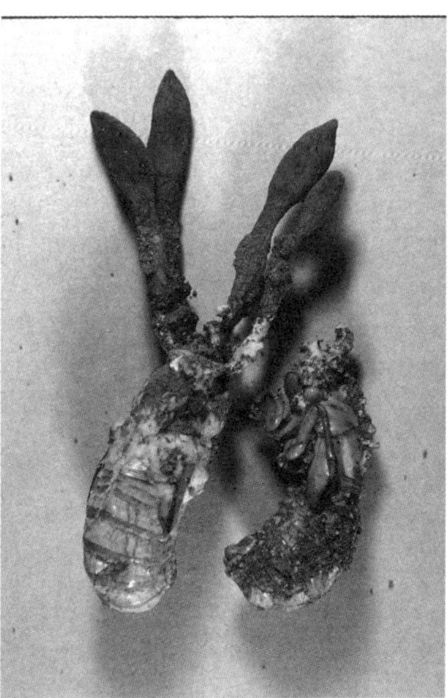

Figure 8.9 Ascocarps (fruiting bodies) of *Cordyceps* sp. growing out of mummified cicada nymphs. Many rhizophagous insects, such as the immature cicadas, are exposed to the infection of many entomopathogens whose spores remain for a long time in the soil.

pest control, normally also an exotic pest. This strategy is based on the hypothesis that the exotic pest is efficiently controlled by one or several natural enemies in its place of origin. Therefore, by introducing such a natural enemy (*e.g.* a fungus), it will spread out by itself in the pest population, causing severe epizootics and the depletion of the pest density. It is also expected that no further releases of the fungus are required. Before the release of the exotic fungus, a series of safety tests are required to minimize the possibility of any damaging effects on non-target species.

An example of classical biological control is the recent introduction of *E. maimaiga* into natural populations of gypsy moth in Bulgaria, where this pathogen has been successfully established.[119] Another example is the introduction of an Israeli strain of *Zoophthora radicans* into exotic populations of the clover aphid, *Therioaphis trifolii*, in Australia.[120] The spread of the fungus in the aphid population was corroborated. In spite of this being a very attractive strategy, it is very difficult to find the right situation where the introduced fungus is not only able to get established in a new environment, but also to become the key biotic factor to maintain the pest population under control. Still, it may happen and all depends on a thorough knowledge on the biology of both pest and pathogen, as well as their interaction.

8.4.5 Conclusions

Although the use of fungi as biological control agents has been established worldwide, there are still some limitations that should be overcome in order to maximize their potential. There is a need for a better understanding of the diversity within a given species, as some exhibit strains with a high degree of specificity and others with a wider host range. Selection of genomically homogeneous lines may help to improve pathogenicity and hence efficiency. A better knowledge of the optimum growth conditions for each strain will also help improve production conditions and ease commercialization. Additionally, there is a great potential for improving the pathogenicity of strains by genetic manipulation. A great number of heterologous genes are still there, waiting to be tested, including those of peptide toxins already found in Hyphomycete species. Finally, and very importantly, production protocols must be strictly adhered to in order to avoid the commercialization of poorly standardized products. Mycoinsecticides, as with many other bioinsecticides, are still struggling to be widely accepted by consumers, and so production of high quality products is a must.

8.5 Future Perspectives on the Use of Microbial Insecticides

As mentioned in the introduction, there is an increased concern of society regarding the use of chemical insecticides for insect control in agriculture and for vectors of important human diseases, due to the environmental and health problems these chemicals can cause.[1] Nevertheless, there is a great dependence on chemical insecticides on insect control due to their proven efficacy. The use of environmental friendly microbial insecticides depends greatly on enhancing their efficacy in field conditions and lowering their production costs. In the case of bacterial toxins, their expression in transgenic crops has enhanced their efficacy in the field resulting in a lower use of chemical insecticides. For the acceptance of microbial insecticides as spray or wettable formulations, however, there is still much to be done to produce microbial insecticides that show a similar efficacies and production costs as their chemical counterparts.

References

1. G. J. Devine and M. J. Furlong, *Agric. Human Values,* 2007, **24**, 281.
2. A. Bravo, S. S. Gill and M. Soberón, in *Comprehensive Molecular Insect Science,* ed. L. I. Gilbert, K. Iatrou and S. S. Gill, Elsevier B.V., The Netherlands, 2005, p. 175.
3. R. H. ffrench-Constant, A. Dowling and N. R. Waterfield, *Toxicon,* 2007, **49**, 436.
4. M. Gohar, K. Faegri, S. Perchat, S. Ravnum, O. A. Økstad, M. Gominet, A. B. Kolstø and D. Lereclus, *PLoS ONE,* 2008, **3**, e2793.
5. A. Bravo, S. S. Gill and M. Soberón, *Toxicon,* 2007, **49**, 423.
6. P. Christou, T. Capell, A. Kohli, J. A. Gatehouse and M. R. Gatehouse, *Trends Plant Sci.,* 2006, **11**, 302.
7. J. B. Zhang, T. C. Hodgman, L. Krieger, W. Schnetter and H. U. Schairer, *J. Bacteriol.,* 1997, **179**, 4336.
8. F. Barloy, M. M. Lecadet and A. Delecluse, *Curr. Microbiol.,* 1998, **36**, 232.
9. J. Y. Roh, J. Y. Choi, M. S. Li, B. R. Jin and Y. H. Je, *J. Microbiol. Biotechnol.,* 2007, **17**, 547.
10. K. van Frankenhuyzen, in *Entomopathogenic Bacteria, from Laboratory to Field Application,* ed. J. F. Charles, A. Delécluse and C. Nielsen-LeRoux, Kluwer Academic Publishers, Dordrecht, The Netherlands, 2000, p. 371.
11. S. C. MacIntosh, T. B. Stone, S. R. Sims, P. L. Hunst, J. T. Greenplata, P. G. Marrone, F. J. Perlak, D. A. Fischhoff and R. L. Fuchs, *J. Invertebr. Pathol.,* 1990, **56**, 258–66.
12. P. Guillet, D. C. Kurstak, B. Philippon and R. Meyer, in *Bacterial Control of Mosquitoes and Blackflies,* ed. H. de Barjac and D. J. Sutherland, Rutgers University Press, Piscataway, NJ, USA, 1990, p. 187.
13. N. Becker, in *Entomopathogenic Bacteria, from Laboratory to Field Application,* ed. J. F. Charles, A. Delécluse and C. Nielsen-LeRoux, Kluwer Academic Publishers, Dordrecht, The Netherlands, 2000, p. 383.
14. C. James, Global status of commercialized biotech/GM crops, *ISAAA Briefs,* 2008, **35**, 1.
15. M. Qaim and D. Zilberman, *Science,* 2003, **299**, 900.

16. G. H. Toenniessen, J. C. O'Toole and J. deVries, *Curr. Opin. Plant Biol.*, 2003, **6**, 191.

17. S. Guo, S. Ye, Y. Liu, L. Wei, J. Xue, H. Wu, F. Song, J. Zhang, X. Wu, D. Huang and Z. Rao, *J. Struct. Biol.*, 2009, **168**, 259.

18. A. Bravo, S. S. Gill and M. Soberón, in *Insect Control*, ed. L. I. Gilbert and S. S. Gill, Elsevier B.V., London, UK, 2010, 278.

19. S. Pacheco, I. Gómez, I. Arenas, G. Saab-Rincon, C. Rodríguez-Almazán, S. S. Gill, A. Bravo and M. Soberón, *J. Biol. Chem.*, 2009, **284**, 32750.

20. I. Arenas, A. Bravo, M. Soberón and I. Gómez, *J. Biol. Chem.*, 2010, **285**, 12497.

21. D. Bradley, M. A. Harkey, M.-K. Kim, M. K. D. Biever and L. S. Bauer, *J. Invertebr. Pathol.* 1995, **65**, 162.

22. M. Abdullah, O. Alzate, M. Mohammad, R. McNall, M. Adang and D. H. Dean, *Appl. Environ. Microbiol.*, 2003, **69**, 5343.

23. S. L. Xinyan and D. H. Dean, *Protein Eng. Design Select.*, 2006, **19**, 107.

24. R. A. de Maagd, M. Weemen-Hendriks, W. Stiekema and D. Bosch, *Appl. Environ. Microbiol.*, 2000, **66**, 1559.

25. F. S. Walters, C. M. deFontes, H. Hart, G. W. Warren and J. S. Chen, *Appl. Environ. Microbiol.*, 2010, **76**, 3082.

26. A. Bravo, *J. Bacteriol.* 1997, **179**, 2793.

27. J. M. Scriber, *Proc. Natl. Acad. Sci. U. S. A.*, 2001, **98**, 12328.

28. J. E. Losey, L. S. Rayor and M. E. Carter, *Nature.* 1999, **399**, 214.

29. E. J. Rosi-Marshall, J. L. Tank, T. V. Royer, M. R. Whiles, M. Evans-White, C. Chambers, N. A. Griffiths, J. Pkelsek and M. L. Stephen, *Proc. Natl. Acad. Sci. U. S. A.*, 2007, **104**, 16204.

30. M. Mendelson, J. Kough, Z. Vaituzis and K. Matthews, *Nature Biotechnol.*, 2003, **21**, 1003.

31. A. M. R. Gatehouse, N. Ferry and R. J. M. Raemaekers, *Trends Genet.*, 2002, **18**, 249.

32. P. D. Jensen, G. P. Dively, C. M. Swan and W. O. Lamp, *Environ. Entomol.*, 2010, **39**, 707.

33. C. R. Rodríguez-Almazán, L. E. Zavala, C. Muñoz-Garay, N. Jiménez-Juárez, S. Pacheco, L. Masson, M. Soberón and A. Bravo, *PLoS ONE*, 2009, **4**, e5545.

34. J.-F. Charles, I. Darboux, D. Pauron, and C. Nielsen-Leroux, in *Comprehensive Molecular Insect Science*, ed. L. I. Gilbert, K. Iatrou and S. S. Gill, Elsevier B.V., Oxford, UK, 2005, p. 207.

35. J.-F. Charles, *Ann. Inst. Pasteur/Microbiol.*, 1987, **138**, 471.

36. A. H. Broadwell and P. Baumann, *Appl. Environ. Microbiol.*, 1987, **53**, 1333–1337.

37. C. Oei, J. Hindley and C. Berry, *J. Gen. Microbiol.*, 1992, **138**, 1515.

38. I. Darboux, C. Nielsen-LeRoux, J.-F. Charles and D. Pauron, *Insect Biochem. Mol. Biol.*, 2001, **31**, 981–990.

39. T. P. Romão, K. D. de Melo Chalegre, S. Key, C. F. Ayres, C. M. Fontes de Oliveira, O. P. de-Melo-Neto and M. H. Silva-Filha, *FEBS J.*, 2006, **273**, 1556.

40. O. Opota J.-F. Charles, S. Warot, D. Pauron and I. Darboux, *Comp. Biochem. Physiol., B: Biochem. Mol. Biol.*, 2008, **149B**, 419.

41. J.-L. Schwartz, L. Potvin, F. Coux, J.-F. Charles, C. Berry, M. J. Humphreys, A. F. Jones, I. Bernhart, M. Dalla Serra and G. Menestrina, *J. Membr. Biol.* 2001, **184**, 171.

42. I. Darboux, Y. Pauchet, C. Castella, M. H. Silva-Filha, C. Nielsen-LeRoux, J.-F. Charles and D. Pauron, *Proc. Natl. Acad. Sci. U. S. A.*, 2002, **99**, 5830.

43. J. Schirmer, I. Just and K. Aktories, *J. Biol. Chem.*, 2002, **277**, 11941.

44. J. Schirmer, H.-J. Wieden, V. Rodnina Marina and K. Aktories, *Appl. Environ. Microbiol.*, 2002, **68**, 894.

45. C. Berry, and M. H. N. L. Silva Filha, in *Insect Control*, ed. L. I. Gilbert and S. S. Gill, Elsevier B.V., Dordrecht, The Netherlands, 2010, p. 308.

46. X. Hu, W. Fan, B. Han, H. Liu, D. Zheng, Q. Li, W. Dong, J. Yan, M. Gao, C. Berry and Z. Yuan, *J. Bacteriol.*, 2008, **190**, 2892.

47. H. Nishiwaki, K. Nakashima, C. Ishida, T. Kawamura and K. Matsuda, *Appl. Environ. Microbiol.*, 2007, **73**, 3404.

48. G. W. Jones, C. Nielsen-Leroux, Y. Yang, Z. Yuan, V. F. Dumas, R. G. Monnerat and C. Berry, *FASEB J.*, 2007, **21**, 4112.

49. T. A. Jackson, D. G. Boucias and J. O. Thaler, *J. Invertebr. Pathol.*, 2001, **4**, 232.

50. M. R. Hurst, T. R. Glare, T. A. Jackson and C. W. Ronson, *J. Bacteriol.* 2000, **182**, 5127.

51. M. R. Hurst, T. R. Glare and T. A. Jackson, *J. Bacteriol.* 2004, **186**, 5116.

52. D. Bowen, T. A. Rocheleau, M. Blackburn, O. Andreev, E. Golubeva, R. Bhartia and R. H. ffrench-Constant, *Science*, 1998, **280**, 2129.

53. M. Sergeant, P. Jarrett, M. Ousley and J. A. W. Morgan, *Appl. Environ. Microbiol.*, 2003, **69**, 3344.

54. D. Liu, S. Burton, T. Glancy, Z. S. Li, R. Hampton, T. Meade and D. J. Merlo, *Nat. Biotechnol.*, 2003, **21**, 1307.

55. N. R. Waterfield, A. Dowling, S. Sharma, P. J. Daborn, U. Potter and R. H. ffrench-Constant, *Appl. Environ. Microbiol.*, 2001, **67**, 5017.

56. A. J. Dowling, P. A. Wilkinson and R. H. ffrench-Constant, in *Insect Control*, ed. L. I. Gilbert and S. S. Gill, Elsevier B.V., Dordrecht, The Netherlands, 2010, p. 328.

57. G. Yang, A. J. Dowling, U. Gerike, R. H. ffrench-Constant and N. R. Waterfield, *J. Bacteriol.*, 2006, **188**, 2254.

58. A. J. Dowling, N. R. Waterfield, M. C. Hares, G. Le Goff, C. H. Streuli and R. H. ffrench-Constant, *Cell. Microbiol.*, 2007 **9**, 2470.

59. B. E. Tabashnik, *Proc. Natl. Acad. Sci. U. S. A.*, 2008, **105**, 19029.

60. A. Bravo and M. Soberón, *Trends Biotechnol.*, 2008, **26**, 573.

61. M. Soberón, L. Pardo-López, I. López, I. Gómez, B. E. Tabashnik and A. Bravo, *Science*, 2007, **318**, 1640.

62. M. Kos, J. J. A. van Loon, M. Dicke and L. E. M. Vet, *Trends Biotechnol.*, 2009, **27**, 621.

63. T. W. Tinsley and D.C. Kelly, in *Viral Insecticides for Biological Control*, ed. K. Maramorosh and K. E. Sherman, Academic Press, Orlando, FL, USA, 1985, p. 3.

64. C. M. Fauquet, M. A. Mayo, J. Maniloff, U. Desselberger and L. A. Ball, *Virus Taxonomy. Eighth Report of the International Committee on Taxonomy of Viruses*, Elsevier Academic Press, San Diego, CA, USA, 2005.

65. S. G. Kamita, K.-D. Kang, A. B. Inceoglu and B. D. Hammock, in *Insect Control*, ed. L. I. Gilbert and S. S. Gill, Elsevier B.V., Dordrecht, The Netherlands, 2010, p. 383.

66. P. R. Hughes, H. A. Wood, J. P. Breen, S. F. Simpson, A. J. Duggan and J.A. Dybas, *J. Invertebr. Pathol.*, 1997, **69**, 112.

67. L. F. Carbonell, M. R Hodge, M. D Tomalski and L. K. Miller, *Gene*, 1988, **73**, 409.

68. L. M. D. Stewart, M. Hirst, M. López Ferber, A. T. Merryweather, P. J. Cayley and R. D. Possee, *Nature*, 1991, **352**, 85.

69. X. Sun, H. Wang, X. Sun, X. Chen, C. Peng, D. Pan, J. A. Jehle, W. van der Werf, J. M. Vlak and Z. Hu, *Biol. Control*, 2004, **2**, 124.

70. M. D. Tomalski and L. K. Miller, *Nature*, 1991, **352**, 82.

71. G. G. Prikhod'ko, M. Robson, J. W. Warmke, C. J. Cohen, M. M. Smith, P. Wang, V. Warren, V. Kaczorowski, L. H. T. Van der Ploeg and L. K. Miller, *Biol. Control*, 1996, **7**, 236.

72. B. D. Hammock, B. C. Bonning, R. D. Possee, T. N. Hanzlik and S. Maeda, *Nature*, 1990, **344**, 458.

73. J. W. M Martens, G. Honee, D. Zuidema, J. W. M. Van Lent, B. Visser and J. M. Vlak, *Appl. Environ. Microbiol.*, 1990, **56**, 2764.

74. A. T. Merryweather, U. Weyer, M. P. G. Harris, M. Hirst, T. Booth and R. D. Possee, *J. Gen. Virol.*, 1990, **71**, 1535.

75. M. C. Del Rincon-Castro and J. E. Ibarra, *Biocontrol Sci. Technol.*, 2005, **15**, 701.

76. Q. Li, L. Li, K. Moore, C. Donly, D. A. Theilmann and M. Erlandson, *J. Gen. Virol.*, 2003, **84**, 123.

77. N. van Beek, A. Lu, J. Presnail, D. Davis, C. Greenamoyer, K. Joraski, L. MoorePierson, R. Herrmann, L. Flexner, J. Foster, A. Van, J. Wong, D. Jarvis, G. Hollingshaus and B. McCutchen, *Biol. Control*, 2003, **27**, 53.

78. J. A. Ashour, D. A Ragheb, E. A El-Sheikh, E. A Gomaa, S. G Kamita and B. D. Hammock, *Int. J. Environ. Res. Public Health*, 2007, **4**, 111.

79. F. Coulibaly, E. Chiu, K. Ikeda, S. Gutmann, P. W. Haebel and S. B. Clemens, *Nature*, 2007, **446**, 97.

80. P. P. C. Mertens, S. Rao and H. Zhou, in *Virus Taxonomy: Eighth Report of the International Committee on Taxonomy of Viruses*, ed. C. M. Fauquet, M. A. Mayo, J. Maniloff, U. Desselberger and L. A. Ball, Elsevier/Academic Press, London, UK, 2004, p. 522.

81. R. Goorha and K. G. Murti, *Proc. Natl. Acad. Sci. U. S. A.*, 1982, **79**, 248.

82. G. Delhon, D. R. Tulman, C. L. Afonso, Z. Lu, J. J. Becnel, B. A. Moser, G. F. Kutish and D. L. Rock, *J. Virol.*, 2006, **80**, 8439.

83. R. R. Granados, *Biotechnol. Bioeng.*, 1980, **XXII**, 1377.

84. R. R. Granados and K. A. Williams, in *The Biology of Baculoviruses*, ed. R. R. Granados and B. A. Federici, CRC Press, Boca Raton, FL, USA, 1986, p. 89.

85. M. Erlandson, *J. Invertebr. Pathol.*, 1991, **57**, 255.

86. J. J. Becnel, S. E. White and A. M. Shapiro, *J. Invertebr. Pathol.*, 2003, **83**, 181.

87. R. Goorha, *J. Virol.*, 1982, **43**, 519.

88. J. D. Podgwaite, in *Viral Insecticides for Biological Control,* ed. K. Maramorosch and K. E. Sherman, Academic Press, Orlando, FL, USA, 1985, p. 775.

89. J. Huber, in *The Biology of Baculoviruses*, ed. R. R. Granados and B. A. Federici, CRC Press, Boca Raton, FL, USA, 1986, p. 181.

90. J. J. L. Velasco de Castro Oliveira, A. Caldas Wolff, A. Garcia-Maruniak, B. Morais Ribeiro, M. E. Batista de Castro, M. Lobo de Souza, F. Moscardi, J. E. Maruniak and P. M. de Andrade Zanotto, *J. Gen. Virology*, 2006, **87**, 3233.

91. K. V. Raman, J. Alcazar and A. Valdez, *Boletín de Capacitación CIP-2*, Centro Internacional de la Papa, Lima, Peru, 1992.

92. J. B. Carter, in *Biotechnology and Genetic Engineering Reviews*, ed. W. Russell, Nottingham University Press, Nottingham, UK, 1984, p. 375.

93. J. Cisneros, J. A. Pérez, D. I. Penagos, J. Ruiz V., D. Goulson, P. Caballero, R. D. Cave and T. Williams, *Biol. Control* 2002, **23**, 87.

94. L. J. Copping, *The Biopesticide Manual*, British Crop Protection Council. Farnham, UK, 2001.

95. E. A. Steinhaus, *Disease in a Minor Chord*, Ohio State University Press, Columbus, OH, USA, 1975.

96. G. Zimmermann, B. Papierok and T. Glare, *Biocont. Sci. Technol.*, 1995, **5**, 527.

97. P. Keeling, B. S. Leander, A. Simpson, in *The Tree of Life Web Project*, http://tolweb.org/2009 (Version 28 October 2009) http://tolweb.org/Eukaryotes/3/2009.10.28.

98. D. S. Hibbett, M. Binder, J. F. Bischoff, M. Blackwell, P. F. Cannon, O. Eriksson, S. Huhndorf, T. James, P. M. Kirk, R. Lücking, T. Lumbsch, F. Lutzoni, P. B. Matheny, D. J. McLaughlin, M. J. Powell, S. Redhead, C. L. Schoch, J. W. Spatafora, J. A. Stalpers, R. Vilgalys, M. C. Aime, A. Aptroot, R. Bauer, D. Begerow, G. L. Benny, L. A. Castlebury, P. W. Crous, Y.-C. Dai, W. Gams, D. M. Geiser, G. W. Griffith, C. Gueidan, D. L. Hawksworth, G. Hestmark, K. Hosaka, R. A. Humber, K. Hyde, U. Köljalb, C. P. Kurtzman, K.-H. Larsson, R. Lichtwardt, J. Longcore, J. Miadlikowska, A. Miller, J.-M. Moncalvo, S. Mozley-Standridge, F. Oberwinkler, R. Parmasto, V. Reeb, J. D. Rogers, C. Roux, L. Ryvarden, J. P. Sampaio, A. Schuessler, J. Sugiyama, R. G. Thorn, L. Tibell, W. A. Untereiner, C. Walker, A. Wang, A. Weir, M. Weiss, M. White, K. Winka, Y.-J. Yao and N. Zhang, *Mycolog. Res.*, 2007, **111**, 509.

99. T. Y. James, F. Kauff, C. Schoch, P. B. Matheny, V. Hofstetter, C. J. Cox, G. Celio, C. Geuidan, E. Fraker, J. Miadlikowska, H. T. Lumbsch,

A. Rauhut, V. Reeb, A. E. Arnold, A. Amtoft, J. E. Stajich, K. Hosaka, G.-H. Sung, D. Johnson, B. O'Rourke, M. Crockett, M. Binder, J. M. Curtis, J. C. Slot, Z. Wang, A. W. Wilson, A. Schüßler, J. E. Longcore, K. O'Donnell, S. Mozley-Standridge, D. Porter, P. M. Letcher, M. J. Powell, J. W. Taylor, M. M. White, G. W. Griffith, D. R. Davies, R. A. Humber, J. B. Morton, J. Sugiyama, A. Rossman, J. D. Rogers, D. H. Pfister, D. Hewitt, K. Hansen, S. Hambleton, R. A. Shoemaker, J. Kohlmeyer, B. Volkmann-Kohlmeyer, R. A. Spotts, M. Serdani, P. W. Crous, K. W. Hughes, K. Matsuura, E. Langer, G. Langer, W. A. Untereiner, R. Lücking, B. Büdel, D. M. Geiser, A. Aptroot, P. Diederich, I. Schmitt, M. Schultz, R. Yahr, D. S. Hibbett, F. Lutzoni, D. J. McLaughlin, J. W. Spatafora and R. Vilgalys, *Nature*, 2006, **443**, 818.

100. B. A. Federici and C. J. Lucarotti, *J. Invertebr. Pathol.* 1986, **48**, 259.

101. J. L. Fetter-Lasko and R. K. Washino, *Environ. Entomol.* 1983, **12**, 635.

102. J. K. Pell, J. Eilenberg, A. E. Hajek and D. C. Steinkraus, in *Fungi as Biocontrol Agents*, ed. T. M. Butt, C. Jackson and N. Magan, CABI Publishing, Wallingford, UK, 2001, p. 71.

103. B. Huang, C. R. Li, Z. G. Li, M. Z. Fan and Z. Z. Li, *Mycotaxon*, 2002, **81**, 229.

104. A. E. Hajek and R. J. St. Leger, *Annu. Rev. Entomol.* 1994, **39**, 293.

105. A. Vey, R. E. Hoagland and T. M. Butt, in *Fungi as Biocontrol Agents*, ed. T. M. Butt, C. Jackson and N. Magan, CABI Publishing, Wallingford, UK, 2001, p. 311.

106. C. A. Bradley, W. E. Black, R. Kearns and P. Wood, in *Frontiers in Industrial Mycology*, ed. G. E. Leatham, Chapman and Hall, New York, USA, 1992, p. 160.

107. N. E. Jenkins, G. Heviefo, J. Langewald, A. J. Cherry and C. J. Lomer, *Biocontrol News Informat.*, 1998, **19**, 21N.

108. M. S. Goettel, J. Eilenberg and T. R. Glare, in *Comprehensive Molecular Insect Science*, ed. L. O. Gilbert, K. Iatrou and S. Gill, Elsevier B.V., Oxford, UK, 2005, p. 361.

109. G. D. Inglis, M. S. Goettel, T. M. Butt and H. Strasser, in *Fungi as Biocontrol Agents* ed. T. M. Butt, C. Jackson and N. Magan, CABI Publishing, Wallingford, UK, 2001, p. 23.

110. S. L. Elliot, G. J. De Moraes, I. Delalibera, C. A. Da Silva, M. A. Tamai and J. D. Mumford, *Bull. Entomol. Res.*, 2000, **90**, 191.

111. J. Harris and D. Dent, *Priorities in Biopesticide Research and Development in Developing Countries*, Biopesticides Series 2, CABI Publishing, New York, USA, 2000.

112. R. St. Leger and S. Screen, in *Fungi as Biocontrol Agents*, ed. T. M. Butt, C. Jackson and N. Magan, CABI Publishing, Wallingford, UK, 2001, p. 219.

113. M. Pava-Ripoll, F. J. Posada, B. Momen, C. Wang and R. St. Leger, *J. Invertebr. Pathol.*, 2008, **99**, 220.

114. D. Lu, M. Pava-Ripoll, Z. Li and C. Wang, *Appl. Microbiol. Biotechnol.* 2008, **81**, 515.
115. W. Fang, J. Feng, Y. Fan, Y. Zhang, M. J. Bidochka, R. J. St. Leger and Y. Pei, *J. Invertebr. Pathol.*, 2009, **102**, 155.
116. J. Ibarra, unpublished results.
117. A. E. Hajek and R. E. Webb, *Biol. Control,* 1999, **14**, 11.
118. J. Celaya-Gortari, M. I. Sotelo, E. P. Delgado and B. Arias, in *Proceedings of the XXVII National Meeting on Biological Control*, Mexican Society for Biological Control, Los Mochis, Mexico, 2004 p. 101.
119. D. Pilarska, M. McManus, A. E. Hajek, F. Herard, F. E. Vega, P. Pilarski and G. Markova, *J. Pest Sci.,* 2000, **73**, 125.
120. R. J. Milner, R. S. Soper and G. G. Lutton, *J. Austral. Entomol. Soc,.* 1982, **21**, 113.

The Challenge of Green in a Pesticide-Dominant IPM (Integrated Pest Management) World

S. J. CASTLE*[1] AND N. PRABHAKER[2]

[1] USDA-ARS, Arid-Land Agricultural Research Center, 21881 N. Cardon Lane, Maricopa, AZ 85138, USA; [2] Department of Entomology, University of California, Riverside, CA 92521, USA

9.1   Introduction

The success of modern agriculture in sustaining food production at levels necessary to meet the demands of a rapidly increasing human population is one of the great achievements of the past half century. One may recall that human numbers had doubled from three to six billion people in just 40 years – from 1960 to 1999.[1] At the beginning of this period there was deep skepticism about the capacity of earth's resources to sustain the human population at its current rate of increase, much of it elicited by prognostications of severe food shortages and famines that would occur late in the 20th century due to overpopulation.[2] These projections ultimately proved unfounded as the Green Revolution and other advances in modern agriculture combined to increase the world's food supply and meet the accelerating demand. Gross world food production (cereals, coarse grains, roots and tubers, pulses and oil crops) grew from

RSC Green Chemistry No. 11
Green Trends in Insect Control
Edited by Óscar López and José G. Fernández-Bolaños
© Royal Society of Chemistry 2011
Published by the Royal Society of Chemistry, www.rsc.org

1.84 billion tons in 1961 to 4.38 billion tons in 2007, an increase of 138%.[3] *Per capita* agricultural production outpaced population growth with a theoretical 29% more food available *per capita* today than in 1960.[3] Despite these optimistic trends, imbalances in agricultural production, arable land resources, wealth and stability of countries, as well as other factors have impeded a more equitable distribution of food,[4,5] causing an estimated 1.02 billion people to remain undernourished.[6] Another critical period of 40 years lies ahead for agriculture as the global population is projected to rise to nine billion by 2050, with anticipated food production needing to increase by 70% over current levels to meet the demand.[7]

One of the key advances in modern agriculture that contributed to the success of the Green Revolution was the development of pest and disease management technologies for combating our principal food competitors. Indeed, genetic resistance to wheat rusts was a key attribute of new varieties introduced in 1961 that enabled cereal production to increase dramatically over the next few decades.[8,9] The overall challenge, however, became much greater as production agriculture was transformed into more favorable environments for numerous herbivorous pests and weed species. Dramatic increases in the use of irrigation and synthetic fertilizers[10] generated faster growing and more enriched crops containing higher levels of nitrogen, often a growth-limiting element for herbivores in natural plant communities or subsistence agriculture.[11] Intensification of agriculture increased substantially over this period[12] and was supported by greater mechanization on the farm as the agronomic capacity to produce crops grew, but also in the supply channels as global demand increased and markets developed to accommodate the production and distribution of food. Intensified agriculture also provided more constant food resources for many crop pests, especially polyphagous species capable of feeding on diverse crop types. Despite the elevated pest challenges that came with high production agriculture, pest populations were for the most part kept at bay through a continually evolving array of technologies ranging from biological to chemical, biotechnological to cultural. The tactics required to effectively implement these diverse approaches also continued to develop through improved knowledge, both basic and applied.

In this chapter, we will consider the role that pest management has played in the success of modern agriculture and evaluate the challenges ahead. Before examining specific tactics and considering their potential role in a greener IPM (Integrated Pest Management), we will attempt to gain perspective on what it means to be green in agriculture and pest management and how these ideals fit with a world in demand for food. We will then question why agricultural systems present unique environments to which pest species are well adapted and continuously try to exploit. Discussion of the various pest management approaches that have developed over the past 50 years will be limited to those that protect crops from arthropod pests. This is not to ignore other pressing topics such as the growing use of herbicides in connection with transgenic crops engineered for resistance to a specific herbicide, or the questionable practices that are used in rearing farm animals. These are issues that, along with the management of arthropods in crops, require much discussion regarding the impact they have on the environment and on supporting growth of the food supply.

9.2 The Concept of Green

The inordinate dependency on pesticides from the 1950s onwards tended to relegate many age-old cultural and biological control tactics to minimal roles in pest management. The rapid and continuous expansion of the pesticide market over succeeding decades helped to create a perpetual demand for chemical control. Despite frequent problems of resistance and secondary pest outbreaks, a new compound always seemed to become available for combating recalcitrant pests. A number of field entomologists recognized the dire warning signs that had become increasingly apparent in crop production where pesticides were being overused.[13–15] Acting on this concern, Stern *et al.*[16] developed the integrated control concept that advocated allowing natural control to operate up to a threshold level where crop injury could occur, and only then apply an insecticide to reduce the pest population and avoid economic damage. However, it was the contribution of a non-entomologist, Rachel Carson, through her book *Silent Spring*[17] that engaged mass attention concerning the heavy use of insecticides for protecting crops and the destructive effects they were having in and out of fields. The enduring impact of this single book cannot be underestimated, especially with respect to the deep suspicion of agrochemicals and pest control technologies that was crystallized in the minds of consumers.

The raising of environmental consciousness that began with Silent Spring had a profound effect on pest control practices across the globe. The earnest response to the depiction of an Earth poisoned by pesticides ignited a grass-roots concern about the environment that eventually led to the creation of government bodies for regulating the use of pesticides. It also helped to promote IPM as a rational alternative to intensive chemical control. Moreover, the increased awareness of pesticide abuses stimulated the formation and growth of environmental groups that have played a prominent role in shaping cultural movements and opinion. The popularization of "green" as a term that expresses environmental conscientiousness and stewardship of the earth is one example of the cultural impact of environmentalism. While there is little controversy about striving for green as an ideal, questions arise concerning how best to balance preservation of natural systems with the technological innovation that serves humanity.

In the case of agriculture, natural environments have been modified through the millennia to allow crops to be cultivated and livestock to be grazed – our present numbers are a testimonial to our success at altering landscapes to produce food. Although agriculture has a major impact on the environment at both local and global scales, there are few who object in principle to the practice of agriculture as a basic human endeavor for securing a food supply, or to the concept of pest management for defending that food supply. This consensus

Table 9.1 Precepts of green in agriculture.

1. Conserve natural resources
2. Preserve ecosystem stability
3. Maintain biodiversity
4. Enhance food security and sustainability
5. Preserve life and well being

becomes weaker, however, when consideration is given to the manner in which particular aspects of agriculture are conducted, especially with regard to how pest management is implemented. Fundamental questions challenge societies and policy makers over issues, such as the degree to which agriculture should remain agrarian rather than industrialized, or whether pest management should incorporate biotechnological advances to a greater degree,[18] in the process potentially deemphasizing centuries-old cultural practices. How society responds to these questions has tremendous relevance to food security, health and well being of humans, and environmental conservation.

We shall approach the general topic of green trends in pest control from the perspective that all technologies have a potential role to play ensuring food security over the next 40 years. Determining which pest management practices are deserving of the label "green" is a matter of judgment, since there is no strict understanding of what it means to be green in agriculture or any other realm. At the present time, "being green" or "going green" are comparative ideas where an activity is judged by its perceived impact on the environment relative to an existing activity. In a broader sense, those activities that conserve natural resources, preserve ecosystem stability, maintain biodiversity, enhance food security and sustainable management, and preserve life and well being may very well fall under the umbrella of green (see Table 9.1), although a given practice or activity may not meet all of these criteria. The challenge of green in an IPM world will be to attain as many of these ideals as possible, but without sacrificing productivity over the long term. Faced with the evidence that agricultural output has begun to plateau in the last few years,[6] there is simply too much at stake in terms of potential human suffering to be anything but pragmatic about implementing pest control technologies that best ensure food security.

9.3 Modern Agriculture and Pest Forcing

9.3.1 The Pesticide Connection

Estimates of crop losses due to all pests (animals, weeds, pathogens) is rather staggering when considered in the context of the powerful management tools that are available to fight back. Global losses of soybean, wheat and cotton are estimated to be 26–29%, but are even higher for maize, rice, and potatoes at 31, 37 and 40%, respectively.[19] Such losses demonstrate not only the tremendous diversity of organisms that compete with humans for food resources, but the impracticality of devising management solutions for every species deriving nutrition and shelter from a crop. Differentiating between primary and secondary pests has been an essential point of distinction for allocating finite pest management resources. Targeting primary pests over and over again with pesticides, however, brought out another aspect of pest organisms – their resiliency. Resistance to pesticides has evolved across the pest spectrum and has exacerbated pest management costs, by increasing both the number of treatments and concentration of active ingredients necessary to manage resistant pest populations. Increasing pesticide use also led to gross imbalances in crop systems by depleting natural enemies of pest species, to such a degree that normally insignificant species would occasionally be released from natural control and increase to pestiferous levels. It was the careless treatment with pesticides of one pest problem that too often would lead to a second pest problem, followed by additional treatments that eventually resulted in a pesticide treadmill.[20] Such scenarios became increasingly common in the 1950s and thereafter and became another motivating reason for Stern *et al.*[16] to develop the integrated control approach and lessen dependency on pesticides. However, it was many years later before integrated pest management became the definitive term that embodied the principles within the integrated control concept, and still more years before it attained the institutional level that it now holds.[21] In the meantime, the bad habits acquired early in the era of proliferating new pesticides proved hard to break as pesticides dominated pest management approaches.

Following the publication of *Silent Spring*, there was considerable examination by entomologists into the role that pesticides had played in crop failures and ecological crises worldwide. Cotton in particular came under scrutiny as pesticides were implicated in multiple destructive pest outbreaks. The pattern of pesticide-related pest problems in cotton led Smith[22] to identify five phases of cotton production that had been observed on different continents, all ending disastrously. The pattern described began with low input cotton production during "the subsistence phase", but eventually proceeded to "the exploitative phase" through intensification measures, such as the addition of fertilizers or development of irrigation systems for constant water delivery. As Smith[22] saw it, the investment of capital during the exploitative phase required protection from pests that would otherwise diminish yield and cut into profits. Thus, the intensive use of pesticides to the virtual exclusion of all other control methods brought about the third phase that he called "the crisis phase". It was the crisis phase that was critical to the predominant thinking that eventually developed concerning the primary role of pesticides in pest outbreaks in agriculture. Once the situation had reached the crisis phase, pest problems would continue to mount due to a combination of pesticide resistance, pest resurgence, and induction of secondary pests to major pest status that altogether would bring on "the disaster phase". At this point, pest problems would be so rampant that cotton would no longer be profitable to grow, leading in some cases to its abandonment as purportedly occurred in some regions.[20,23,24] An alternative to abandonment was to reform pest management practices through adoption of IPM methods that emphasized biological and cultural controls to return to a more natural balance. This would initiate the final part of the five-part sequence known as "the recovery phase".

The pattern of pesticide-induced disasters depicted by Smith[22] and the supporting examples he presented were reiterated numerous times by various authors.[25–29] These crisis-in-cotton examples came to represent the paradigm for pest control failures in general where pesticides were involved. It was not necessary that a major crop failure occur, only that unsatisfactory control leading to economic damage be realized that, more often than not, would elicit an explanation that pesticides were to blame for the resistance they had caused in the target population, or that natural enemies that had been eliminated had led to a resurgence of the target pest or an increase in secondary pests. While well-documented examples exist for each one of these phenomena, specific evidence in the vast majority of cases was rarely presented. For example, speculation made about the increased pest status of the whitefly *Bemisia tabaci* stated "Once whiteflies are established in an area, we believe that the disruptive influence of pesticides, coupled with resistance, is the largest factor responsible for the crisis situations often associated with population outbreaks".[30] Prior to this statement, there had been recent outbreaks of *B. tabaci* in the Sudan Gezira that had been attributed to pesticides either because of severe disruption of natural enemies that eliminated natural regulation of *B. tabaci*,[29] or resistance to organophosphates which arose in *B. tabaci* and hormoligosis to DDT residues occurred.[31] While both of these publications acknowledged that pesticides were behind the outbreaks of *B. tabaci*, they did not agree how pesticides exerted their disruptive influence. An earlier report about outbreaks of *B. tabaci* in the Imperial Valley of California[32] also implicated pesticides, in this case an upset of the natural balance caused by the recently commercialized pyrethroids. So there was general support for the brief summary about *B. tabaci* outbreaks referred to earlier concerning the "disruptive influence of pesticides",[30] but a lack of certainty and confirmatory data concerning the specific ways that pesticides had induced these outbreaks. The same is true of other pest species for which investigative studies were rarely done to identify underlying mechanisms leading to outbreaks. However, the ubiquitous use of pesticides during the second half of the 20th century and the scorn they received following *Silent Spring* made them an easy target for blame whenever pest populations were not adequately controlled.

Despite their unsavory reputation, pesticides continue to play a major role in pest management and are one of the critical considerations, perhaps impediments, to adopting greener IPM programs. Accurately assessing the role pesticides have played in managing pests, or perhaps exacerbating pests in some instances, is crucial to designing better IPM programs that will not have to rely as heavily on pesticides as in the past. Some would argue that reliance on pesticides in the first place was misguided and that dependency grew over time due to the careless way they were implemented into pest management programs.[33] An alternative view (see Figure 9.1b) is that intensification measures that were the hallmark of the Green Revolution contributed to the

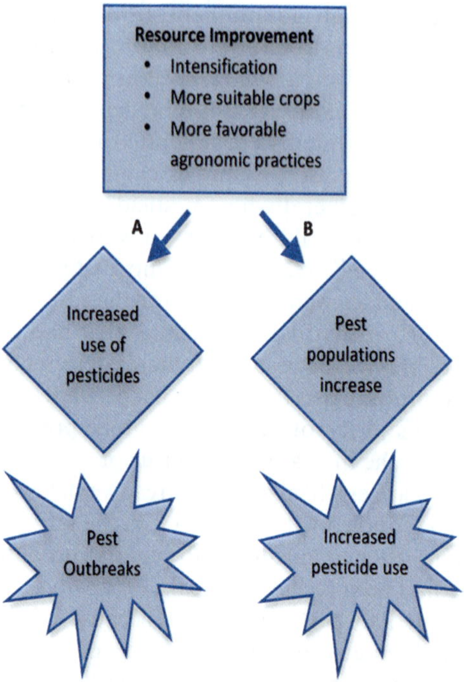

Resource Improvement
- Intensification
- More suitable crops
- More favorable agronomic practices

A

B

Increased use of pesticides

Pest populations increase

Pest Outbreaks

Increased pesticide use

Figure 9.1 Two scenarios describing increased pest problems following resource improvement associated with agricultural development. The predominant view (A) expressed by Smith[22] and others suggests that pesticides used to protect crops cause resistance to pesticides and resurgences in pests resulting in outbreaks. An alternative view (B) suggests that resident pest populations respond with growth to improved resources brought on by intensification measures, and increased pest pressure is met by stepped-up pest management responses including greater pesticide use.

forcing of pest populations by creating a more enriched environment for the pest species. That is, the natural increases in pest populations responding to superior environments led to greater pest pressure and ultimately forced heavier pesticide use. This is a sharp distinction from the predominant view (see Figure 9.1A) that pesticide abuses reduced natural enemies to ineffectual densities, and caused resistance and resurgences of pest populations leading to economic damage and in some cases field failures.

9.3.2 Agricultural Intensification

Looking further back than the 1950s, we can find well-documented cases of *B. tabaci* outbreaks prior to the advent of synthetic organic pesticides. Nymphal densities as high as 150 cm^{-2} were reported in cotton during the early 1940s in the Sudan Gezira.[34] Even higher densities were reported from cotton grown in India during the late 1920s and early 1930s[35] following large-scale increases in area planted to so-called "American" cotton varieties newly introduced to India at that time. Destructive infestations of *B. tabaci* in vegetables were also described in Palestine during the 1930s.[36] What these early records indicate is that *B. tabaci* was already a serious pest of agriculture before synthetic pesticides were developed. There were clearly circumstances involving various crops and locations in which *B. tabaci* populations developed to high densities that would be considered outbreak levels by today's standards. The changeover to American cotton that occurred in parts of the Punjab region of India involved not just a shift to a new variety, but actually to a different species of cotton. Cotton production in India at the beginning of the 20th century depended exclusively on locally adapted varieties belonging to *Gossypium arboreum* L. Between 1911–13, *ca.* 6000 ha were planted to *G. hirsutum*, called American cotton at that time in India, but now more commonly known as "upland cotton". Another 111 694 ha of *G. hirsutum* were planted between 1917–18,[37] greatly expanding the scale of American cotton that turned out to be a more susceptible cotton species to *B. tabaci*.[35] It was soon after this transformation that whitefly numbers began to increase and cotton crop failures were reported.[37] However, there was considerable uncertainty about the cause of the cotton crop failures, since there was no precedent for a minute insect such as *B. tabaci* causing destruction without ever chewing holes in leaves. It was the thorough research of Husain and Trehan[35] that finally resolved the causal nature of the crop failures and by doing so identified a destructive new pest species.

The change to a higher yielding and more desirable cotton fiber that took place in India represented a form of agricultural intensification, not unlike the higher yielding plant varieties introduced at the start of the Green Revolution. Improved quality and yield were the goals in both cases, but with some uncertainty in terms of how respective pest complexes might respond. Two lines of evidence, *i.e.* the improved performance of *B. tabaci* on *G. hirsutum versus* the indigenous *G. arboreum* as determined in controlled studies,[35] and the rise in *B. tabaci* infestations reaching outbreak status that coincided with the wide-scale adoption of *G. hirsutum* in India, together suggest that the alteration in cotton agriculture itself may have been largely responsible for the increased pest problem in Indian cotton during the 1920s. In the Sudan Gezira, there was a 2.3-fold increase in cotton acreage that occurred between 1956 and 1967, going from 103 062 to 232 545 ha in just 11 years. During the same period, the area planted to groundnuts increased 106-fold, from 403 to 42 698 ha. This coincided with the time period of the late 1950s to mid 1960s that Eveleens[29] claimed was the time frame when *B. tabaci* attained the status of a serious pest. The additional acreage for both crops greatly expanded the resource base, but also improved the quality of that base since cotton and groundnuts are favored crops of *B. tabaci*. With the example from the Imperial Valley of California *ca.* 1980, there also was a major expansion of cotton acreage that had taken place over the previous 12–15 years that peaked in 1977 at 58 275 ha. Cotton acreage remained high in the Imperial Valley through 1980, the year of a major outbreak of *B. tabaci*.[32] Although it was the role of pesticides only that was implicated in the outbreaks of *B. tabaci* in Sudan[29,31] and California,[32] the simultaneous expansion of favored crops cannot be ignored given the historical examples of *B. tabaci* outbreaks prior to the era of synthetic organic insecticides.

In addition to the two aspects of agricultural intensification identified from the historical examples for *B. tabaci*, *i.e.* (1) the expansion of favorable crop acreage in Sudan and California, and (2) the replacement of an indigenous cotton species with an exotic, more susceptible cotton species in India that allowed higher colonization rates by *B. tabaci*, still other forms of intensification occurred in the Sudan during the putative emergence of *B. tabaci* as a primary pest.[31] One involved a shift in cultural practices made in the 1960s with the adoption of a longer season, more vegetatively vigorous cotton variety. The sowing date for cotton in the Gezira Scheme had originally been at the end of the rainy season so as to avoid a rain-dispersed bacterial blight.[38] The new variety 'Barakat' was resistant to the causal bacterium, *Xanthomonas malvacearum* (Downson), and so could be sowed at an earlier time, thereby extending the cotton season and enabling additional generations of *B. tabaci* to develop. Another critical development was the expansion of the Gezira Scheme through construction of a new irrigation project. Lying between the White and Blue Nile rivers, the Sudan Gezira region was supplied with irrigation water beginning in the 1920s after construction of a dam on the Blue Nile at Sennar. The success of the Gezira Cotton Scheme from the 1930s–50s provided impetus for further expansion of the irrigation infrastructure in the Gezira region. A project called the Managil Extension opened up another 336 000 ha of irrigated cropland in the 1960s and virtually doubled the cultivated area. The traditional fallowing of one-half of the cropland area had been gradually reduced in the established areas of the Gezira, with the same limited fallowing observed for the new areas under the Managil Extension.[39] A push to diversify the agriculture by adding new crops, while also increasing areas planted to traditional crops, greatly intensified agriculture in the expanded and fully irrigated Gezira Scheme through the 1960s–70s.

9.3.3   Fertilizer Effects

Increased fertilization of crops was a major force behind agricultural intensification and the Green Revolution. It was only after the synthesis of ammonia from atmospheric nitrogen was discovered in 1909 by Fritz Haber, and scaled up to industrial production in 1913 by Carl Bosch that nitrogen could be produced as urea and various nitrate forms of fertilizer.[40] This did not happen at a global level until after two world wars and many years of agronomic research had determined what rates of fertilizer to apply for optimal yields. Nitrogenous fertilizer has been analogized to fossil fuel where food is to nitrogen as energy is to carbon.[41] Plant growth rates had always been limited by nitrogen availability as natural sources of nitrogen, biofixation by bacteria symbiotic with legumes, atmospheric deposition and recycling of plant residues and manure, could only supply half of the global demand for

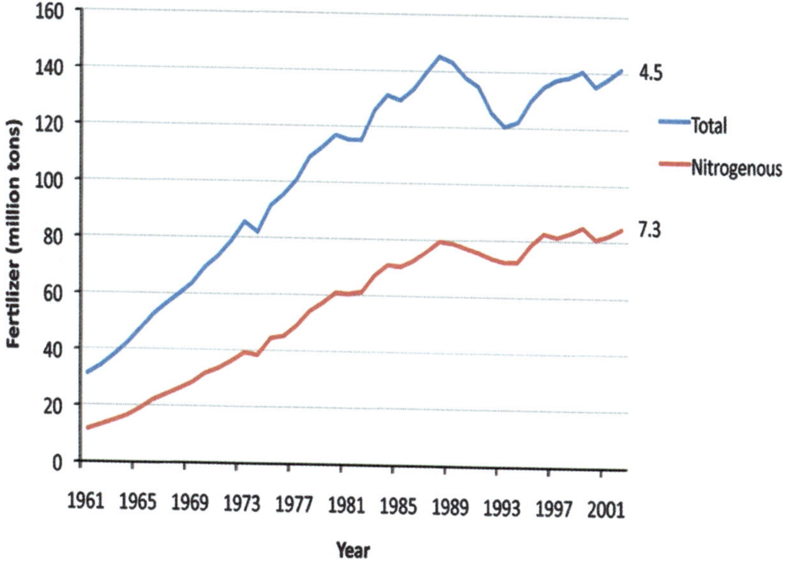

Figure 9.2 World consumption of total and nitrogenous fertilizers in agriculture. Values adjacent to each trace indicate the fold increases in usage between 1961 and 2002.[43]

nitrogen.[42] The ready availability of nitrogen to crops through supplied fertilizer beginning in the late 1940s and 1950s is what drove the tremendous increases in crop productivity thereafter. The demand also increased for the other two plant macronutrients, phosphorous and potassium, and was met by the mining of rock deposits containing phosphates and potassium. However, the rate of consumption of nitrogenous fertilizers has increased 7.3-fold between 1961 and 2002, compared to a 4.5-fold increase for total fertilizers (see Figure 9.2).[43]

Enhanced nitrogen content of plants has been shown to positively influence growth and development of immature insects.[11,44] Higher nitrogen content in plant tissue often improves assimilation efficiency leading to greater accumulation of body tissue in insects.[45] Although faster growth rates do not always occur with enhanced nitrogen content in plant tissue,[46] in most cases laboratory performance measures do improve while population growth rates in the field also are accelerated.[47–50] Some of the benefit to insects that feed on nitrogen-fertilized plants may result more from accumulation of foliar biomass in a vigorous plant than plant nitrogen concentration.[50] Phloem-feeding insects often respond favorably to higher nitrogen content, in part because they are not exposed to secondary plant compounds, such as alkaloids and tannins, to the same degree as plant-chewing insects and can therefore more readily utilize nitrogen-containing compounds.[44] Various studies have shown positive effects of nitrogen fertilizer added to host plants of *B. tabaci*.[51–54] In the Sudan Gezira, substantially higher densities of *B. tabaci* occurred consistently over three field seasons in nitrogenous fertilized cotton compared to unfertilized cotton.[55] Collectively, these studies all point to the potential benefit to *B. tabaci* populations that nitrogenous fertilization provided to the Sudan Gezira and subsequently to many other locations worldwide where outbreaks of *B. tabaci* have occurred.

9.3.4   Intensification Impact

The intensification measures that we have described specifically for the Sudan Gezira are representative of changes that have taken place in agriculture worldwide since the 1950s. In addition to the 7.4-fold increase in nitrogenous fertilization globally, irrigated agriculture has increased 1.7-fold since 1961,[19] making possible the growing of crops on demand rather than limiting crop production to particular seasons. Irrigation also enables greater diversification of crops that can be grown serially through multiple seasons, especially in temperate and subtropical climate regions with mild winters. In addition to boosting the agricultural output for a given region, continuous cropping seasons also provide a more constant food resource for polyphagous pests that move from one crop to the next in an unbroken cycle of population growth.[56] Some of the most important global pests including *B. tabaci*, *Helicoverpa armigera*, *H. zea*, *Myzus persicae*, *Frankliniella occidentalis* and numerous other species have taken advantage of diversified cropping systems where they can exploit crops belonging to different plant families. However, even in crops where a major pest species is essentially monophagous on that crop, *e.g.* Colorado potato beetle (*Leptinotarsa decemlineata*) on potato, irrigation enables a constant and more robust growth pattern in the crop that benefits a developing insect. High yielding varieties of the major food crops that were introduced at the beginning of the Green Revolution were well suited to the optimal agronomic conditions provided by fertilization and in some instances irrigation. The breeding that produced the hybrids did not, however, always account for potential susceptibility to particular pests and pathogens. For example, certain hybrid rice varieties proved to be highly susceptible to the brown planthopper, *Nilaparvata lugens* (Stål), a monophagous rice pest that developed into a major pest species synchronously with the Green Revolution.[57]

In some locations such as the Gezira Scheme in Sudan, extensification[12] of agricultural land occurred through development of irrigation infrastructure in dryland regions or by clearing forested and savannah regions in tropical to subtropical latitudes, such as the Amazon basin. For the most part, however, increases in agricultural productivity have occurred due to intensification measures adopted on existing croplands. Consequently, new agricultural lands have increased a relatively modest 1.1-fold since the early 1960s.[58] The greater energy, nutrient and irrigation inputs to existing croplands have substantially raised productivity levels, and by doing so have avoided converting natural habitats and further damaging biodiversity. Another positive effect of agricultural intensification on the environment has been a reduction of greenhouse gases. This is based on an analysis that compared actual global inputs over the past 45 years with two hypothetical "alternative world" scenarios, where either agricultural technology and farm practices remained as they were in 1961 but current living standards were maintained by agricultural extensification, or where 1961 living standards were maintained with production levels again met by extensification to meet the needs of the growing population.[12] The study concluded with an estimate that emissions to the atmosphere of up to 161 gigatons of carbon have been avoided since 1961 by using intensification measures that prevented the conversion of natural habitats to farmland. However, intensification enacted over this same period has not been done without exacting serious environmental costs including soil erosion, eutrophication of freshwater and marine habitats, disruption of food webs and local extinctions by pesticides, and contribution to greenhouse gases that could still be improved tremendously by refining approaches and being more precise with chemical inputs.

While the impact of agricultural intensification can be measured in numerous ways from pollution outfall to crop yields, the effect it has had on insect populations is unknown. In thinking intuitively about how agricultural intensification has affected pest populations, it may be worthwhile to consider agriculture at an earlier time when it existed much more commonly as a subsistence pursuit. The rise of insect pests out of natural habitats to cultivated and relatively more uniform plots of grain, root or vegetable crops represented a transition from the less to the more predictable. Cultivated plantings appeared as a rich new resource for those species that were able to colonize and adapt to sowing and harvesting patterns over time. The concentrated food resource contained within the primitive plots was spatially simplified and more apparent[59] for emerging pests compared to their natural habitat. The scale of such early plots was almost certainly quite limited, at least until agriculture became institutionalized and people began to dedicate larger portions of their lifetime pursuits to growing food. One can imagine an alert farmer recognizing the destruction caused by an invading pest population and spending hours and perhaps days to hand pick feeding larvae from the cultivated plants, an effective strategy at a limited scale. As human populations grew and stratified socially, however, a dedicated force of agriculturists would need to spend much longer periods protecting crop plants as the scale of plantings increased. Maintaining control at larger scales would become more unpredictable, even with a full complement of natural enemies working unhindered from the pesticide applications of the future. An increase in stochastic variability as some function of increasing scale would make it more likely that pest species immigrating into the field could escape in time and space, both from natural enemies and the eyes of early farmers. This kind of uncertainty would have been a normal feature of agriculture through the ages, as would people's less squeamish attitudes regarding the occasional, but inevitable encounter with insects while consuming produced or stored grain. The expectation of perfect produce in today's markets still drives a large portion of pesticide use.[60]

It is not necessary to go back to the dawn of agriculture to imagine how an insect species that had evolved in a natural habitat could make the transition to a habitat altered by agriculture. There are abundant records that describe particular species collected from their native habitat, but that become better known later on for particular crops they infest. For example, Smith and Allen[14] described how the alfalfa caterpillar, *Colias philodice eurytheme* (Boisduval), also known as the orange sulfur butterfly, lived on scattered native legumes in California that was part of its native range as a western nearctic species. In the time period around 1850, its population levels were much lower than the present, reaching its greatest numbers during the spring when its host plants were most abundant, but then retreating in range and numbers to mountainous riparian areas during the summer. The cultivation of alfalfa began in northern California in 1853 or 1854,[61] and spread eastward carrying along the orange sulfur butterfly that is now ubiquitous in North America.[14] It was the scaling up of alfalfa production, however, that so favored the orange sulfur butterfly and turned it into a common pest species. As Smith and Allen[14] noted, the concentrated cultivation of alfalfa favors their reproduction, and "butterflies emerging from one field have a good chance of finding a nearby field which is in the proper stage for egg-laying. Thus a moderate outbreak in one field can result in a severe outbreak in another". The example of the orange sulfur butterfly is not unlike that of a more notorious pest species, the Colorado potato beetle. It was only after potato had been brought under widespread cultivation in the United States that this insect very quickly became an important pest. Using this example, Stern *et al.*[16] described how "man has created conditions that permit certain species to increase their population densities". The rise to pest status of both of these insect species occurred without a pesticide forcer – it was simply the changes that occurred in agriculture from the late 19th to the early 20th century in North America and elsewhere in the world where expansion was taking place.

The above examples really point to the expansion of agriculture behind changes that occurred in pest status and how new, modified environments were created for particular species that could move out of the obscurity of natural habitats, to a wealth of food resources previously unknown. However, when we consider the intensification efforts that slowly began to take place after the Second World War, reaching full swing by 1960, the examples of a corresponding rise in pest status are not as clear as the previous examples because of a more complex environment that developed with intensification. Most prominently, synthetic organic pesticides were now in the mix and being used with little restraint, making them prime suspects whenever pest management failures occurred. There were also new hybrid plant varieties with high yielding attributes, but also potential susceptibility to particular herbivorous arthropods. Along with increases in fertilization, these were all aspects of intensification changes that occurred in modern agriculture and it is therefore difficult to separate their respective contributions to particular pest problems. Our view is that major pest species have been affected in a significant way and that the cause of the effect is due to the various aspects of intensification that have been discussed, and not just to the monolithic influence of pesticides as so often portrayed. If we are to implement greener and more effective pest management strategies going forward, then we must have a clear understanding of pest ecology and recognize the intrinsic potential within pest species. We have seen historically that species such as the orange sulfur butterfly and Colorado potato beetle have become radically more common in alfalfa and potato fields, respectively, prior to the time of pesticides. We therefore need to follow the trajectory to modern agriculture and seriously consider the possibility that an even richer agriculture, through irrigation and fertilization, has benefitted certain pest species and led to greater population growth and increased pest pressure. If this is the case as we presume, then we must be aware of all pest management approaches and how best to implement each one in order to sustain agricultural productivity.

9.4   The Biotic Challenge

Pest pressure can vary tremendously from one agroecosystem to another, even where the same crops and pests are present. For example, outbreak conditions of *B. tabaci* persisted for many years in the lower desert valleys of California, such as the Imperial Valley, but barely rose to the level of occasional pest in the San Joaquin Valley situated 500 km to the northwest. Both regions grow cotton, vegetables and alfalfa, but the San Joaquin Valley accumulates only half the degree-days and many fewer generations of *B. tabaci* compared to the Imperial Valley. In considering pest pressure in different types of major crops, the analysis of potential crop losses to animal pests[19] revealed the least pressure in wheat at 8.7% and the greatest pressure in cotton at 36.8%. Actual losses for the same two crops were 7.9 and 12.3%, respectively, indicating the value of pest management efforts in cotton to be sure.[19] It is probably not surprising that the warm climate grown cotton faces a greater onslaught of arthropod pests than the higher latitudes and cool season grown wheat. The appropriateness of and demand for different pest defense tactics probably vary considerably between these two crops. Cotton at one time was the biggest user of pesticides, but that designation has now shifted to maize for its tremendous usage of herbicides. However, more insecticides continue to be used in cotton than in other major crops.

The rationale for using insecticides to fight infestations may not always or may rarely be justified, depending on one's frame of reference. The rules of IPM are supposed to provide an objective basis for using an insecticide application or not, depending on whether an economic threshold (ET) has been attained.[16] Too often the problem is that neither the ET nor the economic injury level (EIL) have been determined for a crop, nor has a sampling plan for evaluating pest densities been worked out for particular insect pests in a target crop, relying instead on a nominative ET that is determined arbitrarily. IPM then risks becoming more rhetoric[62] than reality and an overdependence on pesticides continues as in the past. This is not to suggest, however, that there isn't a critical need for the right pesticide at the right moment in a pest population's development. Agriculture is replete with pest species that are superbly adapted at colonizing new fields as progression through the seasons and from one annual crop to the next takes place. Colonization is the establishment of a population of a species in a geographical or ecological space not occupied by that species.[63] So for those itinerant species whose life histories often involve a colonizing episode, the absence of population pressures and natural enemies in a new colony with virtually unlimited resources provides exponential growth opportunities. Such colonizers are able to take advantage of their dispersal capabilities to escape natural enemies that have accumulated at one site, by colonizing a new site that is free of conspecifics and natural enemies, at least temporarily. It is this brief period in enemy free space that enables the colonizing species to stay ahead of its natural enemies, at least until intraspecific competition factors begin to slow population growth rates of the pest species.

A synoptic model of insect population dynamics was developed by Southwood and Comins[64] that expertly accounts for differences in life history patterns and how these influence population growth rates of pest species. This model makes use of the concept of r- or K-selected species, or more realistically the r-K-continuum that distributes species according to their life history traits.[65] The terms r and K are derived from the Verhulst logistic growth equation, with r representing the growth rate of the population and K representing the carrying capacity of the local environment. Many agricultural pests possess life history traits consistent with the concept of the r-strategy, including short lifespan, high fecundity and great dispersal ability. It is these r-strategists that are able to disperse to new crops and establish new colonies that initially see rates of growth that approach r_m.[66] One of the key features of the synoptic population model is the "natural enemy ravine", which is deepest at an intermediate position between two stability points: an upper one determined by competition and a lower one determined by natural enemies.[66] Natural enemies are effective only when pest densities are below a certain point in the ravine according to the model. For insect species that normally remain at low to modest numbers, an upset in the balance of control, perhaps in the form of an insecticide treatment, can cause a release of the population to the "epidemic ridge" depicted in the synoptic population model. For the extreme r-strategists that frequently occupy the epidemic ridge, natural enemies may often be of little consequence due to the invasive and colonizing nature of the r-pests.[66] The habitats they occupy, both natural and agricultural, are frequently characterized by their durational instability. Climate seasonality may drive instability in certain natural habitats in which the availability of herbaceous annuals are tied to a rainy season that soon gives way to warming and drying conditions, leaving an abundant plant biomass to wither and recede. For many r-selected insects that exploit annual plants, the consequence of dry-down is a crash in the population that somehow manages to persist thereafter on a dwindling number of hosts until replenishment returns once again with a new rainy season. In an irrigated, continuous cropping system, however, the bust cycle is much less common while the boom cycle predominates. The insect species that evolved to rapidly exploit ephemeral habitats have now been transposed to a modified environment that remains highly unstable, but in a completely different way from the natural habitat, and one that works to the advantage of the r-selected pest. Much the same as the natural habitat, the agro-habitat features ephemeral stands of herbaceous plants that grow rapidly and then decline, but the interval between growth and decline cycles is much shorter, and may actually be non-existent in certain high production agro-ecosystems. Thus, the r-selected pest does not fade away with a decline in food resources as occurs in the natural habitat, but instead disperses from one crop at the end of its growth cycle into an adjacent crop at the beginning of its cycle. The evolutionarily adroit r-selected pest has literally sensed declining conditions in the previous crop including intraspecific crowding, reduced dietary nutrients and water, and/or an accumulation of secondary defensive compounds within the mature crop plants.[67–69] By dispersing to a new field with a fresh young crop, it gains a tremendous fitness advantage by colonizing vigorously growing plants that are uncrowded with conspecifics and on which prey densities have yet to establish, thus improving chances that lower numbers of natural enemies will be present. For a while at least, the new crop represents an unlimited resource upon which population growth rates of the r-selected pest will be maximized.

It is in agro-habitats where a series of nutritionally high-value crops are grown sequentially through an entire annual cycle, and then repeated each year, that some of the greatest potential for insect outbreaks can occur. This was the nature of agriculture in the Sudan Gezira during the outbreaks of *B. tabaci* in the 1960–70s as intensification measures were implemented. Similarly, temporally overlapping vegetable and field crops that are grown year-round characterize agriculture in the irrigated desert agro-ecosystems of the southwestern USA. This region was invaded by the B-biotype of *B. tabaci* in the late 1980s that continued to build numbers that culminated in unprecedented destruction in 1991.[70,71] The clouds of *B. tabaci* adults that arose from cantaloupe fields in late summer (see Figure 9.3A) were an extreme example of the pest pressure that was present throughout the Imperial Valley of California in 1991 and for years thereafter. Of all the irrigated valleys in this region of the USA, the Imperial Valley has perhaps the largest contiguous area of harvestable land at *ca.* 232 000 ha. It is planted to a large diversity of field crops, including cotton and alfalfa, leafy vegetables such as lettuce and broccoli, and various cucurbits. The sequential planting of these crops beginning with curcurbits in the spring, cotton and alfalfa during the summer, and finally the leafy vegetables and second planting of cucurbits in the autumn into winter encourages increasing numbers of whiteflies (see Figure 9.3B), all the way into autumn until much cooler winter temperatures diminish populations.

The agricultural milieu of the Imperial Valley, California was perfectly tailored for an r-selected pest such as *B. tabaci* in terms of the sequence of suitable crops grown and the hot and arid climate that accelerates growth and reproduction and increases the number of generations per year. It wasn't until the B biotype was positively identified as a new biotype of *B. tabaci*[72] that growers in the Imperial Valley and other parts of North America recognized that they were up against a new pest with superior biotic potential compared with the indigenous A biotype. Numerous management changes have since been

Figure 9.3   Outbreaks of *Bemisia tabaci* that occurred in the southwestern USA during the early 1990s including this decimated cantaloupe field (A) in the Imperial Valley, California, in September 1991 (Reproduced with kind permission from Springer). Despite improvements in management, whitefly pressure remained heavy each year, being most intense during the early autumn vegetable season like this broccoli field (B) in September 2005.

implemented and proven highly effective, yet tremendous pest pressure is still exerted each year in the southwestern USA as evidenced by dispersing adults infesting autumn grown crops (see Figure 9.3B). The most influential changes have involved the introduction of highly effective insecticides representing numerous new modes of action,[73] but also cultural changes in terms of the relative areas of crops planted. In particular, cotton and cucurbit acreages have been severely reduced[74] (see Figure 9.4) as Imperial Valley growers recognized that certain structural changes in the agriculture had to be made to contend with *B. tabaci* biotype B. Most growers recognized that the expense and effort

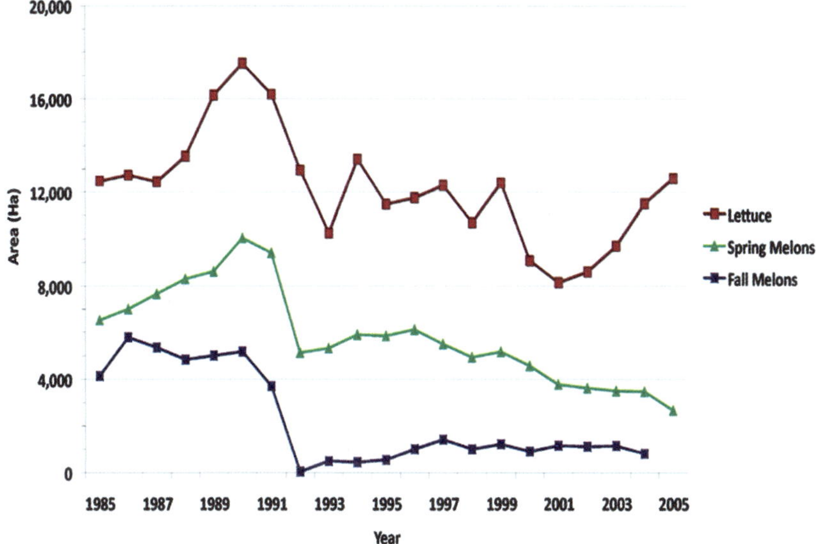

Figure 9.4   A steep decline in area of three favored crops of *Bemisia tabaci* in the Imperial Valley, California (USA) following the destructive outbreak of 1991 (indicated by arrow). Crop areas have remained at reduced levels ever since, due to perennially high population densities of *B. tabaci*.[74]

to fight back a perennial outbreak pest on susceptible crops, like cantaloupes, were too high and the risk too great to continue at the same level prior to the 1991 outbreak.

It was because of pests such as *B. tabaci* and agricultural environments like the Sudan Gezira or the Imperial Valley that Southwood[66] suggested that insecticides would always have an essential role in managing explosive pest species. *B. tabaci* probably experiences at least 13 generations per year in Yuma, Arizona[75] and the same in the Imperial Valley 80 km to the west. The rapid and regular turnover of crops in these regions is ideal for the highly polyphagous *B. tabaci* that is also a very adept disperser among fields. *B. tabaci* effectively rides wind currents between fields, and in a large and contiguous region like the Imperial Valley its chances of finding a suitable host crop are quite good. Its pattern of moving into a new field, often just as plants are breaking through the ground, always puts *B. tabaci* ahead of its natural enemies in terms of population growth and creating a lag period that natural enemies are not able to overcome. It was only through the commercialization of imidacloprid and subsequent neonicotinoid insecticides that retarding *B. tabaci* populations in the Imperial Valley became possible beginning in 1995. Soil-applied treatments of imidacloprid at the time of planting were very effective in suppressing the first couple of generations that normally occurred in the spring cantaloupe crop. This had a tremendous effect on the population dynamics of *B. tabaci* right through the summer and autumn cropping seasons, especially in concert with other modes of action that altogether contributed to the transformation of *B. tabaci* into a reasonably well managed pest.

9.5   Tactics for Defending Against Pests

Some of the tactics used to defend food production from competitor species have been through a technological revolution of their own, while others are as old as agriculture itself. Bioengineered crops have become a powerful new force in pest management, but depending on the situation, the time-tested practice of crop rotation and other cultural practices are no less valuable in terms of minimizing damage caused by pests. All tactics potentially have a role to play in a pest management scenario according to production circumstances and how the cost/benefit relationship plays out. However, the increasing tendency worldwide to practice high production agriculture has led to the adoption of pest defense methods that are expedient and absolute in their action, in turn relying less on strategic cultural maneuvers that can avoid or minimize pest problems over a longer time frame.[55] It was the quick-fix approach that was so subversively appealing back in the early days of the new synthetic pesticides that helped to create the chemical revolution in pest management. The numerous problems associated with an unrestrained chemical approach were soon revealed, however, and eventually led to the adoption of IPM as the pragmatic alternative. Unfortunately IPM has become a catch-all term for all pest management programs, no matter how far they depart from the precepts first laid down in the Integrated Control Concept.[16] The quick-fix mentality still predominates as too often the incentive to implement a complete IPM program is insufficient.[76] Weaning from pesticides in pest management is akin to weaning from petroleum in energy consumption – alternatives exist in both cases, but their implementation requires a different knowledge set and fresh creativity, and usually they are more costly but less definitive in their outcomes.

9.5.1   Cultural Options

So many of the routine operations involved with the cultivation of crops fundamentally affect population dynamics of pests, although the impact that any one procedure or practice has is usually not readily evident. Part of the reason for this is that the scale at which a particular operation is implemented must be considered in the context of the entire landscape that may involve dozens or perhaps hundreds of individual farm sites. The actions taken by a single grower to manage cultural practices at the local level of a single farm may easily be overwhelmed by regional population dynamics and give the appearance of having no effect on pest populations. However, if somehow the contribution of each unit of land to the metapopulation could be measured, clear differences in pest numbers generated among farm sites would likely emerge based on differences in cultural practices alone. Some of these are what might be considered indirect cultural practices, ones that do not explicitly target pest populations but are more customarily considered in the context of their agronomic importance. For example, the amount of nitrogenous fertilizer added to a crop would more normally be considered in terms of what it does for the growth and ultimately the yield of the crop, not how it may affect a colonizing pest population. The underlying research that has been used to set fertilization rates has been performed solely with the intention of identifying the best agronomic practices for producing maximum crop yields. The evidence for negative effects of over fertilization comes from a different set of scientists that are concerned with potential impact on insects in general and pests *per se*. Unfortunately there are few specific crop production guidelines that incorporate findings from both disciplines, or even acknowledge a potential risk of too much fertilizer on a crop from the standpoint of how it may affect a pest population. An analogous situation occurs with frequency of crop irrigation, as the guidelines are predominantly dictated by the agronomic requirements of the crop and not the potential effects on pests, despite recognition of how the water status of plants can affect pest dynamics.[77,78] While there are just enough research findings to suggest that comprehensive improvements could be made that would mutually benefit the goals of crop agronomy and pest management, there is probably insufficient interest in carrying out the demanding field research that could establish new regimens. The environment would be an obvious benefactor of such research if both fertilizer consumption and water use could be reduced without infringing on agricultural productivity.

There are numerous other indirect cultural practices that no doubt influence pest population dynamics, but again are often vague in terms of their effects. Choices that involve planting dates and cultivar selection are most often made with respect to market timing or according to local growing conditions, respectively. However, under the threat of severe pest pressure, growers will sometimes adjust to an earlier or later planting date to try to avoid peak pest populations. Similarly, if supporting information is available, growers may change to a more resistant or tolerant cultivar to better withstand a pest onslaught. This brings up the topic of host plant resistance, sometimes considered under a separate heading within pest management tactics, one that definitely does not fall under the category of indirect cultural practices. Instead, host plant resistance plays a direct role in combating serious pests of major crops such as the Hessian fly, *Mayetiola destructor* (Say), in wheat[79,80] or brown planthopper in rice.[81,82] For these and other highly specialized pests that have a limited host range, identifying resistance genes in germplasm collections for breeding into commercial cultivars is a highly relevant approach towards a greener IPM. Major crops occupy tremendous land area for which any management tactic that can suppress pests without using pesticides would have a significant impact on reducing pesticide consumption. Refraining from pesticide use would also help to avoid resistance to pesticides evolving in target pests, a problem that has made brown planthopper management that much more difficult.[83,84] However, insects also develop resistance to natural plant defenses and both the Hessian fly and the brown planthopper have been notorious for overcoming resistance genes in wheat or rice cultivars. Sixteen Hessian fly biotypes have been identified based on their responses to four resistance genes in wheat.[85] There have also been resistance biotypes to various rice resistance genes that have evolved in brown planthopper in a "gene-for-gene" interaction[86] that occasionally undercut management solutions based on host plant resistance. Currently there are 19 resistance genes in rice to the brown planthopper that have been identified,[82] but modern laboratory methods are improving the mapping of resistance genes and gaining understanding of how their products interact at the molecular level against brown planthopper and other pests and pathogens. Traditional plant breeding for resistance to pests is a long process that requires interdisciplinary collaboration, to first identify resistance traits and then perform the crosses to move the desired traits from one genotype to another without sacrificing other desirable traits. Entomologists are needed to conduct bioassays on target pests to evaluate progress in moving the desired trait, but also to characterize the nature of the resistance mechanism. Plant-based resistance can act against feeding insects on a physiological level, known as "antibiosis", or on a behavioral level, known as "antixenosis". The latter mechanism involves a non-preference response by the herbivore based on whether it accepts or rejects a plant as a feeding host. Although the research challenges of isolating resistance genes and breeding them into crops can be daunting, the effort nevertheless can provide a substantial return on the investment from both monetary and environmental perspectives.[87]

Crop rotation serves many purposes in agriculture and can be extremely important in management of pests that overwinter in the same field or adjacent to it. In most cases, insects that spend winters in diapause will do so as eggs or immature stages in the soil. If they are pests of perennial crops, their emergence from diapause is probably linked to the same environmental clues that break dormancy in plants. With annual crops, however, the rotation of crops among fields each year will act to void the deposit of eggs made at the end of the previous crop season. For crop rotation to be effective, a non-host crop must be planted into the field infested with the overwintering generation of the pest. A classic example of this strategy is the rotation of maize and soybeans as a control tactic for the Western corn rootworm, *Diabrotica virgifera virgifera* (LeConte), and the Northern corn rootworm, *D. barberi* (Smith and Lawrence). Larvae of both species of corn rootworm are obligate feeders of maize roots and must therefore be in close proximity of maize plants when they hatch from eggs. Rotating to soybeans the next season after corn effectively leaves the corn rootworm larvae to starve and eliminates much of the overwintered egg cache. In a perfect example of the adaptability of insects to changing environments, however, a segment of the corn rootworm population began expressing a two-year diapause in the egg stage that effectively skipped the non-host soybean crop before hatching two years later when maize was planted back into the same field.[88] This was not the only adaptation to develop, as a second mechanism appeared in which some portion of adult females began ovipositing in soybean fields at the end of the season, which would then be planted to maize the following spring.[89] These examples of "cultural resistance" are not to be unexpected[90] and point to the necessity to be resourceful and flexible in the application of pest management tactics.

Crop sanitation and early crop destruction following harvest are other cultural practices that can help to limit pest buildup. Managing weeds on the farm and preventing alternative hosts for polyphagous pests can prevent early invasion of a new crop, although immigration from outside the farm can override diligent efforts to maintain a "clean" farm. This illustrates the importance of having good knowledge of what crops are being grown nearby the farm and how release of pests from an earlier crop offsite might influence pest management requirements on the farm. It is the potential for releasing massive numbers of dispersing adults each day following the last harvest that makes early crop destruction an important cultural practice.

Thorough cultural practices on the farm build the foundation for good pest management overall. Much greater emphasis was formerly placed on cultural management out of necessity, prior to the era of synthetic organic insecticides. Greater awareness was required regarding how insect pest problems developed and what could be done culturally to prevent them. There is still a need for these strategies to be implemented in the current era of pest management if progress towards a greener IPM is to be made. A fundamental shift will be required, however, to think about pesticides as a last line of defense rather than as the only defense.

9.5.2 Biocontrol Options

There is perhaps a tendency to think of biological control as a specialized field of entomology, more as an academic pursuit rather than as a tactic to be incorporated into an IPM program. In some respects this is one of the beauties of biocontrol, in that it is already in place and working on behalf of pest management without having to be consciously applied. Unfortunately among non-specialists there is too often a lack of appreciation for the background role that natural enemies play in suppressing pest numbers, or for modifying certain pest management procedures that could augment the existing level of control to help create a greener IPM. It was this biocontrol component of "environmental resistance" that Stern *et al.*[16] was motivated to conserve through development of the integrated control concept. They viewed biocontrol agents as self perpetuating and capable of responding to fluctuations in pest populations. However, they also recognized situations in which environmental resistance, including the biocontrol component, was insufficient to keep pest populations from causing economic damage to the crop. It was only at this point that they could justify an insecticide application to reduce a pest infestation back below the ET level. Even here, however, they saw recourse in the choice of insecticide to be made in terms of selectivity differences among insecticides, and this was back in 1959 when only a few modes of action were recognized. The concern for insecticide selectivity was that an application be the least disruptive to natural enemies so that they could continue to contribute at the highest possible level to environmental resistance against the pest population.

The practice of conservation biological control is one of the three major components of biological control, along with classical and augmentative control. Applications of conservation biocontrol extend from prescribing selective insecticides that are least disruptive of natural enemies to large-scale habitat modification that promotes diversity and food web interactivity. Foundational information in support of a particular application of conservation biocontrol is often lacking, as there are few field-based studies from which a consensus approach can be identified. Variation among types and phenologies of crops and the faunistic composition of agroecosystems makes generalizations difficult about which type of cover or perimeter planting will provide the best refuge for natural enemies. For example, a recent study on various plant mixes associated with a cucurbit host crops identified two out of the four companion plantings as hosting a desirable complement of natural enemies relative to harmful pests.[91] Whether or not the same findings would apply in a different crop, location or time of year is unknown, but is worthy of consideration. The solution may be as simple as performing enough studies under variable circumstances to determine how universally acceptable a particular companion crop may be for a particular pest species. However, it is imperative that such studies be rigorously conducted and critically analyzed, measuring densities not only of various predator and parasitoid species, but also of the major pest species for which the companion crop is intended.[92] The potential for increasing pest densities on a companion crop or in a managed refuge for natural enemies is one of the downsides to the conservation biocontrol approach. This same criticism can be applied at a larger scale where it has been argued that greater plant diversity in an agroecosystem will support ecosystem services that benefit pest control efforts.[93,94] Part of the argument is that with increased services, fewer external inputs are required of agroecosystems that are more ecologically balanced and self-sustaining. What is less evident in such arguments is the level of pest damage or agricultural productivity that is attained in high diversity/low input agroecosystems relative to intensive agriculture. Comparisons have been made that indicate that pest species in general are less abundant in diverse systems compared with simple monocultures,[94] but without indicating respective damage levels in the contrasting environments. One of the risks of reduced productivity is that conversion of additional natural habitats to agricultural land would be required to meet food demands.[12] While decreases in chemical and energy inputs to agriculture might be gained, an increased conversion of natural habitat would threaten to reduce biodiversity.

The inadvertent relocation of insects from their indigenous geographic ranges to new ranges through human travel and commerce continues to be a major challenge to agriculture. Invasive species have caused havoc to crops the world over, arriving into new habitats that are often deficient in co-evolved biocontrol agents characteristic of their home range. Classical biological control is the practice of importing natural enemies found in the indigenous range of the invasive pest and releasing into its new environment. There are many examples where the imported natural enemy has rapidly established in the new territory and driven populations of the pest species down to low, stable equilibrium points, the most classic case being the cottony cushion scale, *Icerya purchasi* (Maskell). The return on the investment for one person to travel to Australia from California, search and locate the vedalia beetle, *Rodolia cardinalis* (Mulsant), then return with a handful of specimens for rearing and release has been tremendous. The outstanding benefit that was gained in the cottony-cushion scale example applies to other classical biological control examples as well,[95] but there are many other cases where imported natural enemies have not made a substantial impact on the target pest population. Moreover, controversy has arisen in recent years about the possibility that classical biological control approaches may, in certain cases, have been harmful to ecosystems.[96,97] The problem is mainly one of spillover of the introduced predator or parasitoid onto species other than the intended target, especially when native species are attacked. Food web dynamics are potentially altered to where decline in certain affected species may occur. Regulations have since been tightened to require impact studies that might avert attack of non-target species and minimize ecological destabilization. It is a valid concern to be sure, but one that must be considered in the context of already altered ecosystems that have been optimized for the mass production of a select group of plant species. In the meantime, classical biological control remains an extremely important approach for combating invasive pests and one that fits well with the goals of a greener IPM.

The commercial mass rearing of beneficial insects for release into both open and closed agricultural systems has turned augmentative biocontrol into a valuable enterprise in pest management. Numerous species of coccinellid beetles, parasitic wasps, predaceous mites and other predators are raised in insectaries and then express shipped to sites of request. Inoculative releases are generally made early in a crop cycle to bolster natural defenses in an open field or to create a line of defense against invading pests into a closed system, *e.g.* greenhouse, whereas inundative releases represent a much greater release rate to stop or slow a pest population approaching a damaging level. In California citrus, augmentative releases of parasitic wasps, *Aphytis melinus* and *Comperiella bifasciata*, have been a critical and long-established component of an IPM program to combat California red scale infestations. Incremental releases of up to 100 000 are recommended in California's San Joaquin Valley.[98] The rapid availability of natural enemies has greatly improved prospects for biologically based management of key pest species, especially whiteflies, aphids, thrips, spider mites and leafminers that tend to be chronic problems in closed systems.[99] The tolerance levels for pest damage tend to be extremely low for vegetables or ornamental plants grown in the greenhouse and often necessitate a pesticide application.[100] However, the capacity to incorporate commercially supplied natural enemies as part of an IPM program for greenhouse pest management has reduced dependency on pesticides and helped produce a greener harvest.

9.5.3 Pesticide Options

For all of their negative connotations, pesticides remain an indispensable part of pest management. Pesticide modes of action have continued to expand despite perennial concerns expressed as early as the 1970s about the finite number of active ingredients remaining to be discovered and the tremendous cost of development of each new insecticide.[66] Nonetheless, whole new insecticide classes have been discovered and commercialized since that time, as insecticides in general have become safer and more effective.[101] The resulting diversification of insecticides has provided the knowledgeable pest manager with many more treatment options than were formerly available with broad-spectrum modes of action. It was not a matter that too few products were available in the early decades of synthetic organic insecticides, but that the modes of action were limited to just two or three target sites in insects. This limitation must have resulted in greater selection pressure being exerted compared with the much larger selection of modes of action that are available in today's pest management. Indeed resistance to insecticides quickly rose to become a major problem in pest management and remains a serious concern as always. With greater awareness of the problem, however, and a much wider selection of products representing many times more modes of action than in the past, the outlook for managing resistance is much brighter.

Pesticides are powerful agents of destruction that are intended for specific pest species but often have much broader impact on non-pest organisms as well. This is a principal reason why they need to be used with the greatest care and restraint, to avoid as much as possible destroying non-target organisms that contribute to a stable equilibrium among pest and non-pest species. In addition to the obvious beneficial insects that include predators and parasitoids of the pest species of concern, non-target also designates all those other arthropods in a crop that generally do not merit concern from a plant injury standpoint, but whose presence may contribute to food web complexity and encourage greater biological control. Insecticides that are physiologically more selective in activity towards a target pest are ones that express higher relative toxicity to the pest species than to non-target species. This is possible because molecular receptors at the target sites where the insecticide molecules interact vary in receptivity according to species level differences or at higher taxonomic categories. Thus, activity spectra among insecticides vary tremendously with those affecting a broader range of vertebrate and invertebrate animal species being considered "broad-spectrum" insecticides. From an environmental and public health perspective, narrow spectrum or higher selectivity insecticides are those that are most toxic to insects only and are relatively non-toxic to vertebrate animals including humans. However, insecticide selectivity extends beyond the vertebrate/invertebrate divide to where particular taxonomic categories of insects are either more or less susceptible to an insecticide. From an IPM perspective, selective insecticides that are more toxic to pest species than to non-target species are ones that are good candidates for integrating into biologically intensive IPM. The overall selectivity of an insecticide treatment can be improved further by timing an application to when a target pest is more exposed and vulnerable, or possibly by reducing the dose that is still effective against the pest without decimating non-target organisms. These latter approaches fall into the category of ecological selectivity and should always be considered prior to an insecticide application so as to minimize exposure to non-target organisms.

Using insecticides that have systemic mobility in plants can further enhance ecological selectivity. While a number of organophosphate and carbamate insecticides have systemic properties, they are more commonly used in spray formulations with the exception of aldicarb, a highly toxic compound that is applied as granules to the soil for plant uptake. Imidacloprid is a much more widely used systemic insecticide that was commercially introduced in the early 1990s. It was the first member of the neonicotinoid group of insecticides[102] that has since expanded to include additional compounds that also can be soil-applied for systemic uptake. The advantage of soil application from an IPM standpoint is that there is no direct contact by foraging beneficial insects with an insecticide applied as a spray to the crop or as a residue after application. Imidacloprid is often applied to the soil ahead of the seed drop at time of planting, especially in situations where there is chronic pest pressure and no uncertainty about whether the pests will invade the crop. Herbivorous insects that feed on treated plants should be the only mechanism by which exposure occurs. In reality, however, the translocation of imidacloprid from the roots through xylem vessels to all parts of the plant results in contamination of nectar and pollen. Honeybees are thus vulnerable along with other nectar and pollen feeders. It is now also known that certain weather and soil moisture conditions that are conducive to guttation occurring in a plant can result in residues of imidacloprid being deposited on leaf surfaces. Despite these drawbacks, soil-applied imidacloprid is much more compatible with biocontrol agents than most foliar insecticides simply because the exposure levels are much lower.

There are still newer groups of insecticides than the neonicotinoids that also exhibit systemicity within plants. One of these is the Insecticide Resistance Action Committee (IRAC)[102] group-28 diamide insecticides that have good activity against lepidopteran pests that also helps fill an important gap in the pest spectrum the neonicotinoids do not cover. A second new insecticide group is the tetronic and tetramic acid derivatives (group 23) from which the compound spirotetramat is both xylem and phloem mobile within plants (see Figure 9.5). This compound is unique in that it is applied to foliage rather than to the soil, yet is translaminar in the leaf where it moves into phloem tissue, translocated to the roots and then back up the plant *via* the xylem. Foliar treatments of spirotetramat defeat the limited exposure concept of soil-applied systemic compounds, but nevertheless have little contact toxicity due to their mode of action that inhibits lipid synthesis and growth regulation.

In addition to conserving natural enemies, a second principal reason for the restrained use of pesticides is to conserve active ingredients by protecting them from resistance. The capacity to diversify insecticide treatments has increased dramatically over the past two decades. For example, at the time of the *B. tabaci* outbreak in the Imperial Valley in 1991 (see Figure 9.3A), only three modes of action were available that included pyrethroids, cyclodiene organochlorines, and the organophosphates and carbamates that share the same target site. The number of modes of action registered for use on crops against

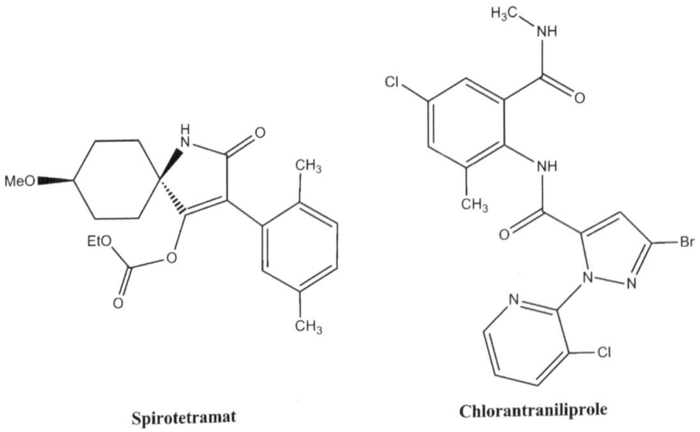

Spirotetramat

Chlorantraniliprole

Figure 9.5 Structures of spirotetramat (Insecticide Resistance Action Committee's group 23: tetronic and tetramic acid derivatives) and chlorantraniliprole (IRAC group 28: diamides).

B. tabaci in the USA is now ten, more than triple the number from two decades ago. These newer products have been indispensable to restoring control of *B. tabaci* in regions where population outbreaks were exacerbated by resistance. Not only are the new products more effective control agents at reduced rates compared to conventional insecticides, they are much less destructive to natural enemies. Many of the diverse modes of action have highly specific activity against certain stages of development and therefore must be used in the proper context according to circumstances in the field. This requires a high level of understanding on the part of pest managers to be able to make correct treatment decisions. Good stewardship of these valuable products requires that they be used knowledgeably and with restraint to ensure food security and sustainability.

9.5.4 Biotechnology Options

The development of biotech crops represents a revolutionary change in pest management, but one that has been restrained by societal concerns from becoming an even larger part of the pest management repertoire. Nevertheless, the trend towards increasing acceptance of biotech crops appears to be accelerating with their adoption having occurred on all continents with the exception of Antarctica. As of 2008, biotech crops were planted in 25 countries on a total of 125 million ha, a 9.4% increase over the previous year.[103] More impressively, when the area planted to biotech crops is considered in terms of acres, 2008 marked the first time that biotech crops exceeded 2 billion acres. The time required to attain the first billion acres was 10 years, from 1996 till 2005, but then only 3 years (2006–2008) were required to accumulate the second billion acres.[103] Herbicide-tolerant biotech crops accounted for 63% (79 million ha) of the total area in 2008 compared to only 15% (19.1 million ha) of insect-resistant crops. However, another 22% (26.9 million ha) consisted of stacked double or triple traits that included at least one insect-resistant trait.[103] There are currently several new biotech crop products under development that feature multiple traits including insect-resistant, herbicide-tolerant and other agronomic traits that are expected to become dominate in the near future.[103]

All insect-resistant biotech crops commercialized thus far have drawn from the family of *cry* genes isolated from the soil bacterium *Bacillus thuringiensis*, commonly known as Bt in crop protection (see Chapter 8). The first insect-resistant biotech crops were transformed by splicing into the crop plant genome just a single *cry* gene, *e.g. cry1Ac* in cotton. The translated Cry proteins from their respective *cry* genes vary in toxicity against several lepidopteran and coleopteran pest species. By pyramiding two or more *cry* genes into a plant, it has been possible to expand the pest spectrum to which a particular biotech cultivar expresses resistance. For example, the original Bollgard® cotton expresses Cry1Ac protein that is highly active against lepidopterans that feed on cotton bolls, such as *Heliothis virescens* and *Pectinophora gossypiella*, and reasonably effective against *Helicoverpa armigera* and *H. zea*.[104] By pyramiding the *cry2Ab* gene along with *cry1Ac*, Bollgard II® was created that offers much improved protection against *H. armigera* and *H. zea*, while retaining outstanding activity against *H. virescens* and *P. gossypiella*. Another example of pyramided genes includes the *cry1F* and *cry1Ac* genes in cotton, which adds protection against *Spodoptera* spp. while retaining efficacy against bollworms. In addition to increasing the breadth of protection against pest species, pyramided cultivars putatively decrease the likelihood of resistance by raising the toxicity to a higher level for a resistance mechanism to overcome. This approach ties into the high dose/refuge strategy of resistance management that posits an inverse relationship between the expression of a toxin and the frequency of a genotype able to survive. The theory behind the refuge part of the strategy is that if, or when, a resistant genotype appears in a population, a surplus of susceptible genotypes maintained in an untreated refuge will greatly outnumber resistant genotypes. The probability of a cross occurring between the rare resistant and abundant susceptible genotypes to produce a heterozygote at the resistance locus will be increased in the presence of a susceptible refuge. Assuming that the inheritance of the resistant allele is recessive, the heterozygote will not survive a high dose with any more likelihood than a fully susceptible genotype. After 15 years of deployment of insect-resistant biotech crops, there have been very few signs of resistance development in insect populations, either in countries with mandated resistance management programs or ones without. Recently, however, populations of *H. zea* in the southeastern USA have shown increased tolerance to Cry1Ac and Cry2Ab in Bt cotton in laboratory bioassays, but without any apparent loss of field efficacy.[105]

The accumulated data since the first deployment of insect-resistant biotech crops present a very positive picture regarding pest management, agronomic and environmental benefits. Significant global declines in insecticide use have been recorded in insect-resistant maize (−35.3%) and cotton (−21.9%) during the period 1996–2008.[106] Other evaluations suggest an even larger reduction in insecticide use in individual countries, *e.g.* 50 and 65% reductions in India and China, respectively.[107] In China and perhaps elsewhere, the impact of insect-resistant crops has been even greater as decreased insecticide use has been measured in other crops (corn, peanuts, soybeans, vegetables) across regions where Bt cotton is grown and *H. armigera* is a primary pest.[108] This is an example where Bt cotton has served as a trap crop[109] for all other crops in the region that also faced infestation by the polyphagous *H. armigera*, effectively becoming a dead-end for any mated female that deposited eggs in the Bt cotton. In China and elsewhere, farmers have benefitted not only from reduced insecticide costs, but also by gains in yield and gross profit margins. This is believed to be a major reason why the rate of adoption of biotech insect-resistant crops has been so rapid, especially in developing countries.[107]

There are a number of metrics that have been developed to estimate the environmental impact that has occurred through adoption of biotech crops. Consider for a minute the amount of fossil fuel required and the emissions generated to fly crop-duster airplanes over fields or to drive tractors up and down rows to spray pesticides. Then there is the release of volatile organic compounds into the atmosphere, the runoff into aquatic systems and contamination of ground water, the toxicity to birds and fish and mammals, health effects on humans, *etc.*, all from the millions of tons of pesticides applied to the earth each year. Researchers at Cornell University developed an environmental impact quotient (EIQ) that incorporates three principal components that include farm worker, consumer and ecological components, each one consisting of multiple rating factors.[110] An EIQ equation is then used to distill all of the factors for each pesticide into a single EIQ value. In a counterfactual analysis using data obtained from conventional crops, Brookes and Barfoote[106] estimated the pesticide applications that were not made as a result of cultivating all biotech crops, herbicide tolerant and insect resistant. They determined that a 16.3% change in reduced environmental impact has occurred globally from 1996–2008 as a result of biotech crops. The two biggest changes occurred in insect-resistant maize (–29.4%) and insect-resistant cotton (–24.8%). There have also been enormous savings in fuel consumption and carbon dioxide (CO_2) emissions from the cultivation of every one of the biotech crops.[106] For example, the fuel savings in insect-resistant cotton (assuming four tractor passes per ha, 1.045 l ha^{-1} fuel consumed per pass) over the 1996–2008 period amounts to 124.99 million liters, with 343.75 million kilograms of CO_2 emissions also saved.[106]

The most apparent negative effect from growing insect-resistant biotech cotton has been an increase in various hemipteran species, most probably due to a decline in insecticide use. With the bollworms and other foliage feeding Lepidoptera kept in check by the Bt proteins expressed in the insect-resistant biotech cotton plants, plant bugs and stinkbugs have become more of a focus of pest management based on a slight increase in insecticides targeting these pests.[111] In northern China, mirid bugs have increased progressively in conjunction with Bt cotton adoption. Cotton under conventional management acts as a sink for mirid bugs, but the decrease in insecticide sprays in Bt cotton has now made it a source.[112]

9.5.5 Push-Pull Options

The "push-pull" strategy is a means by which behavior-modifying stimuli can be used to manipulate pest and/or beneficial insect populations in a way that improves pest management.[113] An alternative terminology is "stimulo-deterrent diversion" or SDD that was developed independently[114] but with overlap to the approaches described originally in the push-pull strategy.[115,116] The terms "stimulo" and "deterrent" are a bit more intuitive in the sense of communicating activities that promote or depress activities such as feeding and oviposition, yet "push-pull" appears to be the more prevalent terminology. Whatever the case, both terminologies employ a combination of approaches for drawing away harmful insects, or alternatively drawing in beneficial insects to promote biological control. Semiochemicals play a major role in manipulating the behaviors of beneficial and destructive insects and can be used in both push and pull applications. One of the major advantages of using semiochemicals in IPM is that they work by non-toxic modes of action and are often naturally derived, thus making at a minimum those approaches that rely upon habitat manipulation compatible with low-input farming.[117] A few example applications of semiochemicals to "push" away from the crop include: (1) masking of host attraction; (2) repellents, antifeedents and oviposition deterrents; and (3) attractants for predators and parasitoids.[116] On the other side of the equation are examples to "pull" into traps or trap crops and include: (1) host attractants; (2) aggregation, sex and oviposition pheromones; and (3) visual cues.[118] Of these many and diverse approaches, pheromones have been used with the most success and to the greatest degree in IPM. Pheromones have been incorporated into IPM programs in various ways including: monitoring established populations, detection and surveys for invasive species, mass trapping, and mating disruption.[119] The incorporation of synthetic female sex pheromones in mating disruption programs for the codling moth, *Cydia pomonella* (L.), and oriental fruit moth, *Grapholita molesta* (Busck), have become a major part of IPM programs in fruit orchards in the western United States.[120] This process has been aided by pressure from regulatory agencies to reduce organophosphate insecticide use. Despite the effectiveness of pheromones in mating disruption of various pests in different cropping systems, the wider adoption of pheromone applications is frequently inhibited by their higher costs compared with chemical control.[119]

Habitat manipulation is another means of employing push-pull strategies to manage pests and in a certain sense can be viewed as a form of cultural control. Greater diversity of plant types in cropping systems using intercropping techniques and other methods often enhances food web complexity and improves biological control.[93,94] Another benefit of diverse plant stands may be an increase in plant volatiles that act to push and/or pull various pest and beneficial species. The influence of plant volatiles on insect/plant interactions has only recently begun to be explored but may result in the identification of plant semiochemicals that enhance pest control when deployed effectively. One prominent example has been the role that methyl salicylate plays in attracting various predator species to plants under attack by different pest species, including soybean aphid, *Aphis glycines* (Matsumura),[121] and the two-spotted spider mite, *Tetranychus urticae* (Koch). Various synthetic plant volatile compounds, including methyl salicylate, are the subject of field research in a number of crops to evaluate whether predatory insect species can be drawn into fields in sufficient numbers to effectively suppress pest populations.[122] While synthetic formulations of semiochemicals hold promise in protecting crops from pests, low input farmers in developing countries could also benefit through cultural push-pull approaches that utilize multi-cropping systems[123] or wild hosts, as has been demonstrated for cereal stem borers in Kenya.[124,125] Such information could be of great value to low input farmers that have to rely upon their wits to grow crops in a manner that avoids pest infestations through various methods of habitat management.

Trap crops can be highly effective at pulling in a pest species in the protection of a main crop. Herbivorous insects respond differentially to host plants of various types as food sources and when presented with a choice will settle on a preferred host. The selection of an appropriate trap crop has to come from the range of hosts that an herbivore uses and ideally be significantly more attractive than the main crop. If there is not a substantial differential in attractiveness, then there is the possibility of using a semiochemical attractant to pull to the trap crop or a deterrent to push away from the main crop. However, the agronomic logistics must also be considered in terms of the feasibility of cultivating two different crops simultaneously and in close proximity to one another. These are tricky issues to work out, and in fact many of the push-pull possibilities that work on paper are not executable in the field because of costs and logistical issues. It is the pragmatic factor that is crucial in modern agriculture, with only so much leeway that can be expected from a grower that must meet his/her commitments like any other business.

9.6   An IPM Synthesis

The practice of IPM over the years has generated substantial criticism for its continued heavy reliance on pesticides to curtail pest infestations. New paradigms have been proposed in an effort to correct the persistent skew towards chemical control. For example, "biologically intensive IPM" that relies on host-plant resistance, biological control and cultural control was introduced as an alternative to chemically intensive IPM.[126] A similar emphasis was expressed in "ecologically based pest management" that emerged from a committee convened by the National Research Council's Board of Agriculture, but one that was criticized[21] for having essentially the same goals of IPM, but with a loftily stated focus on safety, cost-effectiveness, sustainability and ecosystems. There have been other elaborations of IPM, *e.g.* "IPM implementation at three levels of integration"[21] that really only obfuscate the simple elegance contained in the principles of the integrated control concept.[16] These principles have endured at the very core of IPM because they provide a robust methodology for determining if and when a pesticide application should be made. It was the solid linking of this methodology to the economic bottom line of the grower that makes the integrated control concept so powerful, so easily adoptable. It also included, however, the compelling argument regarding the potential of environmental resistance to limit pest populations, and how efforts should be made to preserve biological control and not upset it. It is these underlying values of the integrated control concept, and by extension IPM, that get lost in the more familiar and oft-quoted instructions to apply pesticide when the ET has been reached, which if practiced in the context of the whole of the integrated control concept is laudable. Where it too often comes up short, however, is that the ET has almost certainly been arbitrarily and comfortably defined on the side of safe rather than sorry.

There are very few examples of IPM programs that are true to the fundamentals of IPM that include the empirical determination of ET and EIL, the development of a statistically based sampling program, or the incorporation of selective insecticides to better conserve natural enemies. It can require many years and much effort to develop each one of these components to a level where they can be brought together into a cohesive and complete IPM program. An outstanding example of tying all components together into a comprehensive IPM program is one that was developed in Arizona for control of *B. tabaci* in cotton. Research had begun in the early 1990s at a time when perennial outbreaks of *B. tabaci* were severely disrupting agricultural production in the southwestern USA. Early studies were concerned with identifying spatial distributions of *B. tabaci* on a cotton plant to identify a representative leaf position to serve as a sample unit.[127–129] In addition, insect growth regulator (IGR) insecticides not yet registered for commercial use were being evaluated for their efficacies and selectivity towards natural enemies.[130,131] Action thresholds were determined empirically[132,133] and refined over time to configure to different treatment regimens. The compelling event that brought it all together was a resistance crisis that occurred during the 1995 cotton season.[134] Putting together an IPM program in time for the 1996 season required pushing through emergency registrations on the two IGRs, buprofezin and pyriproxyfen (see Figure 9.6), to the U.S. Environmental Protection Agency, and then training growers in workshops conducted around the state on how to most effectively

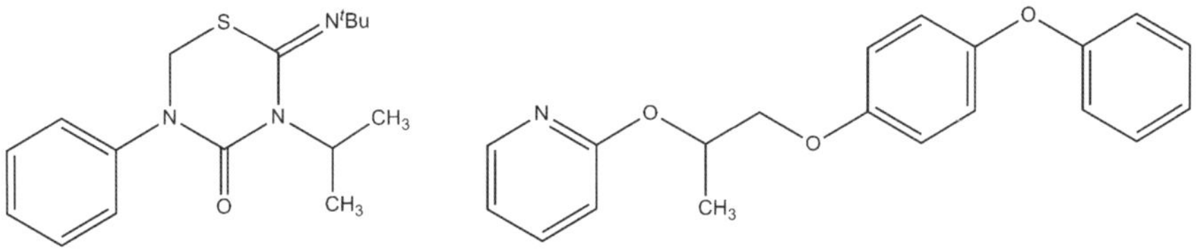

Figure 9.6 Structures of buprofezin (IRAC group 16: Inhibitors of chitin biosynthesis, type 1) and pyriproxyfen (IRAC group 7C: juvenile hormone mimics).

use the IGRs. The training was required as a condition of the emergency registration and provided the opportunity to familiarize growers with the different elements of the IPM program including how to sample for whiteflies and how to know when an economic threshold (ET) had been reached.

Implementation of the new IPM program beginning in 1996 produced almost immediate results, as whitefly populations were kept under control with fewer insecticide treatments. This began a pattern of decreasing insecticide use each year until a low point was reached in 2001, after which other new insecticides became available that tended to disrupt the integrity of the IGR-based IPM program. In the meantime, further research that examined differences in fields where conventional broad-spectrum insecticides were sprayed to those that were treated with the IGR insecticides began to reveal a crucial component of the program's success. Life table studies performed in the field were able to partition and identify the different sources of whitefly mortality in the conventional *versus* IGR-treated fields. The findings that were most evident and consistent through all comparisons were that treatments with either IGR supported more natural enemies than the conventional treatments, and that a single application of a broad spectrum insecticide could significantly affect the beneficial arthropod community for up to 7 weeks.[135] In contrast, predator : prey ratios in the IGR-treated cotton recovered relatively quickly following applications, as the beneficial insects continued to supply a significantly larger component of mortality relative to the conventional-treated cotton. This phenomenon of a decline in immature whitefly numbers soon after a treatment with an IGR, but then persisting at depressed levels beyond the time when a chemical residual would still be effective, gave rise to the concept of bioresidual activity as the beneficial insects rapidly re-colonized and continued to suppress whitefly numbers. The bioresidual finding is a powerful example of how chemical and biological controls can be integrated to produce a stable and sustainable IPM. Since 1996, a 70% reduction in foliar insecticides has been realized along with a > $200 million saving in control costs.[135]

The whitefly IPM program in Arizona demonstrates that when the core principles of IPM are observed and put into practice, a reduction in insecticide use and an increase in natural enemy effectiveness are not only possible for a single season, but can become a stable and persistent outcome. Another important change that occurred in Arizona cotton was the large-scale adoption of Bt cotton for control of the pink bollworm that began in 1996 and increased to 60–80% of total cotton acreage over the next decade.[135] The reduction in insecticide applications against the pink bollworm also benefitted the natural enemy populations in cotton, possibly advancing the breadth of biocontrol in conjunction with the IGRs used against whiteflies. Perhaps the most impressive aspect of this IPM success story is that we have a thorough understanding of why it has been successful. The research that first went towards the development of the IPM methods, and then into the determination of what was working and why, is both exceptional and exemplary of what can be achieved in pest management. Further understanding and knowledge like that gained from the

Arizona experience will be needed to support sustainable intensification[3] and increase food production to the levels required to meet the growing human demand.

These lessons may also prove valuable in the fight against insects that serve as vectors of infectious agents that cause diseases such as malaria in humans. Managing mosquitoes and other disease vectors is a challenging endeavor due to the involvement of aquatic habitats that are often intricate parts of natural habitats. Maintaining eco-balance to encourage natural control of both immature and adult stages is of obvious importance for avoiding vector outbreaks and spillover to populated areas. The use of IPM methodologies against vector species is more realistic in populated areas, where avoidance measures must be emphasized to curtail breeding sites for mosquitoes across the landscape. However, the much more dispersed nature of micro-breeding habitats for mosquitoes is a significant departure from the concentrated nature of crop pests that can be easily targeted using biological or chemical control tactics. Vector control agencies play vital roles in managing mosquitoes in aquatic habitats that are dispersed through rural and urban settings, but also rely on an educated public to contribute to the management effort by preventing ephemeral breeding habitats from forming during rainfall events. As with crop protection, effective management of vector insects requires an informed workforce to be aware of the pest and disease potential and to take appropriate action when required. More than just the farmer that faces the consequences of insufficient pest management in crops, every person within endemic disease zones potentially pays the price for inadequate vector control.

References

1. Population Reference Bureau, http://www.prb.org/Publications/GraphicsBank/PopulationTrends.aspx (last accessed August 10th, 2010).
2. P. R. Erhlich, *The Population Bomb,* Ballantine, New York, USA, 1968.
3. Royal Society of London, *Reaping the Benefits: Science and the Sustainable Intensification of Global Agriculture*, Royal Society, London, UK, 2009.
4. P. Hazell and S. Wood, *Philos. Trans. R. Soc. London,Ser. B*, 2008, **363**, 495.
5. D. G. Johnson, *Proc. Natl. Acad. Sci. U. S. A.*, 1999, **96**, 5915.
6. FAO, *The State of Food Insecurity in the World,* http://www.fao.org/docrep/012/i0876e/i0876e00.htm (last accessed August 2nd, 2010).
7. T. J. A. Bruce, *Food Security,* 2010, **2**, 133.
8. M. S. Hovmøller, S. Walter and A. F. Justesen, *Science*, 2010, **329**, 369.
9. E. Stokstad, *Science*, 2009, **324**, 710.
10. V. Smil, *Feeding the World: A Challenge for the Twenty-First Century*, MIT Press, MA, USA, 2000.
11. D. R. Strong, J. H. Lawton, T. R. E. Southwood, *Insects on Plants*, Harvard Press, MA, USA, 1984.

12. J. A. Burney, S. J. Davis and D. B. Lobell, *Proc. Nat. Acad. Sci. U. S. A.*, 2010, **107**, 12052.
13. P. DeBach and B. Bartlett, *J. Economic Entomol.*, 1951, **44**, 372.
14. R. F. Smith and W. W. Allen, *Sci. Am.*, 1954, **190**, 38.
15. W. E. Ripper, *Annu. Rev. Entomol.*, 1956, **1**, 403.
16. V. M. Stern, R. F. Smith, R. van den Bosch and K. S. Hagen, *Hilgardia*, 1959, **29**, 81.
17. R. Carson, *Silent Spring*, Fawcett Publications, Greenwich, CT, USA, 1962.
18. R. Paarlberg, *Society*, 2009, **46**, 4.
19. E.-C. Oerke, *J. Agric. Sci.*, 2006, **144**, 31.
20. R. van den Bosch, *The Pesticide Conspiracy,*, University of California Press, CA, USA, 1978.
21. M. Kogan, *Annu. Rev. Entomol.*, 1998, **43**, 243.
22. R. F. Smith, *Proc. Accad. Nazion. Lincei*, 1969, **366**, 21.
23. P. L. Adkisson, *Connecticut Agric. Exp. Stn. Bull.*, 1969, **708**, 155.
24. R. F. Luck, R. van den Bosch and R. Garcia, *Bioscience*, 1977, **27**, 606.
25. R. L. Doutt and R. F. Smith, in *Biological Control*, ed., C. B. Huffaker, Plenum, New York, USA, 1971, p. 3.
26. R. F. Smith and H. T. Reynolds in *The Careless Technology. Ecology and International Development*, ed. M. T. Farvar and J. P. Milton, The Natural History Press, New York, USA, 1972, p. 373.
27. P. DeBach, *Biological Control by Natural Enemies*, Cambridge University Press, London, UK, 1974.
28. D. G. Bottrell and P. L. Adkisson, *Annu. Rev. Entomol.*, 1977, **22**, 451.
29. K. G. Eveleens, *Crop Protect.*, 1983, **2**, 273.
30. D. N. Byrne, T. S. Bellows Jr. and M. P. Parrella, in *Whiteflies: their Bionomics, Pest Status and Management,* ed. D. Gerling, Intercept, Andover, UK, 1990, p. 227.
31. V. Dittrich, S. C. Hassan and G. H. Ernst, *Crop Protect.*, 1985, **4**, 161.
32. M. W. Johnson, N. C. Toscano, H. T. Reynolds, E. S. Sylvester, K. Kido and E. T. Natwick, *California Agric.*, 1982, **36**, 24.
33. M. T. Farvar and J. P. Milton, *The Careless Technology. Ecology and International Development*, The Natural History Press, New York, USA, 1972.
34. J. D. Tothill, *Agriculture in the Sudan*, Oxford University Press, Oxford, UK, 1948.
35. M. F. Husain and K. K. Trehan, *Indian J. Agric. Sci.*, 1933, **3**, 701.
36. Z. Avidov, *Ktavim* (English edition), 1956, **7**, 25.
37. W. Roberts, *Empire Cotton Grow. Rev.*, 1930, **7**, 181.
38. W. T. Griffiths, MSc thesis, University of London, 1984.
39. H. R. J. Davies, *The Agriculture of the Sudan,* ed. G. M. Craig, Oxford University Press, Oxford, UK, 1991, p. 339.
40. V. Smil, *Nature*, 1999, **400**, 415.
41. R. H. Socolow, *Proc. Nat. Acad. Sci. U. S. A.*, 1999, **96**, 6001.
42. V. Smil, *Enriching the Earth*, MIT Press, MA, USA, 2001.

43. FAOSTAT, http://faostat.fao.org/default.aspx (last accessed August 16th, 2010).

44. L. M. Schoonhoven, T. Jermy and J. J. A. van Loon, *Insect–Plant Biology*, Chapman & Hall, London, UK, 1998.

45. S. McNeill and T. R. E Southwood in *Biochemical Aspects of Plant and Animal Coevolution*, ed. J. B. Harborne, Academic Press, London, UK, 1978, p. 77.

46. J. M. Scriber and F. Slansky Jr., *Annu. Rev. Entomol.*, 1981, **26**, 183.

47. C. P. Onuf, in *The Ecology of Arboreal Folivores*, ed. G. G. Montgomery Smithsonian Institution, Washington, D. C., USA, 1978, p. 85.

48. R. A. Prestidge, *J. Appl. Ecol.*, 1982, **19**, 735.

49. E. A. Heinrichs and F. G. Medrano, *Int. Rice Res. Newsletter*, 1985, **10**, 20.

50. R. K. Jansson, G. L. Leibee, C. A. Sanchez and S. H. Lecrone, *Entomol. Exp. Appl.*, 1991, **61**, 7.

51. M. J. Blua and N. C. Toscano, *Environ. Entomol.*, 1994, **23**, 316.

52. J. Bentz, J. Reeves, P. Barbosa and B. Francis, *Environ. Entomol.*, 1995, **24**, 271.

53. J. L. Bi, G. R. Ballmer, D. L. Hendrix, T. J. Henneberry and N. C. Toscano, *Entomol. Exp. Appl.*, 2001, **99**, 25.

54. J. L. Bi, D. M. Lin, K. S. Lii and N. C. Toscano, *Insect Sci.*, 2005, **12**, 31.

55. J. E. Jackson, H. O. Burhan and H. M. Hassan, *J. Agric. Sci.*, 1973, **81**, 491.

56. J. R. Bradley Jr., in *Evolution of Insect Pests*, ed. K. C. Kim and B. A. McPheron, John Wiley & Sons Inc., New York, USA, 1993, p. 375.

57. K. Kiritani, *Annu. Rev. Entomol.*, 1979, **24**, 279.

58. D. Tilman, *Proc. Natl. Acad. Sci. U. S. A.*, 1999, **96**, 5995.

59. P. Feeny, *Recent Adv. Phytochem.*, 1976, **10**, 1.

60. J. C. Palumbo and S. J. Castle, *Pest Manag. Sci.*, 2009, **65**, 1311.

61. F. D. Coburn, *The Book of Alfalfa*, Orange Judd Co., New York, USA, 1907.

62. V. W. Ruttan, *Proc. Natl. Acad. Sci. U. S. A.*, 1999, **96**, 5960.

63. R. C. Lewontin, in *The Genetics of Colonizing Species*, ed. H. G. Baker and G. L. Stebbins, Academic Press, New York, USA, 1965.

64. T. R. E. Southwood and H. N. Comins, *J. Anim. Ecol.*, 1976, **65**, 949.

65. R. H. MacArthur, E. O. Wilson, *The Theory of Island Biogeography*, Princeton University Press, NJ, USA, 1967.

66. T. R. E. Southwood, *Am. Sci.*, 1977, **65**, 30.

67. R. Karban and G. M. English-Loeb, *Exp. Appl. Acarol.*, 1988, **4**, 225.

68. R. Karban, *Oikos*, 1990, **59**, 27.

69. J. Li and D. C. Margolies, *Entomol. Exp. Appl.*, 1993, **68**, 79.

70. T. M. Perring, A. Cooper, D. J. Cooper, C. Shields and J. Shields, *California Agric.*, 1991, **45**, 10.

71. T. M. Perring, A. D. Cooper, R. J. Rodriguez, C. A. Farrar and T. S. Bellows, *Science*, 1993, **259**, 74.

72. H. S. Costa and J. K. Brown, *Entomol. Exp. Appl.*, 1991, **61**, 211.

73. P. C. Ellsworth and J. L. Martinez-Carrillo, *Crop Protect.*, 2001, **20**, 853.

74. http://www.co.imperial.ca.us/ag/ (last accessed August19th, 2010).

75. J. C. Palumbo, A. R. Horowitz and N. Prabhaker, *Crop Protect.*, 2001, **20**, 739.

76. J. T. Trumble, *Integrated Pest Manag. Rev.*, 1998, **3**, 195.

77. T. C. R. White, *Oecologia*, 1984, **63**, 90.

78. H. M. Flint, S. E. Naranjo, J. E. Leggett and T. J. Henneberry, *J. Economic Entomol.*, 1996, **89**, 1288.

79. R. H. Ratcliffe, H. W. Ohm, F. L. Patterson, S. E. Cambron and G. G. Safranski, *J. Economic Entomol.*, 1996, **89**, 1309.

80. R. H. Ratcliffe, S. E. Cambron, K. L. Flanders, N. A. Bosque-Perez, S. L. Clement and H. W. Ohm, *J. Economic Entomol.*, 2000, **93**, 1319.

81. M. B. Cohen, S. N. Alam, E. B. Medina and C. C. Bernal, *Entomol. Exp. Appl.*, 1997, **85**, 221.

82. B. Du, W. Zhang, B. Liu, J. Hu, Z. Wei, Z. Shi, Z. He, L. Zhu, R. Chen, B. Han and G. He, *Proc. Natl. Acad. Sci. U. S. A.*, 2009, **106**, 22163.

83. P. Jeschke and R. Nauen, *Pest Manag. Sci.*, 2008, **64**, 1084.

84. K. Gorman, Z. Liu, I. Denholm, K.-U. Bruggen and R. Nauen, *Pest Manag. Sci.*, 2008, **64**, 1122.

85. R. H. Ratcliffe in *Radcliffe's IPM World Textbook*, ed. E. B. Radcliffe and W. D. Hutchison, University of Minnesota, St. Paul, MN, http://ipmworld.umn.edu (last accessed August 14th, 2010).

86. J. Hollander and P. K. Pathak, *Entomol. Exp. Appl.*, 1981, **29**, 76.

87. C. M. Smith, in *Biological and Biotechnological Control of Insect Pests*, ed. J. E. Rechcigl and N. A. Rechcigl, CRC Press, Boca Raton, FL, USA, 2000, p. 171.

88. J. L. Krysan, D. E. Foster, T. F. Branson, K. R. Ostlie and W. S. Cranshaw, *Bull. Entomol. Soc. Am.*, 1986, **32**, 250.

89. E. Levine, J. L. Spencer, S. A. Isard, D. W. Onstad and M. E. Gray, *Am. Entomologist*, 2002, **48**, 94.

90. F. Gould, *Am. Sci.*, 1991, **79**, 496.

91. S. A. Qureshi, D. J. Midmore, S. S. Syeda and D. J. Reid, *Bull. Entomol. Res.*, 2010, **100**, 67.

92. M. Jonsson, S. D. Wratten, D. A. Landis and G. M. Gurr, *Biol. Cont.*, 2008, **45**, 172.

93. M. A. Altieri, *Agric. Ecosys. Environ.*, 1999, **74**, 19.

94. D. A. Andow, *Annu. Rev. Entomol.*, 1991, **36**, 561.

95. L. E. Caltagirone, *Annu. Rev. Entomol.*, 1981, **26**, 213.

96. D. Simberloff and P. Stiling, *Ecology*, 1996, **77**, 1965.

97. D. R. Strong, *Science*, 1997, **277**, 1058.

98. University of California IPM Online, http://www.ipm.ucdavis.edu/PMG/r107301111.html (last accessed September 2nd, 2010).

99. P. A. Stansly and E. T. Natwick, in *Bemisia: Bionomics and Management of a Global Pest*, ed. P. A. Stansly and S. E. Naranjo, Springer, Dordrecht, The Netherlands, 2010, p. 467.

100. P. A. Stansly, P. A. Sanchez, J. M. Rodriguez, F. Canizares, A. Nieto, M. J. Lopez-Leyva, M. Fajardo, V. Suarez and A. Urbaneja, *Crop Protect.*, 2004, **23**, 701.

101. J. E. Casida and G. B. Quistad, in *Modern Agriculture and the Environment*, ed. D. Rosen, E. Tel-Or, Y. Hadar and Y. Chen, Kluwer Academic Publishers, Dordrecht, The Netherlands, 1997, p. 3.

102. Insecticide Resistance Action Committee (IRAC) Mode of Action Classification, http://www.irac-online.org (last accessed September 2nd 2010).

103. C. James, *ISAAA Brief No. 39*, International Service for the Acquisition of Agri-Biotech Applications, Ithaca, NY, USA, 2008.

104. J. Ferré, J. Van Rie and S. C. MacIntosh, in *Integration of Insect-Resistant Genetically Modified Crops within IPM Programs*, ed. J. Romeis, A. M. Shelton and G. G. Kennedy, Springer Science, Dordrecht, The Netherlands, 2008, p. 41.

105. B. E. Tabashnik, A. J. Gassmann, D. W. Crowder and Y. Carrière, *Nature Biotechnol.*, 2008, **26**, 1074.

106. G. Brookes and P. Barfoot, *AgBioForum*, 2010, **13**, 76.

107. M. Qaim, C. E. Pray and D. Zilberman, in *Integration of Insect-Resistant Genetically Modified Crops within IPM Programs*, ed. J. Romeis, A. M. Shelton, G. G. Kennedy, Springer Science, Dordrecht, The Netherlands, 2008, pp. 329–356.

108. K. M. Wu, Y. H. Lu, H. Q. Feng, Y. Y. Jiang and J. Z. Zhao, *Science*, 2008, **321**, 1676.

109. A. M. Shelton and F. R. Badenes-Perez, *Annu. Rev. Entomol.*, 2006, **51**, 285.

110. J. Kovach, C. Petzoldt, J. Degni and J. Tette, A method to measure the environmental impact of pesticides, *New York's Food and Life Sciences Bulletin*, NYS Agricultural Experiment Station, Cornell University, Geneva, NY, USA, 1992, http://www.nysipm.cornell.edu/publications/ EIQ.html (last accessed August 25th, 2010).

111. S. E. Naranjo, J. R. Ruberson, H. C. Sharma, L. Wilson and K. M. Wu, *Integration of Insect-Resistant Genetically Modified Crops within IPM Programs*, ed. J. Romeis, A. M. Shelton and G. G. Kennedy, Springer Science, Dordrecht, The Netherlands, 2008, p. 159.

112. Y. Lu, K. Wu, Y. Jiang, B. Xia, P. Li, H. Feng, K. A. G. Wyckhuys and Y. Guo, *Science*, 2010, **328**, 1151.

113. S. M. Cook, Z. R. Khan and J. A. Pickett, *Annu. Rev. Entomol.*, 2007, **52**, 375.

114. J. R. Miller and R. S. Cowles, *J. Chem. Ecol.*, 1990, **16**, 197.

115. M. Rice, in *Proc. Heliothis Ecol. Workshop 1985*, ed. M. P. Zaluki and P. G. Twine, Dept. Primary Ind., Brisbane, QLD, Australia, 1986, p. 27.

116. B. Pyke, M. Rice, B. Sabine and M. P. Zalucki, *Australia Cotton Growers*, 1987, May–July, 7.

117. A. Hassanali, H. Herren, Z. R. Khan, J. A. Pickett and C. M. Woodcock, *Philos. Trans. R. Soc. London, Ser. B*, 2008, **363**, 611.

118. N. Agelopoulos, M. A. Birkett, A. J. Hick, A. M. Hooper and J. A. Pickett, *Pestic. Sci.*, 1999, **55**, 225.
119. T. C. Baker, in *Integrated Pest Management*, ed. T. Radcliffe and B. Hutchison, Cambridge University Press, UK, 2008, p. 273.
120. V. P. Jones, T. R. Unruh, D. R. Horton, N. J. Mills, J. F. Brunner, E. H. Beers and P. W. Shearer, *Pest Manag. Sci.*, 2009, **65**, 1305.
121. J. Zhu and K.-C. Park, *J. Chem. Ecol.*, 2005, **31**, 1733.
122. D. G. James, *Environ. Entomol.*, 2003, **32**, 977.
123. Z. R. Khan, *Nature*, 1997, **388**, 631.
124. Z. R. Khan, P. Chiliswa, K. Ampong-Nyarko, L. E. Smart, A. Polaszek, J. Wandera and M. A. Mulaa, *Insect Sci. Appl.*, 1997, **17**, 143.
125. Z. R. Khan, J. A. Pickett, L. J. Wadhams and F. Muyekho, *Insect Sci. Appl.*, 2001, **21**, 375.
126. R. E. Frisbie and J. W. Smith Jr., in *Progress and Perspectives for the 21st Century*, presented at the Entomological Society of America Centennial Symposium, Lanham, MD, USA, ed. J. J. Menn and A. L. Steinhauer, 1991, p. 151.
127. S. E. Naranjo and H. M. Flint, *Environ. Entomol.*, 1995, **24**, 261.
128. S. E. Naranjo, H. M. Flint and T. J. Henneberry, *J. Economic Entomol.*, 1995, **88**, 1666.
129. S. E. Naranjo, H. M. Flint and T. J. Henneberry, *Entomol. Exp. Appl.*, 1996, **80**, 343.
130. P. C. Ellsworth and D. L. Meade, in *Cotton, a College of Agriculture, Report Series P-94*, University of Arizona, Tucson, AZ, USA, 1993, p. 280.
131. P. Ellsworth and D. Meade, in *Silverleaf Whitefly (Formerly Sweetpotato Whitefly Strain B): 1995 Supplement to the 5-Year National Research and Action Plan*, US Dept. Agric., *Agric. Res. Serv.*, 1995, p. 72.
132. R. Nichols, P. Ellsworth and T. Dennehy, in *Proceedings of the Beltwide Cotton Conference*, ed. P. Dugger and D. A. Richter, National Cotton Council, Memphis, TN, USA, 1996, p. 153.
133. S. E. Naranjo, P. C. Ellsworth, C. C. Chu, T. J. Henneberry, D. G. Riley, T. F. Watson and R. L. Nichols, *J. Economic Entomol.*, 1998, **91**, 1415.
134. T. J. Dennehy and L. Williams, *Pestic. Sci.*, 1997, **51**, 398.
135. S. E. Naranjo and P. C. Ellsworth, *Pest Manag. Sci.*, 2009, **65**, 1267.

Subject Index